Rudolf Berghammer Bernhard Möller
Georg Struth (Eds.)

Relational and Kleene-Algebraic Methods in Computer Science

7th International Seminar on Relational Methods in Computer Science
and 2nd International Workshop on Applications of Kleene Algebra
Bad Malente, Germany, May 12-17, 2003
Revised Selected Papers

Springer

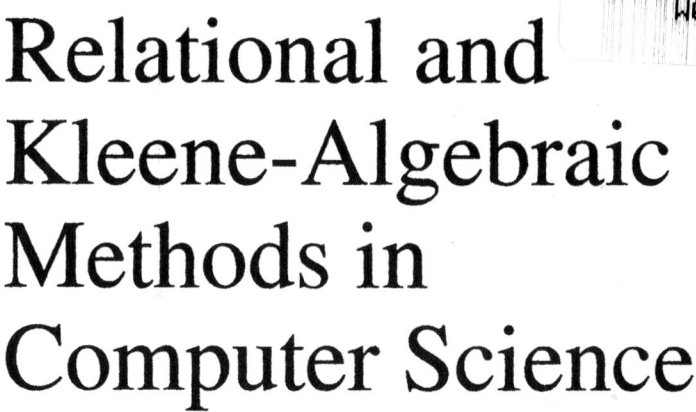

Volume Editors

Rudolf Berghammer
Christian-Albrechts-Universität zu Kiel
Institut für Informatik und Praktische Mathematik
Olshausenstr. 40, 24098 Kiel, Germany
E-mail: rub@informatik.uni-kiel.de

Bernhard Möller
Georg Struth
Universität Augsburg, Institut für Informatik
Universitätsstr. 14, 86135 Augsburg, Germany
E-mail: {bernhard.moeller, georg.struth}@informatik.uni-augsburg.de

Library of Congress Control Number: 2004106383

CR Subject Classification (1998): F.4, D.2.4, F.3, I.1, I.2.3, G.2

ISSN 0302-9743
ISBN 3-540-22145-X Springer-Verlag Berlin Heidelberg New York

Springer-Verlag is a part of Springer Science+Business Media

springeronline.com

Printed in Germany

Typesetting: Camera-ready by author, data conversion by DA-TeX Gerd Blumenstein
Printed on acid-free paper SPIN: 11011163 06/3142 5 4 3 2 1 0

Lecture Notes in Computer Science

Commenced Publication in 1973
Founding and Former Series Editors:
Gerhard Goos, Juris Hartmanis, and Jan van Leeuwen

Springer
Berlin
Heidelberg
New York
Hong Kong
London
Milan
Paris
Tokyo

In Memoriam

Armando Haeberer

(1947—2003)

Preface

This volume contains the proceedings of the 7th *International Seminar on Relational Methods in Computer Science* (RelMiCS 7) and the 2nd *International Workshop on Applications of Kleene Algebra*. The common meeting took place in Bad Malente (near Kiel), Germany, from May May 12–17, 2003. Its purpose was to bring together researchers from various subdisciplines of Computer Science, Mathematics and related fields who use the calculi of relations and/or Kleene algebra as methodological and conceptual tools in their work.

This meeting is the joint continuation of two different series of meetings. Previous RelMiCS seminars were held in Schloss Dagstuhl (Germany) in January 1994, Parati (Brazil) in July 1995, Hammamet (Tunisia) in January 1997, Warsaw (Poland) in September 1998, Quebec (Canada) in January 2000, and Oisterwijk (The Netherlands) in October 2001. The first workshop on applications of Kleene algebra was also held in Schloss Dagstuhl in February 2001. To join these two events in a common meeting was mainly motivated by the substantial common interests and overlap of the two communities. We hope that this leads to fruitful interactions and opens new and interesting research directions.

This volume contains 23 contributions by researchers from all over the world: 21 regular papers and two invited papers *Choice Procedures in Pairwise Comparison of Multiple-Attribute Decision Making Methods* by Raymond Bisdorff and Marc Roubens and *Kleene Algebra with Relations* by Jules Desharnais. The papers show that relational algebra and Kleene algebra have wide-ranging diversity and applicability in theory and practice. Just to give an (incomplete) overview, the papers deal with problems appearing in software technology and program verification and analysis, the formal treatment of pointer algorithms and of algorithms for many problems on discrete structures, applications of relations in combination with fixed points to investigate games, questions arising in the context of databases and data mining, the relational modeling of real-world situations, many topics from artificial intelligence such as knowledge representation and acquisition, preference modeling and scaling methods, and, finally, the use of tools for prototyping and programming with relations and for relational reasoning.

We are very grateful to the members of the program committee and the external referees for their care and diligence in reviewing the submitted papers. We also want to thank Ulrike Pollakowski-Geuther, Ulf Milanese, and Frank Neumann for their assistance; they made organizing this meeting a pleasant experience. Finally, we want to thank Günther Gediga and Gunther Schmidt for their help.

March 2004

Rudolf Berghammer
Bernhard Möller
Georg Struth

Program Committee

Roland Backhouse	(U. Nottingham, UK)
Rudolf Berghammer	(U. Kiel, Germany)
Richard Bird	(Oxford U., UK)
Jules Desharnais	(U. Laval, Canada)
Ivo Düntsch	(Brock U., Canada)
Marcelo Frías	(U. Buenos Aires, Argentina)
Ali Jaoua	(U. Qatar, Qatar)
Dexter Kozen	(Cornell U., USA)
Bernhard Möller	(U. Augsburg, Germany)
Oege de Moor	(Oxford U., UK)
Ewa Orłowska	(U. Warsaw, Poland)
Gunther Schmidt	(U. Armed Forces, Munich, Germany)
Harrie de Swart	(U. Tilburg, The Netherlands)
Joakim von Wright	(Åbo Akademi U., Finland)

External Referees

Hans Bherer	Claude Bolduc	Ernst-Erich Doberkat
Woitek Dzik	Thorsten Ehm	Alexander Fronk
Ali Jaoua	Wolfram Kahl	Michiel van Lambalgen
Vincent Mathieu	Damian Niwinski	Eric Offermann
Dominik Slezak	Andrzej Slezak	Georg Struth
Michael Winter		

Sponsoring Institutions

The generous support of the following institutions and companies is gratefully acknowledged:

Deutsche Forschungsgemeinschaft
EU COST Action 274 TARSKI
Faculty of Engineering of Kiel University
CrossSoft (Kiel)
Lufthansa Revenue Services (Hamburg)

Table of Contents

Invited Papers

Contributed Papers

Choice Procedures in Pairwise Comparison Multiple-Attribute Decision Making Methods

Raymond Bisdorff[1] and Marc Roubens[2]

[1] Department of Management and Informatics
University Center of Luxembourg
Bisdorff@cu.lu
[2] Department of Mathematics
University of Liege
M.Roubens@ulg.ac.be

Abstract. We consider extensions of some classical rational axioms introduced in conventional choice theory to valued preference relations. The concept of kernel is revisited using two ways : one proposes to determine kernels with a degree of qualification and the other presents a fuzzy kernel where every element of the support belongs to the rational choice set with a membership degree. Links between the two approaches is emphasized. We exploit these results in Multiple-attribute Decision Aid to determine the good and bad choices. All the results are valid if the valued preference relations are evaluated on a finite ordinal scale.

1 Introduction

We consider a pair wise comparison multiple-attribute decision making procedure that assigns to each ordered pair $(x, y), x, y \in A$ (the set of alternatives) a global degree of preference $R(x, y)$. $R(x, y)$ represents the degree to which x is weakly preferred to y.

We suppose that $R(x, y)$ belongs to a finite set $L : \{c_0, c_1, \ldots, c_m, \ldots, c_{2m}\}$ that constitutes a $(2m + 1)$-element chain $\{c_0, c_1, \ldots, c_{2m}\}$. $R(x, y)$ may be understood as the level of credibility that "a is at least as good as b". The set L is built using the values of R taking into consideration an antitone unary contradiction operator $\neg$ such that $\neg c_i = c_{(2m-i)}$ for $i = 0, \ldots, 2m$.

If $R(x, y)$ is one of the elements of L, then automatically $\neg R(x, y)$ belongs to L. We call such a relation an L-valued binary relation.

We denote $L^{\succ m} : \{c_{m+1}, \ldots, c_{2m}\}$ and $L^{\prec m} : \{c_0, \ldots, c_{m-1}\}$.

If $R(x, y) \in L^{\succ m}$, we say that the proposition "$(x, y) \in R$" is L-true. If however $R(x, y) \in L^{\prec m}$, we say that the proposition is L-false. If $R(x, y) = c_m$, the median level (a fix point of the negation operator) then the proposition "$(x, y) \in R$" is L-undetermined. If $R(a, b) = c_r$ and $R(c, d) = c_s, c_r < c_s$, it means that the proposition "a is at least as good as b" is less credible than "c is at least as good as d".

In the classical case where R is a crisp binary relation ($m = 2$, and $R(x, y)$ is never rated c_1; $R(x, y) = c_2 = 1$ is denoted xRy and $R(x, y) = c_0 = 0$

R. Berghammer et al. (Eds.): RelMiCS/Kleene-Algebra Ws 2003, LNCS 3051, pp. 1–7, 2004.

corresponds to $\neg xRy$, we define a digraph $G(A,R)$ with vertex set A and arc family R. A choice in $G(A,R)$ is a non empty set Y of A.

R can be represented by a Boolean matrix and the choice Y can be defined with the use of a subset characteristic row vector $Y(.) = (\ldots, Y(x), \ldots, Y(y), \ldots)$ where

$$Y(x) = \begin{cases} 1 \text{ if } x \in Y \\ 0 \text{ otherwise,} \end{cases} \quad \text{for all } x \in A.$$

The subset characteristic vector of the successors of the elements of the vertex set $Y : \{x \in A \mid \exists\, y \in Y, yRx\}$ is denoted $Y \circ R$ and is obtained using the Boolean composition

$$(Y \circ R)(x) = \vee_{y \neq x}(Y(y) \wedge R(y,x)) \tag{1}$$

where $\vee$ and $\wedge$ represent respectively "disjunction" and "conjunction" for the 2-element Boolean lattice $\mathbf{B} = \{0,1\}$.

The choice Y should satisfy some of the following rationality axioms ($\bar{Y}$ represents the complement of Y in A) :

- *Inaccessibility of* Y (or GOCHA rule, cf.[5], [10])
  $\forall y \in Y, \forall x \in \bar{Y}, \neg xRy$
  $\bar{Y} \circ R \subseteq \bar{Y}$, "the successors of $\bar{Y}$ are inside $\bar{Y}$".
- *Stability of* Y (see [9], [11])
  $\forall y \in Y, \forall x \in Y, \neg yRx$
  $Y \circ R \subseteq \bar{Y}$, "the successors of Y are inside $\bar{Y}$".
- *Dominance of* Y (or external stability, see [9],[11])
  $\forall x \in \bar{Y}, \exists\, y \in Y, yRx$
  $\bar{Y} \subseteq Y \circ R$, "the successors of Y contain $\bar{Y}$".
- *Strong dominance of* Y (or GETCHA rule, cf. [5], [10])
  $\forall y \in Y, \forall x \in \bar{Y}, yRx \equiv \neg yR^d x$
  (R^d is the dual relation, i.e. the transpose of the complement of R)
  $\bar{Y} \circ R^d \subseteq \bar{Y}$.

The maximal set of all non-dominated alternatives (inaccessibility and stability are satisfied) is called the *core* of Y and the internally and externally stable set corresponds to the *kernel*. The GETCHA set is such that the strong dominance rule applies.

No specific property like acyclicity or antisymmetry will be assumed in the sequel. The core guarantees a rather small choice but is often empty. The GETCHA set corresponds to a rather large set and, in this general framework, the kernel (see [5], [8]) seems to be the best compromise. However its existence or uniqueness cannot be guaranteed. . It has been mentioned in [5] that for random graphs – with probability .5 – a kernel almost certainly exists and that in a Moon-Moser graph with n nodes the number of kernels is around $3^{n/3}$.

In order to illustrate all these concepts, we consider a small example.

Table 1. Boolean matrix R and scores

R	a	b	c	d	e	f	g	h	$S(+)$
a	·	1	1	1	0	0	0	0	3
b	1	·	1	1	1	1	1	1	7
c	1	1	·	1	1	0	1	1	6
d	1	1	1	·	1	0	1	1	6
e	0	1	1	1	·	0	1	1	5
f	0	1	1	1	1	·	1	1	6
g	0	1	1	1	1	1	·	1	6
h	0	1	1	1	1	0	0	·	4
$S(-)$	3	7	7	7	6	2	5	6	

Example 1. Consider the following example $A : \{a, b, c, d, e, f, g, h\}$ with 8 alternatives. The Boolean matrix R together with the outgoing and ingoing scores $S(+)$ and $S(-)$ are presented in Table 1.

Core (non dominated elements) : empty set.
Kernels (maximal stable and minimal dominant sets) : $\{b\}, \{a, f\}, \{a, g\}$.
Minimal GETCHA sets : $\{b\}, \{a, e, f, g, h\}$.

We may define generalizations of the previous crisp concepts in the valued case in two different ways :

(i) Starting from the definition of a rational choice in terms of logical predicates, one might consider that every subset of A is a rational choice with a given qualification and determine those sets with a sufficient degree of qualification.
(ii) One might also extend the algebraic definition of a rational choice. In that case, there is a need to define proper extensions of composition law $\circ$ and inclusion $\subseteq$.

Solutions that correspond to this approach give a fuzzy rational set $\tilde{Y}$, each element of A belonging to A to a certain degree (membership function).

It should be interesting to stress the correspondence between these two approaches. The choice of the operators is closely related to the type of scale that is used to quantify the valued binary relation R, i.e. an ordinal scale.

2 Qualification of Crisp Kernels in the Valued Ordinal Context

We now denote $G^L = G^L(A, R)$ a digraph with vertices set A and a valued arc family that corresponds to the L-valued binary relation R . This graph is often called *outranking graph* in the context of multi-attribute decision making.

We define the level of stability qualification of subset Y of X as

$$\Delta^{sta}(Y) = \begin{cases} c_{2m} & \text{if } Y \text{ is a singleton,} \\ \min\limits_{y \neq x} \min\limits_{x \neq y} \{\neg R(x, y)\} & \text{otherwise} \end{cases}$$

Table 2. Outranking relation related to eight cars

R	a	b	c	d	e	f	g	h
a	1	.75	.70	.62	0	0	0	0
b	.76	1	.90	.82	.82	.82	.82	.80
c	.70	.86	1	1	1	.46	.80	.91
d	.64	.65	.94	1	.88	.22	.94	.74
e	.33	.57	.93	1	1	0	.80	.86
f	0	.73	.64	.92	.76	1	.96	.80
g	0	.63	.73	.85	.82	.70	1	.81
h	0	.60	.64	.60	.77	0	0	1

and the level of dominance qualification of Y as

$$\Delta^{dom}(Y) = \begin{cases} c_{2m} & \text{if } Y = A, \\ \min\limits_{x \notin Y} \max\limits_{y \in Y} R(y,x) & \text{otherwise.} \end{cases}$$

Y is considered to be an L-good choice, i.e L-stable and L-dominant, if $\Delta^{sta}(Y) \in L^{\succ m}$ and $\Delta^{dom}(Y) \in L^{\succ m}$. Its qualification corresponds to

$$Q^{good}(L) = \min(\Delta^{sta}(Y), \Delta^{dom}(Y)) \in L^{\succ m}.$$

We denote $C^{good}(G^L)$ the possibly empty set of L-good choices in G^L.

The determination of this set is an NP-complete problem even if, following a result of Kitainik [5], we do not have to enumerate the elements of the power set of A but only have to consider the kernels of the corresponding crisp strict median-level cut relation $R^{\succ m}$ associated to R, i.e. $(x,y) \in R^{\succ m}$ if $R(x,y) \in L^{\succ m}$.

As the kernel in $G(X, R^{\succ m})$ is by definition a stable and dominant crisp subset of A, we consider the possibly empty set of kernels of $G^{\succ m} = G(A, R^{\succ m})$ which we denote $C^{good}(G^{\succ m})$.

Kitainik proved that

$$C^{good}(G^L) \subseteq C^{good}(G^{\succ m}).$$

The determination of crisp kernels has been extensively described in the literature (see, for example [9]) and the definition of $C^{good}(G^L)$ is reduced to the enumeration of the elements of $C^{good}(G^{\succ m})$ and the calculation of their qualification.

Example 2. We now consider the comparison of 8 cars (a,b,c,d,e,f,g) on the basis of maximum speed, volume, price and consumption. Data and aggregation procedure will not be presented here (for more details, see [2]). The related outranking relation is presented in Table 2.

We will consider only the ordinal content of that outranking relation and we transpose the data on a L-scale with $c_0 = 0, c_{2m} = 1, m = 27$ and $c_m = .5$.

The strict median-cut relation $R^{\succ m}$ corresponds to data of Table 1. The set $C^{good}(G^{\succ m})$ corresponds to $(\{b\}, \{a, f\}, \{a, g\})$ with the following qualifications :

$$Q^{good}(\{b\}) = .76, \quad Q^{good}(\{a, f\}) = Q^{good}(\{a, g\}) = .70.$$

3 Fuzzy Kernels

A second approach to the problem of determining a good choice is to consider the valued extension of the Boolean system of equations (1).

If $\tilde{Y}(.) = (\ldots, Y(x), Y(y), \ldots)$, where $\tilde{Y}(x)$ belongs to L for every $x \in A$ is the characteristic vector of a fuzzy choice and indicates the credibility level of the assertion that "x is part of the choice $\tilde{Y}$", we have to solve the following system of equations :

$$(\tilde{Y} \circ R)(x) = \max_{y \neq x}[\min(\tilde{Y}(y), R(y, x))] = \neg\tilde{Y}(x), \quad \forall x, y \in A. \tag{2}$$

The set of solutions to the system of equations (2) is called $\tilde{Y}^{dom}(G^L)$.

In order to compare these fuzzy solutions to the solutions in $C^{good}(G^L)$, we define the crisp choice

$$K_{\tilde{Y}} \subset A \begin{cases} x \in K_{\tilde{Y}} \text{ if } \tilde{Y}(x) \in L^{\succ m} \\ x \notin K_{\tilde{Y}} \text{ otherwise} \end{cases} \tag{3}$$

and we consider a partial order on the elements of $\tilde{Y}^{dom}(G^L)$: $\tilde{Y}$ is sharper than $\tilde{Y}'$, noted $\tilde{Y}' \preceq \tilde{Y}$, iff $\forall x \in A$: either $\tilde{Y}(x) \leq \tilde{Y}'(x) \leq c_m$, either $c_m \leq \tilde{Y}'(x) \leq \tilde{Y}(x)$.

The subset of the sharpest solutions in $\tilde{Y}^{dom}(G^L)$ is called $F^{dom}(G^L)$.

Bisdorff and Roubens have proved that the set of crisp choices constructed from $F^{dom}(G^L)$ using (3) and denoted $K(F^{dom}(G^L))$ coincides with $C^{dom}(G^L)$.

Coming back to Example 2, we obtain 3 sharpest solutions to equation (2)

$$\begin{aligned} \tilde{Y}_{\{b\}} &= (.24, .76, .24, .24, .24, .24, .24, .24) \\ \tilde{Y}_{\{a,f\}} &= (.70, .30, .30, .30, .30, .70, .30, .30) \\ \tilde{Y}_{\{a,g\}} &= (.70, .30, .30, .30, .30, .30, .30, .70). \end{aligned}$$

In this particular case, we obtain only Q^{good} and $\neg Q^{good}$ as components of the $\tilde{Y}$'s but this is not true in general.

4 Good and Bad Choices in Multi-attribute Decision Making

In the framework of decision making procedures, it is often interesting to determine choice sets that correspond to bad choices. These bad choices should be ideally different from the good choices. To clarify this point, let us first consider the crisp Boolean case and define the rationality axiom of

- *Absorbance* of Y (see [10])
  $\forall x \in \bar{Y}, \exists\, y \in Y, xRy = yR^t x$
  $\bar{Y} \subseteq Y \circ R^t$, "the predecessors of Y contain $\bar{Y}$".

As the stability property can be rewritten as $Y \circ R^t \subseteq \bar{Y}$, we immediately obtain the Boolean equation that determines the absorbent kernel (stable and absorbent choice) :

$$\bar{Y} = Y \circ R^t.$$

We notice that for some digraphs (dominant) kernels and absorbent kernels may coincide (consider a digraph $G(A, R)$ with vertices $A : \{a, b, c, d\}$ and four arcs $(a, b), (b, c), (c, d), (d, a)$. $\{a, c\}$ as well as $\{b, d\}$ are dominant and absorbent kernels or good and bad choices).

This last concept can be easily extended in the valued case. Consider the valued graph G^L introduced in Section 2. We define the level of absorbance qualification of Y as

$$\Delta^{abs}(Y) = \begin{cases} c_{2m} & \text{if } Y = A, \\ \min\limits_{x \notin Y} \max\limits_{y \in Y} R(x, y) & \text{otherwise.} \end{cases}$$

The qualification of Y being a bad choice corresponds to

$$Q^{bad}(Y) = \min(\Delta^{sta}(Y), \Delta^{abs}(Y)) > c_m.$$

If $Q^{bad}(Y) \leq c_m$, Y is not considered to be a bad choice.

A fuzzy absorbent kernel is a solution of equation

$$(\tilde{Y} \circ R^t)(x) = \max_{y \neq x} \min(\tilde{Y}(y), R^t(y, x)) = \neg\tilde{Y}(x), \quad \forall x \in A. \tag{4}$$

The set of solutions of equations (4) denoted $\tilde{Y}^{abs}(G^L)$ can be handled in the same way as done in Section 3 for $\tilde{Y}^{dom}(G^L)$ and creates a link between these solutions (4) and subsets of Y being qualified as bad choices.

Reconsidering Example 2, we observe that $\{b\}, \{c\}, \{d\}, \{a, e\}$ and $\{a, h\}$ are absorbent kernels in $G(A, R^{\succ m})$. Qualification can be easily obtained and we get $Q^{bad}(\{a, c\}) = .76, Q^{bad}(\{a, h\}) = .74, Q^{bad}(\{c\}) = .64, Q^{bad}(\{d\}) = .60$, $Q^{bad}(\{b\}) = .57$.

We finally decide to keep car b as the best solution noticing however that it is a bad choice. Going back to digraph $G(A, R^{\succ m})$, we see that b is at the same time dominating and dominated by all the other elements. Car b is indifferent to all the other cars which is not true for a, c, d, e, f, g, h, since indifference is not transitive in this example.

References

[1] Bisdorff, R., Roubens, M. : On defining and computing fuzzy kernels from L-valued simple graphs. In : Da Ruan et al. (eds.) : Intelligent Systems and Soft Computing for Nuclear science and Industry, FLINS'96 Workshop. World Scientific Publishers, Singapore (1996) 113-123

[2] Fodor, J., Roubens, M. : Fuzzy Preference Modelling and Multi-criteria Decision Support. Kluwer Academic publishers, Dordrecht Boston London (1994) 4
[3] Fodor, J. C., Perny, P., Roubens, M. : Decision Making and Optimization. In : Ruspini, E., Bonissone, P., Pedrycz, W. (eds.) : Handbook of Fuzzy Computation, Institute of Physics Publications and Oxford University Press, Bristol (1998) F.5.1 : 1-14
[4] Fodor, J., Orlovski S. A., Perny, P., Roubens, M. : The use of fuzzy preference models in multiple criteria : choice, ranking and sorting. In : Dubois, D., Prade, H. (eds.) : Handbooks and of Fuzzy Sets, Vol. 5 (Operations Research and Statistics), Kluwer Academic Publishers, Dordrecht Boston London (1998) 69-101
[5] Kitainik, L. : Fuzzy Decision Procedures with Binary Relations : towards an unified Theory. Kluwer Academic Publishers, Dordrecht Boston London (1993) 2, 4
[6] Marichal, J.-L. : Aggregation of interacting criteria by means of the discrete Choquet integral. In : Calvo,T., Mayor,G., Mesiar R. (eds.) : Aggregation operators : new trends and applications. Series : Studies in Fuzziness and Soft Computing Vol. 97, Physica-Verlag, Heidelberg (2002) 224-244
[7] Perny, P., Roubens, M. : Fuzzy Relational Preference Modelling. In : Dubois, D,. and Prade, H. (eds.) : Handbooks of Fuzzy Sets, Vol. 5 (Operations Research and Statistics). Kluwer Academic Publishers, Dordrecht Boston London (1998) 3-30
[8] Roy, B. : Algèbre moderne et théorie des graphes. Dunod, Paris (1969) 2
[9] Schmidt, G., StrÜhlein, T. : Relations and Graphs; Discrete mathematics for Computer Scientists. Springer-Verlag, Berlin Heidelberg New York (1991) 2, 4
[10] Schwartz, T. : The logic of Collective Choice, Columbia Univer Press, New York (1986) 2, 6
[11] von Neumann, J., Morgenstern, O. : Theory of Games and Economic Behaviour. Princeton University Press, New York (1953) 2

Kleene Algebra with Relations*

Jules Desharnais

Département d'informatique et de génie logiciel
Université Laval, Québec, QC, G1K 7P4 Canada
Jules.Desharnais@ift.ulaval.ca

Abstract. Matrices over a Kleene algebra with tests themselves form a Kleene algebra. The matrices whose entries are tests form an algebra of relations if the converse of a matrix is defined as its transpose. Abstracting from this concrete setting yields the concept of Kleene algebra with relations.

1 Introduction

It is well known [4, 13] that matrices over a Kleene algebra (KA in the sequel), i.e., matrices whose entries belong to a KA, again form a KA (a heterogeneous KA if matrices with different sizes are allowed). Such matrices can be used to represent automata or programs by suitably choosing the underlying KA (algebra of languages, algebra of relations, ...). Every KA has an element 0 (e.g., the empty language, the empty relation) and an element 1 (e.g., the language containing only the empty sequence, the identity relation). Now, matrices filled with 0's and 1's are again matrices over the given KA, but, in addition, they are relations satisfying the usual properties of relations. Hence, the set of $n \times n$ matrices over a given KA is a KA with relations.

Using this simple remark, we abstract from the concrete world of matrices and define the concept of KA with relations. We also give examples showing the interest of the concept.

In Sect. 2, we give the definition of Kleene algebra. In Sect. 3, we introduce matrices over a KA and describe how the concept of KA with relations may arise. Section 4 defines abstract KAs with relations and gives examples. Section 5 briefly discusses additional axioms and representability. Section 6 is a short section on projections, direct products and unsharpness in KAs with relations.

2 Kleene Algebra

There are some variants of KA around [4, 6, 13, 14]. We use Kozen's first-order axiomatization [14], because this is the least constraining one and it can be used as a basis for the other definitions.

---

* This research is supported by NSERC (Natural Sciences and Engineering Research Council of Canada).

R. Berghammer et al. (Eds.): RelMiCS/Kleene-Algebra Ws 2003, LNCS 3051, pp. 8–20, 2004.

Definition 1. *A* Kleene algebra *is a structure* $\mathcal{K} = (K, +, \cdot, ^*, 0, 1)$ *such that* $(K, +, 0)$ *is a commutative monoid,* $(K, \cdot, 1)$ *is a monoid, and the following laws hold:*

$$\begin{array}{ll} a + a = a, & a \cdot (a + b) = a \cdot a + a \cdot b, \\ a \cdot 0 = 0 \cdot a = 0, & (a + b) \cdot c = a \cdot c + b \cdot c, \\ 1 + a \cdot a^* = a^*, & b + a \cdot c \leq c \Rightarrow a^* \cdot b \leq c, \\ 1 + a^* \cdot a = a^*, & b + c \cdot a \leq c \Rightarrow b \cdot a^* \leq c, \end{array}$$

where $\leq$ *is the partial order induced by* $+$*, that is,*

$$a \leq b \Leftrightarrow a + b = b \ .$$

A KA is Boolean *if there is a complementation operation* $^-$ *such that* $(K, +, ^-, 0)$ *is a Boolean lattice. The meet* $\sqcap$ *of this lattice satisfies* $a \sqcap c = \overline{\overline{a} + \overline{c}}$ *and there is a top element* $\top = \overline{0}$.

A Kleene algebra with tests *[14] is a two-sorted algebra* $(K, T, +, \cdot, ^*, 0, 1, \neg)$ *such that* $(K, +, \cdot, ^*, 0, 1)$ *is a Kleene algebra and* $(T, +, \cdot, \neg, 0, 1)$ *is a Boolean algebra, where* $T \subseteq K$ *and* $\neg$ *is a unary operator defined only on* T.

Operator precedence, from lowest to highest, is $(+, \sqcap), (\cdot), (^-, ^*, \neg)$.

It is immediate from the definition that $t \leq 1$ for any test $t \in T$. The meet of two tests $t, u \in T$ is their product $t \cdot u$. Note that every KA can be made into a KA with tests, by taking $\{0, 1\}$ as the set of tests.

Models of KAs include the following:

1. Algebras of languages: $(2^{\Sigma^*}, \cup, \bullet, ^*, \emptyset, \{\epsilon\})$, where Σ is an alphabet, Σ^* is the set of all finite sequences over Σ, $\bullet$ denotes concatenation, extended pointwise from sequences to sets of sequences, * is the union of iterated concatenations, and ϵ is the empty sequence. The unique set of tests is $\{\emptyset, \{\epsilon\}\}$.
2. Algebras of path sets in a directed graph [20]: $(2^{\Sigma^*}, \cup, \bowtie, ^*, \emptyset, \Sigma \cup \{\epsilon\})$, where Σ is a set of labels (of vertices) and $\bowtie$ denotes concatenation, extended pointwise from paths to sets of paths. Path concatenation is defined as $\epsilon \bowtie \epsilon = \epsilon$, $sa \bowtie at = sat$, for all $a \in \Sigma$ and all paths s, t, and is undefined otherwise. The * operator is again the union of iterated concatenations. The largest possible set of tests is $2^{\Sigma \cup \{\epsilon\}}$, i.e., the set of all subidentities.
3. Algebras of relations over a set S: $(2^{S \times S}, \cup, ;, ^*, \emptyset, I)$, where ; is relational composition, * is reflexive-transitive closure and I is the identity relation. The largest possible set of tests is 2^I, i.e., the set of all subidentities.
4. Abstract relation algebras with transitive closure [21, 22]: $(A, +, ;, ^-, ^\smile, ^*, \mathbb{I})$, where the listed operations are join, composition, complementation, converse, transitive closure and identity relation, in this order. The largest possible set of tests is the set of all subidentities (relations below $\mathbb{I}$).

3 Matrices Over a Kleene Algebra

A (finite) matrix over a KA $(K, +, \cdot, ^*, 0, 1)$ is a function

$$\mathbf{A}_{mn} : \{1, \ldots, m\} \times \{1, \ldots, n\} \to K \ ,$$

where $m, n \in \mathbb{N}$. When no confusion arises, we simply write $\mathbf{A}$ instead of $\mathbf{A}_{mn}$.

We use the following notation for matrices.

$$\mathbf{0} : \text{matrix whose entries are all 0, i.e., } \mathbf{0}[i,j] = 0,$$

$$\mathbf{1} : \text{identity matrix (square), i.e., } \mathbf{1}[i,j] = \begin{cases} 1 & \text{if } i = j \\ 0 & \text{if } i \neq j, \end{cases}$$

$$\mathsf{T} : \text{matrix whose entries are all } \top, \text{ i.e., } \mathsf{T}[i,j] = \top$$
(if K has a greatest element $\top$).

The sum $\mathbf{A} + \mathbf{B}$, product $\mathbf{A} \cdot \mathbf{B}$ and comparison $\mathbf{A} \leq \mathbf{B}$ are defined in the standard fashion, provided that the usual constraints on the size of matrices hold:

$$\begin{aligned} (\mathbf{A}+\mathbf{B})[i,j] &\stackrel{\text{def}}{=} \mathbf{A}[i,j] + \mathbf{B}[i,j], \\ (\mathbf{A}\cdot\mathbf{B})[i,j] &\stackrel{\text{def}}{=} \textstyle\sum(k \mid: \mathbf{A}[i,k]\cdot\mathbf{B}[k,j]), \\ \mathbf{A} \leq \mathbf{B} &\stackrel{\text{def}}{\Leftrightarrow} \mathbf{A}+\mathbf{B} = \mathbf{B} \Leftrightarrow \forall(i,j \mid: \mathbf{A}[i,j] \leq \mathbf{B}[i,j]). \end{aligned} \tag{1}$$

The Kleene star of a square matrix is defined recursively. If $\mathbf{A} = (\,a\,)$, for some $a \in K$, then $\mathbf{A}^* \stackrel{\text{def}}{=} (\,a^*\,)$. If

$$\mathbf{A} = \begin{pmatrix} a & b \\ c & d \end{pmatrix}$$ (with graphic representation 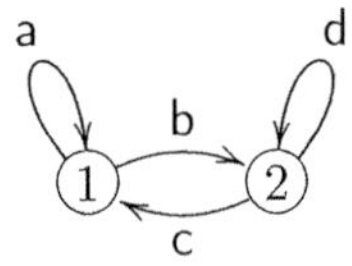

),

for some $a, b, c, d \in K$, then

$$\mathbf{A}^* \stackrel{\text{def}}{=} \begin{pmatrix} f^* & f^*\cdot b\cdot d^* \\ d^*\cdot c\cdot f^* & d^* + d^*\cdot c\cdot f^*\cdot b\cdot d^* \end{pmatrix}, \tag{2}$$

where $f = a + b\cdot d^*\cdot c$; the automaton corresponding to $\mathbf{A}$ helps understand that f corresponds to paths from state 1 to state 1. If $\mathbf{A}$ is a larger matrix, it is decomposed as a 2×2 matrix of submatrices $\mathbf{A} = \begin{pmatrix} \mathbf{B} & \mathbf{C} \\ \mathbf{D} & \mathbf{E} \end{pmatrix}$, where $\mathbf{B}$ and $\mathbf{E}$ are square. Then $\mathbf{A}^*$ is calculated recursively using (2).

Let $\mathcal{M}(\mathcal{K}, m, n)$ be the set of matrices of size $m \times n$ over a KA $\mathcal{K}$. Using the operations defined above, it can be shown that for all n,

$$(\mathcal{M}(\mathcal{K}, n, n), +, \cdot, {}^*, \mathbf{0}_{nn}, \mathbf{1}_{nn})$$

is a KA. See [13] for the details. By setting up an appropriate type discipline, one can define *heterogeneous Kleene algebras* as is done for heterogeneous relation algebras [15, 24]. The set of matrices $\mathcal{M}(\mathcal{K}, m, n)$, for $m, n \in \mathbb{N}$, is such a heterogeneous KA.

Now assume a KA with tests $(K, T, +, \cdot, {}^*, 0, 1, \neg)$ is given. We call *matrix relations* (*relations* for short) those matrices $\mathbf{R}$ whose entries are tests, i.e., $\mathbf{R}[i,j] \in T$ for all i, j. Let $\mathbf{Q}$ and $\mathbf{R}$ be relations. We define the (relational) *converse, meet, top* and *complementation* operations as follows:

$$\begin{array}{lll}(\mathbf{Q}^{\smile})[i,j] \overset{\text{def}}{=} \mathbf{Q}[j,i] & \text{converse (which is the transpose),} & \\ (\mathbf{Q} \sqcap \mathbf{R})[i,j] \overset{\text{def}}{=} \mathbf{Q}[i,j] \cdot \mathbf{R}[i,j] & \text{meet,} & \\ \mathbb{T}[i,j] \overset{\text{def}}{=} 1 & \text{relational top,} & (3) \\ \overline{\mathbf{Q}}^{\mathbb{T}}[i,j] \overset{\text{def}}{=} \neg\mathbf{Q}[i,j] & \text{relational complement.} & \end{array}$$

Again, note that these definitions also apply to nonsquare matrices. In particular, there is a relational top $\mathbb{T}_{mn}$ for every m, n.

A (square) matrix $\mathbf{T}$ is a test if $\mathbf{T} \leq \mathbf{1}$. For instance, if t_1, t_2, t_3 are tests,

$$\begin{pmatrix} t_1 & 0 & 0 \\ 0 & t_2 & 0 \\ 0 & 0 & t_3 \end{pmatrix} \text{ is a test and } \neg \begin{pmatrix} t_1 & 0 & 0 \\ 0 & t_2 & 0 \\ 0 & 0 & t_3 \end{pmatrix} = \begin{pmatrix} \neg t_1 & 0 & 0 \\ 0 & \neg t_2 & 0 \\ 0 & 0 & \neg t_3 \end{pmatrix} .$$

Let $\mathcal{MR}(\mathcal{K}, n, n)$ be the set of (matrix) relations of size $n \times n$ over the KA with tests $\mathcal{K} = (K, T, +, \cdot, ^*, 0, 1, \neg)$. It is straightforward to verify that

$$(\mathcal{MR}(\mathcal{K}, n, n), +, \sqcap, ^{-\mathbb{T}}, \cdot, ^{\smile}, \mathbf{0}_{nn}, \mathbf{1}_{nn}, \mathbb{T}_{nn})$$

is a relation algebra [2, 23, 25]. In particular, it satisfies the Dedekind rule

$$\mathbf{P} \cdot \mathbf{Q} \sqcap \mathbf{R} \leq (\mathbf{R} \cdot \mathbf{Q}^{\smile} \sqcap \mathbf{P}) \cdot (\mathbf{P}^{\smile} \cdot \mathbf{R} \sqcap \mathbf{Q})$$

and the Schröder equivalences

$$\mathbf{P} \cdot \mathbf{Q} \sqcap \mathbf{R} \leq \mathbf{0} \Leftrightarrow \mathbf{P}^{\smile} \cdot \mathbf{R} \sqcap \mathbf{Q} \leq \mathbf{0} \Leftrightarrow \mathbf{R} \cdot \mathbf{Q}^{\smile} \sqcap \mathbf{P} \leq \mathbf{0} .$$

We say that $\mathcal{M}(\mathcal{K}, n, n)$ is a *KA with relations* $\mathcal{MR}(\mathcal{K}, n, n)$.

In $\mathcal{M}(\mathcal{K}, n, n)$, more general variants of the above laws hold: for arbitrary matrices $\mathbf{A}$ and $\mathbf{B}$ and an arbitrary relation $\mathbf{R}$,

$$\begin{array}{lll} \text{(a)} & \mathbf{R} \cdot \mathbf{A} \sqcap \mathbf{B} \leq \mathbf{R} \cdot (\mathbf{R}^{\smile} \cdot \mathbf{B} \sqcap \mathbf{A}) , & \\ \text{(b)} & \mathbf{A} \cdot \mathbf{R} \sqcap \mathbf{B} \leq (\mathbf{B} \cdot \mathbf{R}^{\smile} \sqcap \mathbf{A}) \cdot \mathbf{R} , & \\ \text{(c)} & \mathbf{R} \cdot \mathbf{A} \sqcap \mathbf{B} \leq \mathbf{0} \Leftrightarrow \mathbf{R}^{\smile} \cdot \mathbf{B} \sqcap \mathbf{A} \leq \mathbf{0} , & (4) \\ \text{(d)} & \mathbf{A} \cdot \mathbf{R} \sqcap \mathbf{B} \leq \mathbf{0} \Leftrightarrow \mathbf{B} \cdot \mathbf{R}^{\smile} \sqcap \mathbf{A} \leq \mathbf{0} . & \end{array}$$

We show only part (a). The proof of (b) is similar to that of (a) and (c,d) easily follow from (a,b).

$$\begin{array}{ll} & (\mathbf{R} \cdot \mathbf{A} \sqcap \mathbf{B})[i,j] \\ = & (\mathbf{R} \cdot \mathbf{A})[i,j] \sqcap \mathbf{B}[i,j] \\ = & \sum(k \mid: \mathbf{R}[i,k] \cdot \mathbf{A}[k,j]) \sqcap \mathbf{B}[i,j] \\ = & \quad \langle\ k \text{ is not free in “}\mathbf{B}[i,j]\text{”}\ \rangle \\ & \sum(k \mid: \mathbf{R}[i,k] \cdot \mathbf{A}[k,j] \sqcap \mathbf{B}[i,j]) \\ = & \quad \langle\ \mathbf{R}[i,k] \text{ is a test because } \mathbf{R} \text{ is a matrix relation } \& \\ & \quad \text{In a Boolean KA, for any test } t,\ t \cdot a \sqcap b = t \cdot (a \sqcap t \cdot b)\ [6]\ \rangle \end{array}$$

$$
\begin{aligned}
&\textstyle\sum(k \mid: \mathbf{R}[i,k]\cdot(\mathbf{A}[k,j] \sqcap \mathbf{R}[i,k]\cdot\mathbf{B}[i,j]) \\
=\;& \textstyle\sum(k \mid: \mathbf{R}[i,k]\cdot(\mathbf{A}[k,j] \sqcap \mathbf{R}^{\smile}[k,i]\cdot\mathbf{B}[i,j]) \\
\leq\;& \textstyle\sum(k \mid: \mathbf{R}[i,k]\cdot(\mathbf{A}[k,j] \sqcap \sum(l \mid: \mathbf{R}^{\smile}[k,l]\cdot\mathbf{B}[l,j])) \\
=\;& \textstyle\sum(k \mid: \mathbf{R}[i,k]\cdot(\mathbf{A}[k,j] \sqcap (\mathbf{R}^{\smile}\cdot\mathbf{B})[k,j])) \\
=\;& \textstyle\sum(k \mid: \mathbf{R}[i,k]\cdot(\mathbf{A} \sqcap \mathbf{R}^{\smile}\cdot\mathbf{B})[k,j]) \\
=\;& (\mathbf{R}\cdot(\mathbf{A} \sqcap \mathbf{R}^{\smile}\cdot\mathbf{B}))[i,j] \\
=\;& (\mathbf{R}\cdot(\mathbf{R}^{\smile}\cdot\mathbf{B} \sqcap \mathbf{A}))[i,j]
\end{aligned}
$$

4 Kleene Algebra with Relations

We are now ready to abstract from the concrete setting of matrices and define the concept of Kleene algebra with relations.

Definition 2. *A* Kleene algebra with relations (KAR) *is a two-sorted algebra*

$$(K, R, +, \cdot, {}^*, 0, 1, \sqcap, {}^{-\mathbb{T}}, {}^{\smile}, \mathbb{T})$$

such that

$$(K, +, \cdot, {}^*, 0, 1)$$

is a Kleene algebra and

$$(R, +, \sqcap, \cdot, {}^{-\mathbb{T}}, {}^{\smile}, 0, 1, \mathbb{T})$$

is a relation algebra, where $R \subseteq K$, $\sqcap$ is a binary operator defined at least on R, ${}^{\smile}$ is a unary operator defined at least on R, ${}^{-\mathbb{T}}$ is a unary operator defined only on R, and $\mathbb{T} \in K$.

In the sequel, we let $a, b, c, \ldots$ stand for elements of K, and $p, q, r, \phi, \pi, \sigma$ stand for elements of R.

Note that in a Boolean KAR, $\overline{r}^{\mathbb{T}} = \overline{r} \sqcap \mathbb{T}$.

The relation algebra of a KAR inherits the Kleene star operation from the KA and is thus a relation algebra with transitive closure [21]. Using the axioms of a KA (Definition 1), one can prove that $r^{*\smile} = r^{\smile *}$ (see [21]).

Let a KAR $(K, R, +, \cdot, {}^*, 0, 1, \sqcap, {}^{-\mathbb{T}}, {}^{\smile}, \mathbb{T})$ be given. We now present examples of "interactions" between relations in R and arbitrary elements in K.

We recall that a relation $r \in R$ is *functional* (or *deterministic*, or *univalent*) iff $r^{\smile}\cdot r \leq 1$ (equivalently, $r\cdot\overline{1} \leq \overline{r}$) [2, 23]. It is *total* iff $1 \leq r\cdot r^{\smile}$ (equivalently, $r\cdot\mathbb{T} = \mathbb{T}$). A *mapping* is a total functional relation. A mapping r is *bijective* iff $r^{\smile}$ is also a mapping.

In a relational setting, functional relations satisfy additional laws, such as left-distributivity over meets. We have a similar situation here for Boolean KARs.

Proposition 1. *Let $(K, R, +, \cdot, {}^*, 0, 1, \sqcap, {}^{-\mathbb{T}}, {}^{\smile}, \mathbb{T})$ be a Boolean KAR. Then, for all $a, b \in K$ and $r \in R$,*

1. r *functional* $\Rightarrow r\cdot(a \sqcap b) = r\cdot a \;\sqcap r\cdot b$,
2. r *functional* $\Rightarrow r\cdot\overline{a} \leq \overline{r\cdot a}$,
3. r *total* $\Rightarrow \overline{r\cdot a} \leq r\cdot\overline{a}$.

Proof. 1. Assume $r^{\smile}\cdot r \leq 1$.

$$
\begin{array}{ll}
 & r\cdot(a \sqcap b) \\
\leq & \quad \langle \text{ Monotonicity of composition } \rangle \\
 & r\cdot a \;\sqcap r\cdot b \\
= & \quad \langle \text{ For any relation } r,\ r = (1 \sqcap r\cdot r^{\smile})\cdot r \ \ [23] \ \rangle \\
 & (1 \sqcap r\cdot r^{\smile})\cdot r\cdot a \;\sqcap\; (1 \sqcap r\cdot r^{\smile})\cdot r\cdot b \\
= & \quad \langle \text{ In a Boolean KA, } t \leq 1 \Rightarrow t\cdot(a \sqcap b) = t\cdot a \;\sqcap t\cdot b \ \ [6] \ \rangle \\
 & (1 \sqcap r\cdot r^{\smile})\cdot(r\cdot a \;\sqcap\; r\cdot b) \\
\leq & \quad \langle \text{ Monotonicity of composition } \rangle \\
 & r\cdot r^{\smile}\cdot(r\cdot a \;\sqcap\; r\cdot b) \\
\leq & \quad \langle \text{ Monotonicity of composition } \rangle \\
 & r\cdot(r^{\smile}\cdot r\cdot a \;\sqcap\; r^{\smile}\cdot r\cdot b) \\
\leq & \quad \langle \text{ Hypothesis } r^{\smile}\cdot r \leq 1 \text{ and monotonicity of composition } \rangle \\
 & r\cdot(a \sqcap b)
\end{array}
$$

2. It is shown in [6] that $\forall(a, b \mid : c\cdot(a \sqcap b) = c\cdot a \;\sqcap c\cdot b)$ and $\forall(a \mid : c\cdot\overline{a} \leq \overline{c\cdot a})$ are equivalent even when c is an arbitrary element of K. Thus the result follows from item 1.
3. Assume $r\cdot \mathbb{T} = \mathbb{T}$.

$$
\begin{array}{ll}
 & \overline{r\cdot a} \leq r\cdot\overline{a} \\
\Leftrightarrow & \quad \langle \text{ Shunting } \rangle \\
 & \top \leq r\cdot a \;+ r\cdot\overline{a} \\
\Leftrightarrow & \quad \langle \text{ Distributivity } \rangle \\
 & \top \leq r\cdot(a + \overline{a}) \\
\Leftrightarrow & \quad \langle \text{ Boolean law } \rangle \\
 & \top \leq r\cdot\top \\
\Leftrightarrow & \quad \langle\ \mathbb{T}\cdot\top = \top \text{ (follows from } 1 \leq \mathbb{T}) \ \rangle \\
 & \top \leq r\cdot\mathbb{T}\cdot\top \\
\Leftrightarrow & \quad \langle \text{ Hypothesis } r\cdot\mathbb{T} = \mathbb{T} \ \rangle \\
 & \top \leq \mathbb{T}\cdot\top \qquad \text{—This holds, since } \mathbb{T}\cdot\top = \top
\end{array}
$$

□

The result in Proposition 1(1) is quite interesting. The constraint that r is functional can be written as $r\cdot\overline{1} \leq \overline{r}$. This expression does not involve converse and can thus be used to define what it means for any element $a \in K$ to be deterministic in a certain sense ($a \cdot \overline{1} \leq \overline{a}$). It is shown in [6] that this condition is not sufficient to ensure left distributivity of a over meets. Thus, relations are special.

A relation ϕ is a *homomorphism* from a to b iff ϕ is a mapping and $\phi^{\smallsmile} \cdot a \cdot \phi \leq b$. A relation ϕ is an *isomorphism* between a and b iff ϕ is a homomorphism from a to b and $\phi^{\smallsmile}$ is a homomorphism from b to a, which is equivalent to saying that ϕ is a bijective mapping and $\phi^{\smallsmile} \cdot a \cdot \phi = b$. It is easy to see that if ϕ is a mapping,

$$\phi^{\smallsmile} \cdot a \cdot \phi \leq b \;\Leftrightarrow\; a \cdot \phi \leq \phi \cdot b \;\Leftrightarrow\; a \leq \phi \cdot b \cdot \phi^{\smallsmile} \;\Leftrightarrow\; \phi^{\smallsmile} \cdot a \leq b \cdot \phi^{\smallsmile} .$$

And if ϕ is a bijective mapping, then

$$\phi^{\smallsmile} \cdot a \cdot \phi = b \;\Leftrightarrow\; a \cdot \phi = \phi \cdot b \;\Leftrightarrow\; a = \phi \cdot b \cdot \phi^{\smallsmile} \;\Leftrightarrow\; \phi^{\smallsmile} \cdot a = b \cdot \phi^{\smallsmile} .$$

Thus, the formulae are as in a pure relational setting [23], but apply to a wider range of models. Note, e.g., that matrices over a KA can be used to represent the transition structure of automata [4, 13] or, more generally, transition systems with relations labeling the transitions. For instance,

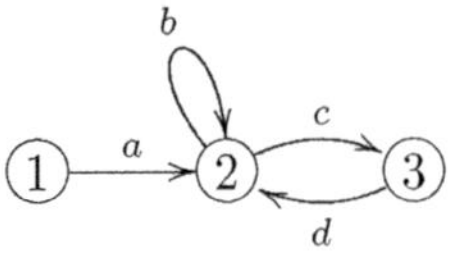

is represented by $\begin{pmatrix} 0 & a & 0 \\ 0 & b & c \\ 0 & d & 0 \end{pmatrix}$,

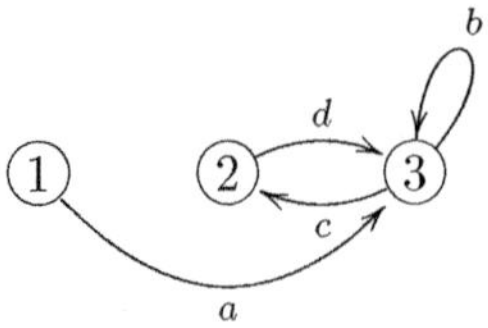

is represented by $\begin{pmatrix} 0 & 0 & a \\ 0 & 0 & d \\ 0 & c & b \end{pmatrix}$,

and

$$\begin{pmatrix} 1 & 0 & 0 \\ 0 & 0 & 1 \\ 0 & 1 & 0 \end{pmatrix} \text{ is an isomorphism between } \begin{pmatrix} 0 & a & 0 \\ 0 & b & c \\ 0 & d & 0 \end{pmatrix} \text{ and } \begin{pmatrix} 0 & 0 & a \\ 0 & 0 & d \\ 0 & c & b \end{pmatrix}.$$

Hence we have a means to describe homomorphisms and isomorphisms between structures *within the same calculus* of Kleene algebra that is used to describe the structures, rather than by external functions.

Other relationships that can be described within the calculus are those of simulation and bisimulation [8, 19]. We say that a relation σ is a *bisimulation* between a and b iff

$$\sigma^{\smallsmile} \cdot a \leq b \cdot \sigma^{\smallsmile} \quad \text{and} \quad \sigma \cdot b \leq a \cdot \sigma \tag{5}$$

(the diagram 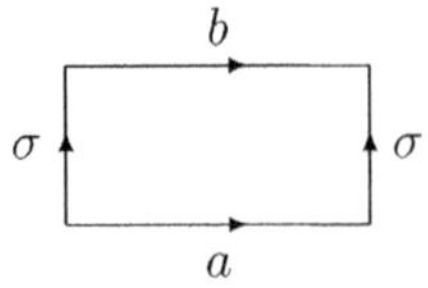

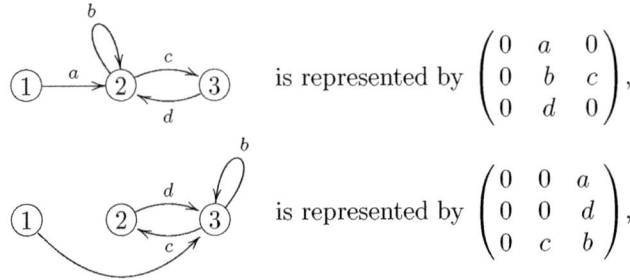

shows how the elements are connected).

Note that this is a standard definition of bisimulation when a, b and σ are relations [7, 8]. The interest here is that it applies to a more general setting.

Since the join of bisimulations is again a bisimulation, there is a largest bisimulation (assuming that arbitrary sums exist in the underlying KA). For instance, consider the following matrices and the graphs (trees) associated to **A** and **B**.

$$\mathbf{A} = \begin{pmatrix} 0 & a & 0 & 0 \\ 0 & 0 & b & c \\ 0 & 0 & 0 & 0 \\ 0 & 0 & 0 & 0 \end{pmatrix}, \quad \mathbf{B} = \begin{pmatrix} 0 & a & a & 0 & 0 \\ 0 & 0 & 0 & b & 0 \\ 0 & 0 & 0 & 0 & c \\ 0 & 0 & 0 & 0 & 0 \\ 0 & 0 & 0 & 0 & 0 \end{pmatrix}, \quad \mathbf{S} = \begin{pmatrix} 0 & 0 & 0 & 0 & 0 \\ 0 & 0 & 0 & 0 & 0 \\ 0 & 0 & 0 & 1 & 1 \\ 0 & 0 & 0 & 1 & 1 \end{pmatrix} .$$

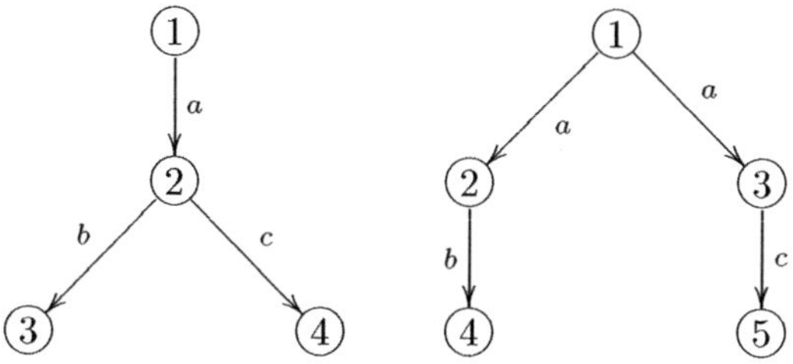

It is a simple task to check that **0** and **S** are bisimulations, no matter what the interpretation of a, b, c is. For instance, if the entries of the matrices come from an algebra of languages over an alphabet $\{\mathsf{a}, \mathsf{b}, \mathsf{c}\}$, we could have

$$a := \{\mathsf{a}\}, \quad b := \{\mathsf{b}\}, \quad c := \{\mathsf{c}\} .$$

In this case, **S** is the largest bisimulation. It shows that the leaves of the trees are bisimilar and that the roots are not (this is the prototypical example of systems that are not bisimilar [18, 19], because $\mathbf{S}[1,1] = 0$).

In an algebra of paths, matrix **S** is still a bisimulation, but it might be possible to find a larger one, because the set of tests is richer than for languages. For instance, with the alphabet $\{\mathsf{a}, \mathsf{b}, \mathsf{c}, \mathsf{d}, \mathsf{e}, \mathsf{f}\}$, and the interpretation

$$a := \{\mathsf{ab}, \mathsf{abc}, \mathsf{bd}\}, \quad b := \{\mathsf{bc}, \mathsf{de}\}, \quad c := \{\mathsf{cd}, \mathsf{df}\} ,$$

one finds that

$$\mathbf{S} = \begin{pmatrix} \{\epsilon, \mathsf{c}, \mathsf{d}, \mathsf{e}, \mathsf{f}\} & 0 & 0 & 0 & 0 \\ 0 & \{\epsilon, \mathsf{a}, \mathsf{b}, \mathsf{e}, \mathsf{f}\} & \{\epsilon, \mathsf{a}, \mathsf{c}, \mathsf{e}, \mathsf{f}\} & 0 & 0 \\ 0 & 0 & 0 & 1 & 1 \\ 0 & 0 & 0 & 1 & 1 \end{pmatrix}$$

is the largest bisimulation.

We now make additional assumptions that will allow us to show how a largest bisimulation can be extracted from (5).

1. We assume a Boolean KAR.
2. We assume (4c), which holds in an algebra of matrices over a Boolean KA with tests, in the form

$$r\cdot a \le b \Leftrightarrow r^{\smile}\cdot\overline{b} \le \overline{a} \ ,$$

where r is a relation and $a, b \in K$ (this does not hold in arbitrary KARs, as shown below). This allows us to rewrite (5) as

$$\sigma\cdot\overline{b\cdot\sigma^{\smile}} \le \overline{a} \quad \text{and} \quad \sigma\cdot b \le a\cdot\sigma \ . \tag{6}$$

3. We assume a complete KAR. This ensures that a left residual operator / can be defined by the Galois connection $a\cdot b \le c \Leftrightarrow a \le c/b$ [1]. We can thus rewrite (6) as

$$\sigma \le \overline{a}/\overline{b\cdot\sigma^{\smile}} \quad \text{and} \quad \sigma \le (a\cdot\sigma)/b \ ,$$

from which we get

$$\sigma \le \overline{a}/\overline{b\cdot\sigma^{\smile}} \sqcap (a\cdot\sigma)/b \ .$$

The function $f(\sigma) \stackrel{\text{def}}{=} \overline{a}/\overline{b\cdot\sigma^{\smile}} \sqcap (a\cdot\sigma)/b$ is monotonic. Due to completeness, a largest solution for σ exists. However, it need not be a relation.
4. We assume that for any $a \in K$, $a \sqcap \mathbb{T} \in R$ (i.e., $a \sqcap \mathbb{T}$ is a relation). With this assumption, we get the largest relation that is a bisimulation as the largest solution of

$$\sigma \le \overline{a}/\overline{b\cdot\sigma^{\smile}} \sqcap (a\cdot\sigma)/b \sqcap \mathbb{T} \ .$$

5 Additional Axioms and Representability

The treatment of bisimulations in the previous section required the introduction of axioms in addition to those provided by Definition 2. So, the question arises as to what is the most suitable set of axioms for Kleene algebra with relations. With a specific intended model in mind, one can be guided in this choice. Here, however, the starting point is that of matrices over arbitrary KAs, and various KAs can be useful, depending on the context. For instance, when describing programs as transition systems using matrices over a KA, the desired degree of precision dictates the type of the underlying KA. If high precision is required, the entries of the matrices are chosen to be relations on the set of states of the program. If a more abstract view is desired, the entries of the matrices can be simple labels naming actions done by the program and the KA is that of languages over these labels.

For many applications in Computer Science, matrices over a Boolean KA are needed (this is the case for the two examples in the previous paragraph). As already noted, these satisfy a form of Schröder equivalences (see (4) above), so that it becomes natural to require

$$r\cdot a \sqcap b \le 0 \Leftrightarrow r^{\smile}\cdot b \sqcap a \le 0 \quad \text{and} \quad a\cdot r \sqcap b \le 0 \Leftrightarrow b\cdot r^{\smile} \sqcap a \le 0 \tag{7}$$

for Boolean KAR. These equivalences do not follow from the definition of a KAR (Definition 2), even when it is Boolean. This is shown by the following example, due to Peter Jipsen.

Let the sets of atoms of K and R be $\{1, \mathsf{r}, \mathsf{a}\}$ and $\{1, \mathsf{r}\}$, respectively. Composition on the atoms of K is defined by the following table. Note that $\mathbb{T} = 1 + \mathsf{r}$ and $\top = 1 + \mathsf{r} + \mathsf{a} = \mathbb{T} + \mathsf{a}$.

$\cdot$	1	r	a
1	1	r	a
r	r	$\mathbb{T}$	$\top$
a	a	$\top$	$\top$

Composition on K and R is obtained by distributivity using this table. The converse operation on R is given by $x^{\smile} = x$, and the Kleene star on K is defined by $0^* = 1$ and $x^* = x^2$ for $x \neq 0$. One can check that with these operations, R is a relation algebra and K is a Boolean KAR. Now,

$$\mathsf{r}^{\smile} \cdot \mathsf{r} \sqcap \mathsf{a} = \mathsf{r} \cdot \mathsf{r} \sqcap \mathsf{a} = \mathbb{T} \sqcap \mathsf{a} \leq 0 \ ,$$

but

$$\mathsf{r} \cdot \mathsf{a} \sqcap \mathsf{r} = \top \sqcap \mathsf{r} = \mathsf{r} \neq 0 \ .$$

Note the following consequence of (7):

$$\mathbb{T} \cdot \overline{\mathbb{T}} \sqcap \mathbb{T} \leq 0 \Leftrightarrow \mathbb{T}^{\smile} \cdot \mathbb{T} \sqcap \overline{\mathbb{T}} \leq 0 \Leftrightarrow \mathbb{T} \sqcap \overline{\mathbb{T}} \leq 0 \Leftrightarrow \mathsf{true} \ .$$

The expression $\mathbb{T} \cdot \overline{\mathbb{T}} \sqcap \mathbb{T}$ is the relational part of the composition $\mathbb{T} \cdot \overline{\mathbb{T}}$. The above result means that the composition of a relation with an element that contains no relational part does not contain any (nonzero) relational part. This is violated in Jipsen's example, since $\mathsf{r} \cdot \mathsf{a} = \top = \mathbb{T} + \mathsf{a}$.

The determination of the intended model is also important in connection with questions about representability, where the goal is to determine whether any algebra satisfying a given set of axioms is isomorphic to a concrete instance of the intended model. As indicated at the beginning of this section, there is no single concrete intended model, since many models may be useful. However, let us say that a KAR $(K, R, +, \cdot, {}^*, 0, 1, \sqcap, {}^{-\mathbb{T}}, {}^{\smile}, \mathbb{T})$ is *representable relatively to a given KA with tests* $(K', T, +, \cdot, {}^*, 0, 1, \neg)$ iff

1. $(K, +, \cdot, {}^*, 0, 1)$ is isomorphic to the set of square matrices of a fixed (finite or infinite[1]) cardinality over K' with the corresponding Kleene operations, and
2. $(R, +, \sqcap, \cdot, {}^{-\mathbb{T}}, {}^{\smile}, 0, 1, \mathbb{T})$ is isomorphic to the subset of these matrices that are matrices over T with the corresponding relational operations.

One can then investigate whether a given set of axioms ensures relative representability with respect to a given KA with tests; this is a topic for future research.

[1] Dealing with infinite matrices is outside the scope of this paper, but this can be done under suitable restrictions (see [16]).

6 Projections and Unsharpness

Projections constitute another example of the use of relations inside a KA. We again assume that the KA is Boolean.

Definition 3. *A pair of relations* (π_1, π_2) *is called a* direct product *iff*

$$\pi_1^{\smile}\cdot\pi_1 = 1, \qquad \pi_2^{\smile}\cdot\pi_2 = 1, \qquad \pi_1\cdot\pi_1^{\smile} \sqcap \pi_2\cdot\pi_2^{\smile} = 1, \qquad \pi_1^{\smile}\cdot\pi_2 = \mathbb{T} \ .$$

The product *of* a_1 *and* a_2 *is* $a_1 \times a_2 \stackrel{\text{def}}{=} \pi_1\cdot a_1\cdot\pi_1^{\smile} \sqcap \pi_2\cdot a_2\cdot\pi_2^{\smile}$. *The relations* π_1 *and* π_2 *are called* projections.

This is the standard definition of projections in a heterogenous setting [23]. However, note that $\mathbb{T}$ need not be the largest element of the algebra, which is $\top$.

Consider the following matrices.

$$\mathbf{P}_1 = \begin{pmatrix} 1 & 0 \\ 1 & 0 \\ 1 & 0 \\ 0 & 1 \\ 0 & 1 \\ 0 & 1 \end{pmatrix} \quad \mathbf{P}_2 = \begin{pmatrix} 1 & 0 & 0 \\ 0 & 1 & 0 \\ 0 & 0 & 1 \\ 1 & 0 & 0 \\ 0 & 1 & 0 \\ 0 & 0 & 1 \end{pmatrix} \quad \mathbf{A}_1 = \begin{pmatrix} a & b \\ c & d \end{pmatrix} \quad \mathbf{A}_2 = \begin{pmatrix} e & f & g \\ h & i & j \\ k & l & n \end{pmatrix}$$

The relations $\mathbf{P}_1$ and $\mathbf{P}_2$ constitute a direct product. The product of $\mathbf{A}_1$ and $\mathbf{A}_2$ is easily calculated and corresponds to the *synchronous product* of the automata or transition systems represented by $\mathbf{A}_1$ and $\mathbf{A}_2$.

$$\mathbf{A}_1 \times \mathbf{A}_2 = \begin{pmatrix} a\sqcap e & a\sqcap f & a\sqcap g & b\sqcap e & b\sqcap f & b\sqcap g \\ a\sqcap h & a\sqcap i & a\sqcap j & b\sqcap h & b\sqcap i & b\sqcap j \\ a\sqcap k & a\sqcap l & a\sqcap n & b\sqcap k & b\sqcap l & b\sqcap n \\ c\sqcap e & c\sqcap f & c\sqcap g & d\sqcap e & d\sqcap f & d\sqcap g \\ c\sqcap h & c\sqcap i & c\sqcap j & d\sqcap h & d\sqcap i & d\sqcap j \\ c\sqcap k & c\sqcap l & c\sqcap n & d\sqcap k & d\sqcap l & d\sqcap n \end{pmatrix} .$$

With direct products in the picture, one naturally wonders what happens to the *unsharpness problem* [3] in this setting. The problem consists in determining whether $(q_1\cdot\pi_1^{\smile} \sqcap q_2\cdot\pi_2^{\smile})\cdot(\pi_1\cdot r_1 \sqcap \pi_2\cdot r_2) = q_1\cdot r_1 \sqcap q_2\cdot r_2$ holds for all relations q_1, q_2, r_1, r_2. It does hold for concrete algebras of relations, but it is shown in [17] that it does not in RA. The counterexample is rather complex. However, the special case $(q_1 \times q_2)\cdot(r_1 \times r_2) = q_1\cdot r_1 \times q_2\cdot r_2$ holds in RA[2] for all relations q_1, q_2, r_1, r_2 [5].

With KAR, it is very simple to exhibit an example illustrating that even the special case $(a_1 \times a_2)\cdot(b_1 \times b_2) = a_1\cdot b_1 \times a_2\cdot b_2$ does not hold. Consider a Boolean KAR K with $\{0, 1\}$ as set of relations (note that $\mathbb{T} = 1$). Let $\pi_1 = 1$ and $\pi_2 = 1$. Then (π_1, π_2) is a direct product. For arbitrary $a, b, c \in K$, we have

$$(a \times a\ \cdot b)\cdot(b\cdot c \times c) = (a \sqcap a\ \cdot b)\cdot(b\cdot c \sqcap c) \ ,$$

[2] Composition $(\cdot)$ has precedence over $\times$.

while

$$a \ \cdot (b \cdot c) \times (a \cdot b) \cdot c = a \ \cdot b \cdot c \ .$$

It is easy to find concrete examples where $(a \sqcap a \ \cdot b) \cdot (b \cdot c \sqcap c) = 0$ and $a \cdot b \cdot c \neq 0$.

In [12], Kempf and Winter create unsharpness in a purely relational setting by requiring $\pi_1^{\smile} \cdot \pi_2 = \mathbb{L} \leq \mathbb{T}$, where $\mathbb{L}$ is the greatest tabular relation instead of the $\mathbb{T}$ relation. This is analogous to the situation with KARs, where $\pi_1^{\smile} \cdot \pi_2 = \mathbb{T} \leq \top$, so that $\pi_1^{\smile} \cdot \pi_2$ need not be the $\top$ element.

7 Conclusion

This paper introduces the concept of Kleene algebra with relations, but only presents basic results and simple motivating applications. There is much more to do, both on the use of the concept for applications and on the development of the theory, in particular on the problem of relative representability. As a conclusion, we note that the idea of finding a relation algebra inside another structure is not new. In [10, 11], von Karger and Hoare introduce *sequential algebras*, which are Boolean KAs with additional laws, but not as constrained as relation algebras; in sequential algebras, a (possibly trivial) subset of the elements behave as relations. Although the approach and motivation are completely different from those presented here, it would be interesting to investigate their relationships, in particular with respect to results on representability [9] versus relative representability as defined in Sect. 5.

Acknowledgements

The author thanks Peter Jipsen for an interesting discussion and for the counterexample presented in Sect. 5 above. He also thanks Rudolf Berghammer, Claude Bolduc, Therrezinha Fernandes, Vincent Mathieu and Bernhard Möller for comments that helped improve the paper.

References

[1] Aarts, C. J.: Galois connections presented calculationally. Technical report, Eindhoven University of Technology, Department of Mathematics and Computer Science (1992) 16

[2] Brink, C., Kahl, W., Schmidt, G., eds.: Relational Methods in Computer Science. Springer-Verlag (1997) 11, 12

[3] Cardoso, R.: Untersuchung paralleler Programme mit relationenalgebraischen Methoden. Diplomarbeit, Institut für Informatik, Technische Universität München (1982) 18

[4] Conway, J. H.: Regular Algebra and Finite Machines. Chapman and Hall, London (1971) 8, 14

[5] Desharnais, J.: Monomorphic characterization of n-ary direct products. Information Sciences – An International Journal **119** (1999) 275–288 18

[6] Desharnais, J., Möller, B.: Characterizing determinacy in Kleene algebras. Information Sciences **139** (2001) 253–273 8, 11, 13, 14
[7] de Roever, W.-P., Engelhardt, K.: Data Refinement: Model-Oriented Proof Methods and their Comparison. Series Cambridge Tracts in Theoretical Computer Science, Cambridge University Press (1998) 14
[8] Hoare, C. A. R., Jifeng, H., Sanders, J. W.: Prespecification in data refinement. Information Processing Letters **25** (1987) 71–76 14
[9] Jipsen, P., Maddux, R.: Nonrepresentable sequential algebras. Logic Journal of the IGPL **5** (1997) 565–574 19
[10] von Karger, B., Hoare, C. A. R.: Sequential calculus. Information Processing Letters **53** (1995) 123–130 19
[11] von Karger, B.: Sequential calculus. Technical Report ProCos II: [Kiel BvK 15/11], Christian-Albrechts Universität zu Kiel (1995) 19
[12] Kempf, P., Winter, M.: Relational unsharpness and processes. In Berghammer, R., Möller, B., eds.: Participant's proceedings of the 7th International Seminar on Relational Methods in Computer Science, in combination with 2nd International Workshop on Applications of Kleene Algebra, Bad Malente (near Kiel), Germany, Institut für Informatik und Praktische Mathematik, Christian-Albrechts-Universität zu Kiel (2003) 270–276 19
[13] Kozen, D.: A completeness theorem for Kleene algebras and the algebra of regular events. Information and Computation **110** (1994) 366–390 8, 10, 14
[14] Kozen, D.: Kleene algebras with tests. ACM Transactions on Programming Languages and Systems **19** (1997) 427–443 8, 9
[15] Kozen, D.: Typed Kleene algebra. Technical Report 98-1669, Computer Science Department, Cornell University (1998) 10
[16] Kozen, D.: Myhill-Nerode relations on automatic systems and the completeness of Kleene algebra. In Ferreira, A., Reichel, H., eds.: 18th Symp. Theoretical Aspects of Computer Science (STACS'01). Volume 2010 of Lecture Notes in Computer Science., Dresden, Germany, Springer-Verlag (2001) 27–38 17
[17] Maddux, R. D.: On the derivation of identities involving projection functions. Technical report, Department of Mathematics, Iowa State University (1993) 18
[18] Milner, R.: A calculus of communicating systems. Volume 92 of Lecture Notes in Computer Science. Springer-Verlag, Berlin (1980) 15
[19] Milner, R.: Communication and Concurrency. Prentice Hall International Series in Computer Science (1989) 14, 15
[20] Möller, B.: Derivation of graph and pointer algorithms. In Möller, B., Partsch, H. A., Schuman, S. A., eds.: Formal Program Development. Volume 755 of Lecture Notes in Computer Science. Springer-Verlag, Berlin (1993) 123–160 9
[21] Ng, K. C.: Relation algebras with transitive closure. PhD thesis, University of California, Berkeley (1984) 9, 12
[22] Ng, K. C., Tarski, A.: Relation algebras with transitive closure. Abstract 742-02-09, Notices of the American Mathematical Society **24** (1977) 9
[23] Schmidt, G., Ströhlein, T.: Relations and Graphs. EATCS Monographs in Computer Science. Springer-Verlag, Berlin (1993) 11, 12, 13, 14, 18
[24] Schmidt, G., Hattensperger, C., Winter, M.: Heterogeneous relation algebra. In Brink, C., Kahl, W., Schmidt, G., eds.: Relational Methods in Computer Science. Springer-Verlag (1997) 10
[25] Tarski, A.: On the calculus of relations. Journal of Symbolic Logic **6** (1941) 73–89 11

Integrating Model Checking and Theorem Proving for Relational Reasoning

Konstantine Arkoudas, Sarfraz Khurshid, Darko Marinov, and Martin Rinard

MIT Laboratory for Computer Science
200 Technology Square
Cambridge, MA 02139 USA
{arkoudas,khurshid,marinov,rinard}@lcs.mit.edu

Abstract. We present `Prioni`, a tool that integrates model checking and theorem proving for relational reasoning. `Prioni` takes as input formulas written in Alloy, a declarative language based on relations. `Prioni` uses the Alloy Analyzer to check the validity of Alloy formulas for a given scope that bounds the universe of discourse. The Alloy Analyzer can refute a formula if a counterexample exists within the given scope, but cannot prove that the formula holds for all scopes. For proofs, `Prioni` uses Athena, a denotational proof language. `Prioni` translates Alloy formulas into Athena proof obligations and uses the Athena tool for proof discovery and checking.

1 Introduction

`Prioni` is a tool that integrates model checking and theorem proving for relational reasoning. `Prioni` takes as input formulas written in the Alloy language [7]. We chose Alloy because it is an increasingly popular notation for the calculus of relations. Alloy is a first-order, declarative language. It was initially developed for expressing and analyzing high-level designs of software systems. It has been successfully applied to several systems, exposing bugs in Microsoft COM [9] and a naming architecture for dynamic networks [10]. It has also been used for software testing [12], as a basis of an annotation language [11], and for checking code conformance [20]. Alloy is gaining popularity mainly for two reasons: it is based on relations, which makes it easy to write specifications about many systems; and properties of Alloy specifications can be automatically analyzed using the Alloy Analyzer (AA) [8].

`Prioni` leverages AA to model-check Alloy specifications. AA finds *instances* of Alloy specifications, i.e., assignments to relations in a specification that make the specification true. AA requires users to provide only a *scope* that bounds the universe of discourse. AA then automatically translates Alloy specifications into boolean satisfiability formulas and uses off-the-shelf SAT solvers to find satisfying assignments to the formulas. A satisfying assignment to a formula that expresses the negation of a property provides a counterexample that illustrates a violation of the property. AA is restricted to finite refutation: if AA does not find a counterexample within some scope, there is no guarantee that no

R. Berghammer et al. (Eds.): RelMiCS/Kleene-Algebra Ws 2003, LNCS 3051, pp. 21–33, 2004.

counterexample exists in a larger scope. Users can increase their confidence by re-running AA for a larger scope, as long as AA completes its checking in a reasonable amount of time.

It is worth noting that a successful exploration of a finite scope may lead to a false sense of security. There is anecdotal evidence of experienced AA users who developed Alloy specifications, checked them for a certain scope, and believed the specifications to hold when in fact they were false. (In particular, this happened to the second author in his earlier work [10].) In some cases, the fallacy is revealed when AA can handle a larger scope, due to advances in hardware, SAT solver technology, or translation of Alloy specifications. In some cases, the fallacy is revealed by a failed attempt to carefully argue the correctness of the specification, even if the goal is not to produce a formal proof of correctness.

`Prioni` integrates AA with a theorem prover that enables the users to prove that their Alloy specifications hold for all scopes. `Prioni` uses Athena for proof representation, discovery, and checking. Athena is a type-ω denotational proof language [2] for polymorphic multi-sorted first-order logic. We chose Athena for several reasons: 1) It uses a natural-deduction style of reasoning based on *assumption bases* that makes it easier to read and write proofs. 2) It offers a strong soundness guarantee. 3) It has a flexible polymorphic sort system with built-in support for structural induction. 4) It offers a high degree of automation through the use of *methods*, which are akin to the tactics and tacticals of HOL [5] and Isabelle [15]. In addition, Athena offers built-in automatic translations from its own notation to languages such as the TPTP standard [1], and can be seamlessly integrated with any automatic theorem prover that accepts inputs in such a language. The use of such provers allows one to skip many tedious steps, focusing instead on the interesting parts of the proof. In this example we used Otter [21]; more recently we have experimented with Vampire [17].

`Prioni` provides two key technologies that enable the effective use of Athena to prove Alloy specifications. First, `Prioni` provides an axiomatization of the calculus of relations in Athena and a library of commonly used lemmas for this calculus. Since this calculus is the foundation of Alloy, the axiomatization and the lemmas together eliminate much of the formalization burden that normally confronts users of theorem provers. Second, `Prioni` provides an automatic translation from Alloy to the Athena relational calculus. This translation eliminates the coding effort and transcription errors that complicate the direct manual use of theorem provers. Finally, we note that since Athena has a formal semantics, the translation also gives a precise semantics to Alloy.

`Prioni` supports the following usage scenario. The user starts from an Alloy specification, model-checks it and potentially changes it until it holds for as big a scope as AA can handle. After eliminating the most obvious errors in this manner, the user may proceed to prove the specification. This attempt may introduce new proof obligations, such as an inductive step. The user can then again use AA to model-check these new formulas to be proved. This way, model checking aids proof engineering. But proving can also help model checking. Even when the user cannot prove that the whole specification is correct, the user may be able to prove that a part of it is. This can make the specification smaller, and AA can then check the new specification in a larger scope than the original specification. Machine-verifiable proofs of key properties greatly increase our trust in the reliability of the system. An additional benefit of having *readable* formal proofs lies in improved documentation: such proofs not only show that the desired properties hold, but also *why* they hold.

2 Model Checking

We next illustrate the use of our `Prioni` prototype on a recursive function that returns the set of all elements in a list. We establish that the result of the function is the same as a simple relational expression that uses transitive closure. The following Alloy specification introduces lists and the function of interest:

```
module List
sig Object {}
sig Node {
  next: option Node,    // next is a~partial function from Node to Node
  data: Object }        // data is a~total function from Node to Object
det fun elms(n: Node): set Object {
  if (no n.next) then result = n.data
  else result = n.data + elms(n.next) }
assert Equivalence { all n: Node | elms(n) = n.*next.data }
check Equivalence for 5
```

The declaration `module` names the specification. The keyword `sig` introduces a *signature*, i.e., a set of indivisible atoms. Each signature can have *field* declarations that introduce relations. By default, fields are total functions; the modifiers `option` and `set` are used for partial functions and general relations, respectively.

The keyword `fun` introduces an Alloy "*function*", i.e., a parametrized formula that can be invoked elsewhere in the specification. In general, an Alloy function denotes a relation between its arguments and the result; the modifier `det` specifies an actual function. The function `elms` has one argument, `n`. Semantically, all variables in Alloy are relations (i.e., sets). Thus, `n` is not a scalar from the set `Node`; `n` is a singleton subset of `Node`. (A general subset is declared with `set`.) In the function body, `result` refers to the result of the function. The intended meaning of `elms` is to return the set of objects in all nodes reachable from `n`. The operator '`.`' represents relational composition; `n.next` is the set of nodes that the relation `next` maps `n` to. Note that the recursive invocation type-checks even when this set is empty, because the type of `n` is essentially a set of `Nodes`.

The keyword `assert` introduces an *assertion*, i.e., a formula to be checked. The prefix operator '`*`' denotes reflexive transitive closure. The expression `n.*next` denotes the set of all nodes reachable from `n`, and `n.*next.data` denotes the set of objects in these nodes. `Equivalence` states that the result of `elms` is exactly the set of all those objects. The command `check` instructs AA to check this for the given *scope*, in this example for all lists with at most five nodes and five objects. AA produces a counterexample, where a list has a cycle. Operationally, `elms` would not terminate if there is a cycle reachable from its argument. In programming language semantics, the least fixed point is taken as the meaning of a recursive function definition. Since Alloy is a declarative, relational language, AA instead considers all functions that satisfy the recursive definition of `elms`.

We can rule out cyclic lists by adding to the above Alloy specification the following: `fact AllAcyclic { all n: Node | n !in n.^next }`. A *fact* is a formula that is assumed to hold, i.e., AA checks if the assertion follows from the conjunction of all facts in the specification. `AllAcyclic` states that there is no node `n` reachable from itself, i.e., no node `n` is in the set `n.^next`; '^' denotes transitive closure. We again use AA to check `Equivalence`, and this time AA produces no counterexample.

3 Athena Overview

Athena is a type-ω denotational proof language [2] for polymorphic multi-sorted first-order logic. This section presents parts of Athena relevant to understanding the example. In Athena, an arbitrary universe of discourse (sort) is introduced with a `domain` declaration, for example:

```
(domain Real)
(domain Person)
```

Function symbols and constants can then be declared on the domains, e.g.:

```
(declare + (-> (Real Real) Real))
(declare joe Person)
(declare pi Real)
```

Relations are functions whose range is the predefined sort `Boolean`, e.g.,

```
(declare < (-> (Real Real) Boolean))
```

Domains can be polymorphic, e.g.,

```
(domain (Set-Of T))
```

and then function symbols declared on such domains can also be polymorphic:

```
(declare insert ((T) -> (T (Set-Of T)) (Set-Of T)))
```

Note that in the declaration of a polymorphic symbol, the relevant sort parameters are listed within parentheses immediately before the arrow `->`. The equality symbol `=` is a predefined relation symbol with sort `((T) -> (T T) Boolean)`.

Inductively generated domains are introduced as *structures*, e.g.,

```
(structure Nat
  zero
  (succ Nat))
```

Here `Nat` is freely generated by the *constructors* `zero` and `succ`. This is equivalent to issuing the declarations `(domain Nat)`, `(declare zero Nat)`, `(declare succ (-> (Nat) Nat))`, and additionally postulating a number of axioms stating that `Nat` is freely generated by `zero` and `succ`. Those axioms along with an appropriate induction principle are automatically generated when the user defines the structure. In this example, the induction principle will allow for proofs of statements of the form $(\forall n : \texttt{Nat})\, P(n)$ by induction on the structure of the number n:

```
(by-induction-on n (P n)
  (zero D1)
  ((succ k) D2))
```

where `D1` is a proof of `(P zero)`—the basis step—and `D2` is a proof of `(P (succ k))` for some fresh variable `k`—the inductive step. The inductive step `D2` is performed under the assumption that `(P k)` holds, which represents the inductive hypothesis. More precisely, `D2` is evaluated in the assumption base $\beta \cup \{(P\texttt{k})\}$, where β is the assumption base in which the entire inductive proof is being evaluated; more on assumption bases below.

Structures can also be polymorphic, e.g.,

```
(structure (List-Of T)
  nil
  (cons T (List-Of T)))
```

and correspondingly polymorphic free-generation axioms and inductive principles are automatically generated.

The basic data values in Athena are terms and propositions. Terms are s-expressions built from declared function symbols such as `+` and `pi`, and from *variables*, written as ?*I* for any identifier *I*. Thus `?x`, `(+ ?foo pi)`, `(+ (+ ?x ?y) ?z)`, are all terms. The (most general) sort of a term is inferred automatically; the user does not have to annotate variables with their sorts. A proposition P is either a term of sort Boolean (say, `(< pi (+ ?x ?y))`); or an expression of the form `(not` P`)` or `(`$\odot$ P_1 P_2`)` for $\odot \in \{\texttt{and}, \texttt{or}, \texttt{if}, \texttt{iff}\}$; or `(`$Q$ $x_1 \cdots x_n$ P`)` where $Q \in \{\texttt{forall}, \texttt{exists}\}$ and each x_i a variable. Athena also checks the sorts of propositions automatically using a Hindley-Milner-like type inference algorithm.

The user interacts with Athena via a read-eval-print loop. Athena displays a prompt `>`, the user enters some input (either a phrase to be evaluated or a top-level directive such as `define`, `assert`, `declare`, etc.), Athena processes the user's input, displays the result, and the loop starts anew.

The most fundamental concept in Athena is the *assumption base*—a finite set of propositions that are assumed to hold, representing our "axiom set" or "knowledge base". Athena starts out with the empty assumption base, which then gets incrementally augmented with the conclusions of the deductions that the user successfully evaluates at the top level of the read-eval-print loop. A proposition can also be explicitly added into the global assumption base with the top-level directive `assert`. (Note that in Athena the keyword `assert` introduces a formula that is supposed to hold, whereas in Alloy `assert` introduces a formula that is to be checked.)

An Athena deduction D is always evaluated in a given assumption base β. Evaluating D in β will either produce a proposition P (the "conclusion" of D in β), or else it will generate an error or will diverge. If D does produce a conclusion P, Athena's semantics guarantee $\beta \models P$, i.e., that P is a logical consequence of β. There are several syntactic forms that can be used for deductions.

The form `pick-any` introduces universal generalizations: `(pick-any` $I_1 \cdots I_n$ D`)` binds the names $I_1 \cdots I_n$ to fresh variables $v_1, \ldots, v_n$

and evaluates D. If D yields a conclusion P, the result returned by the entire `pick-any` is $(\forall v_1, \ldots, v_n) P$.

The form `assume` introduces conditionals: to evaluate `(assume` P D`)` in an assumption base β, we evaluate D in $\beta \cup \{P\}$. If that produces a conclusion Q, the conditional $P \Rightarrow Q$ is returned as the result of the entire `assume`. The form `(assume-let ((`I P`))` D`)` works like `assume`, but also lexically binds the name I to the hypothesis P within D.

The form `(dlet ((`I_1 D_1`)` $\cdots$ `(`I_n D_n`))` D`)` is used for sequencing and naming deductions. To evaluate such a deduction in β, we first evaluate D_1 in β to obtain a conclusion P_1. We then bind I_1 to P_1, insert P_1 into β, and continue with D_2. The conclusions P_i of the various D_i are thus incrementally added to the assumption base, becoming available as lemmas for subsequent use. The body D is then evaluated in $\beta \cup \{P_1, \ldots, P_n\}$, and its conclusion becomes the conclusion of the entire `dlet`.

`Prioni` starts by adding relational calculus axioms and already proved lemmas to the empty assumption base. It then translates the Alloy specification and adds to the assumption base all translated constraints and definitions. Only the translated Alloy assertion is not added to the assumption base; rather, it constitutes the proof obligation.

4 Axiomatization

We next introduce certain key parts of our axiomatization of the calculus of relations in Athena. The axiomatization represents relations as sets of tuples in a typed first-order finite-set theory. Tuples of binary relations (i.e., ordered pairs) are represented with the following polymorphic Athena structure: `(structure (Pair-Of S T) (pair S T))`. `Prioni` introduces similar structures for tuples of greater length as needed.

Sets are polymorphic, their sort being given by a domain constructor: `(domain (Set-Of S))`, and with the membership relation `in` typed as follows:

```
(declare in ((S) -> (S (Set-Of S)) Boolean))
```

Set equality is captured by an extensionality axiom `set-ext`, and set operations are defined as usual. We also introduce a singleton-forming operator:

```
(declare singleton ((T) -> (T) (Set-Of T)))

(define singleton-def
  (forall ?x ?y (iff (in ?x (singleton ?y)) (= ?x ?y))))
```

Relation operations are defined set-theoretically, e.g.:

```
(declare transpose ((T) -> ((Set-Of (Pair-Of T T))) (Set-Of (Pair-Of T T))))

(define transpose-def
  (forall ?R ?x ?y (iff (in (pair ?x ?y) (transpose ?R))
                        (in (pair ?y ?x) ?R))))

(define pow-def-1
  (forall ?R ?x ?y
```

```
    (iff (in (tup [?x ?y]) (pow ?R zero))
         (= ?x ?y))))

(define pow-def-2
  (forall ?R ?k ?x ?y
    (iff (in (tup [?x ?y]) (pow ?R (succ ?k)))
         (exists ?z
           (and (in [?x ?z] ?R)
                (in [?z ?y] (pow ?R ?k)))))))
```

Alloy has one general composition operator '.' that can be applied to two arbitrary relations at least one of which has arity greater than one. Such a general operator could not be typed precisely in a Hindley-Milner-like type system such as that of Athena, and in any event, the general composition operator has a fairly involved definition that would unduly complicate theorem proving. So what our translation does instead is introduce a small number of specialized composition operators `comp-n-m` that compose relations of types $S_1 \times \cdots \times S_n$ and $T_1 \times \cdots \times T_m$, with $S_n = T_1$. Such operators are typed precisely and have straightforward definitions; for instance:

```
(declare comp-2-2 ((S T U) -> ((Set-Of (Pair-Of S T)) (Set-Of (Pair-Of T U)))
                              (Set-Of (Pair-Of S U))))
(forall ?R1 ?R2 ?x ?y
  (iff (in (pair ?x ?y) (comp-2-2 ?R1 ?R2))
       (exists ?z
         (and (in (pair ?x ?z) ?R1)
              (in (pair ?z ?y) ?R2)))))
```

Many Alloy specifications use only `comp-1-2` and `comp-2-2`. In the less common cases, `Prioni` determines the arities at hand and automatically declares and axiomatizes the corresponding composition operators.

Transitive closure is defined in terms of exponentiation. For the latter, we need a minimal theory of natural numbers: their definition as an inductive structure and the primitive recursive definition of addition, in order to be able to prove statements such as $(\forall\, R, n, m)\, R^{n+m} = R^n \,.\, R^m$.

5 Translation

`Prioni` automatically translates any Alloy specification into a corresponding Athena theory. A key aspect of this translation is that it preserves the meaning of the Alloy specification. We next show how `Prioni` translates our example Alloy specification into Athena. Each Alloy signature introduces an Athena domain:

```
(domain Object-Dom)
(domain Node-Dom)
```

Additionally, each Alloy signature or field introduces a constant set of tuples whose elements are drawn from appropriate Athena domains:

```
(declare Object (Set-Of Object-Dom))
(declare Node (Set-Of Node-Dom))
(declare next (Set-Of (Pair-Of Node-Dom Node-Dom)))
(declare data (Set-Of (Pair-Of Node-Dom Object-Dom)))
```

In our example, Alloy field declarations put additional constraints on the relations. The translation adds these constraints into the global assumption base (i.e., a set of propositions that are assumed to hold, as explained in Section 3):

```
(assert (is-fun next))
(assert (is-total-fun Node data))
```

where `is-fun` and `is-total-fun` are defined as expected. Each Alloy "function" introduces an Athena function symbol (which can be actually a relation symbol, i.e., a function to the Athena predefined sort `Boolean`):

```
(declare elms (-> ((Set-Of Node-Dom)) (Set-Of Object-Dom)))

(define elms-def
  (forall ?n ?result
    (iff (= (elms ?n) ?result)
         (and (and (singleton? ?n) (subset ?n Node))
              (and (if (empty? (comp-1-2 ?n next))
                       (= ?result (comp-1-2 ?n data)))
              (if (not (empty? (comp-1-2 ?n next)))
                  (= ?result (union (comp-1-2 ?n data) (elms (comp-1-2 ?n next)))))))))
(assert elms-def)
```

where `empty-def` is as expected. Note that there are essentially two cases in `elms-def`: when `(comp-1-2 ?n next)` is empty, and when it is not. To facilitate theorem proving, we split `elms-def` into two parts, `elms-def-1` and `elms-def-2`, each covering one of these two cases. Both of them are automatically derived from `elms-def`.
Alloy facts are simply translated as formulas and added to the assumption base:

```
(define AllAcyclic
  (forall ?n (not (subset (singleton ?n) (comp-1-2 (singleton ?n) (tc next))))))

(assert AllAcyclic)
```

Finally, the assertion is translated into a proof obligation:

```
(define Equivalence
  (forall ?n (= (elms (singleton ?n))
                (comp-1-2 (comp-1-2 (singleton ?n) (rtc next)) data))))
```

Recall that all values in Alloy are relations. In particular, Alloy blurs the type distinction between scalars and singletons. In our Athena formalization, however, this distinction is explicitly present and can be onerous for the Alloy user. To alleviate this, `Prioni` allows users to intersperse Athena text with expressions and formulas written in an infix Alloy-like notation and enclosed within double quotes. (We will follow that practice in the sequel.) Even though this notation retains the distinction between scalars and singletons, it is nevertheless in the spirit of Alloy and should therefore prove more appealing to Alloy users than Athena's s-expressions. There are some other minor notational differences, e.g., we use '*' as a postfix operator and distinguish between set membership (`in`) and containment (`subset`).

6 Proof

The assertion `Equivalence` is an equality between sets. To prove this equality, we show that `elms` is sound:

$$\texttt{ALL n | elms(\{n\}) subset \{n\}.next*.data} \quad (1)$$

and complete:

$$\texttt{ALL n | \{n\}.next*.data subset elms(\{n\})} \quad (2)$$

The desired equality will then follow from set extensionality.

The proof uses a few simple lemmas from `Prioni`'s library of results frequently used in relational reasoning:

```
(define comp-monotonicity "ALL x y R | y in {x}.R* ==> {y}.R* subset {x}.R*")
(define first-power-lemma "ALL x y R | [x y] in R ==> [x y] in R*")
(define comp-lemma "ALL s1 s2 R | s1 subset s2 ==> s1.R subset s2.R")
(define scalar-lemma "ALL x y R | y in {x}.R <==> [x y] in R")
(define subset-rtc-lemma "ALL n R | {n} subset {n}.R*")
(define fun-lemma "ALL n x R | [n x] in R & is-fun(R) ==> {x} = {n}.R")
(define star-pow-lemma "ALL x n R S | x in ({n}.R*).S ==>
                                      (EXISTS m k | [n m] in R^k & [m x] in S)")
```

and a couple of trivial set-theory lemmas:

```
(define subset-trans "ALL s1 s2 s3 | (s1 subset s2) & (s2 subset s3) ==> s1 subset s3")
(define union-lemma "ALL s1 s2 s | (s1 subset s) & (s2 subset s) ==> (s1 union s2) subset s")
```

We also need the following two lemmas about `next` and `data`:

```
(define elms-lemma-1 "ALL n | {n}.data subset ({n}.next*).data")
(define elms-lemma-2 "ALL n | {n}.data subset elms({n})")
```

The first follows immediately from `comp-lemma` and `subset-rtc-lemma` using the method `prove` (explained below); the second also follows automatically from the definitions of `elms`, `union` and `subset`.

6.1 Soundness

The soundness proof needs an induction principle for Alloy lists. Athena supports inductive reasoning for domains that are generated by a set of free constructors. But Alloy structures are represented here as constant sets of tuples, so we must find an alternative way to perform induction on them. In our list example, an appropriate induction principle is:

$$\frac{(\forall n)\,(\neg(\exists m)\,[n,m] \in \texttt{next}) \Rightarrow P(n) \qquad (\forall n)\,(\forall m)\,[n,m] \in \texttt{next} \Rightarrow P(m) \Rightarrow P(n)}{(\forall n)\,P(n)}$$
$$\text{provided } (\forall n)\, n \notin \{n\}.\texttt{next}^{+}$$

The rule is best read backward: to prove that a property P holds for every node n, we must prove: 1) the left premise, which is the base case: if n does not have a successor, then P must hold for n; and 2) the right premise, which is the inductive step: $P(n)$ must follow from the assumption $P(m)$ whenever m is a successor of n. The proviso $(\forall n)\, n \notin \{n\}.\texttt{next}^{+}$ rules out cycles, which would render the rule unsound.

Athena makes it possible to introduce arbitrary inference rules via *primitive methods*. Unlike regular methods, whose bodies must be deductions, primitive methods are defined by expressions. (The distinction between expressions and deductions plays a key role in type-ω DPLs [2].) A primitive method is thus free to generate any conclusion it wishes by performing an arbitrary computation on its inputs. Since no guarantees can be made about soundness, primitive methods are part of one's trusted base and must be used sparingly.

We have implemented the above induction rule with a primitive method `list-induction` parameterized over the goal property P. P is implemented as an Athena function `goal` that constructs the desired proposition for a given argument. In this case, we have:

```
(define (elms-goal n) "elms({n}) subset {n}.next*.data")
```

The primitive method `list-induction` takes a `goal` as an argument, constructs the two premises from it, checks that they are in the assumption base along with the acyclicity constraint, and if successful, outputs `(forall ?n (goal ?n))`.

The base step is proved automatically:

```
(define base-step
  (!prove "ALL n |~(EXISTS m | [n m] in next) ==> elms({n}) subset ({n}.next*).data"
          [elms-def empty-def scalar-lemma elms-lemma-1]))
```

where `prove` is a binary method.[1] (All method calls in Athena are prefixed with '`!`', which distinguishes them from Athena function calls [2].) A method call (`!prove` P $[P_1 \cdots P_n]$) attempts to derive the conclusion P from the premises $P_1, \ldots, P_n$, which must be in the current assumption base. If a proof is found, the conclusion P is returned. Currently, Otter is used for the proof search. Where deductive forms such as `assume` (and others explained in Section 3) are used to guide the deduction, `prove` is used to skip tedious steps. A call (`!prove` P $[P_1 \cdots P_n]$) essentially says to Athena: "P follows from $P_1, \ldots, P_n$ by standard logical manipulations: universal specializations, modus ponens, etc. There is nothing interesting or deep here—you work out the details." If we are wrong, either because P does not in fact follow from $P_1, \ldots, P_n$ or because it is a non-trivial consequence of them, the method call will fail within a preset maximum time limit (currently 1 min). Otherwise, a proof will invariably be found almost instantaneously and P will be successfully returned.

The proof of the inductive step is more interesting:

```
(pick-any x y
   (assume-let ((hyp "[x y] in next")
                (ihyp (ind-goal y)))
     (dlet ((P1 (!prove "elms({x}) = {x}.data union elms({y})"
                        [elms-def-2 hyp fun-lemma (is-fun next) scalar-lemma empty-def]))
            (P2 (!prove "{y}.next* subset {x}.next*" [hyp comp-monotonicity
                                                        scalar-lemma first-power-lemma]))
            (P3 (!prove "({y}.next*).data subset ({x}.next*).data" [P2 comp-lemma]))
            (P4 (!prove "elms({y}) subset ({x}.next*).data" [P3 ihyp subset-trans])))
       (!prove "elms({x}) subset ({x}.next*).data" [P1 elms-lemma-1 P4 union-lemma]))))
```

[1] Currently, `prove` is a primitive method and thus Otter is part of our trusted base. However, it is not difficult to implement Otter's inference rules (paramodulation, etc.) as Athena methods and then use them to define `prove` as a regular method.

The key constructs of the proof—(`pick-any`, `assume-let`, and `dlet`—are explained in Section 3. At this point, both the base case and the inductive step have been proved and are in the assumption base, so we can now apply `list-induction` to obtain the desired conclusion: `(!list-induction elms-goal)`.

6.2 Completeness

Next we present the completeness proof of the statement `{n}.next*.data subset elms({n})`, for arbitrary `n`. Viewing the transitive closure `next*` as the union of $\texttt{next}^k$ for all k, we proceed by induction on k. Specifically, we prove the following by induction on k:

```
ALL k n m x | [n m] in next^k & [m x] in data ==> x in elms({n})      (3)
```

As before, we first define a function `goal` that constructs the inductive goal for any given k:

```
(define (goal k) "ALL m n x | [n m] in next^k & [m x] in data  ==> x in elms({n})"))
```

The following is the inductive proof of 3:

```
(by-induction-on ?k (goal ?k)
  (zero (!prove (goal zero) [elms-lemma-2 pow-def-1 scalar-lemma subset-def]))
  ((succ k) (pick-any m n x
            (assume-let ((hyp "[n m] in next^k+1 & [m x] in data"))
              (!prove "x in elms({n})" [hyp (goal k) pow-def-2 fun-lemma (is-fun next)
                                       scalar-lemma empty-def elms-def-2 union-def])))))
```

Finally, the completeness proof follows, where `ind-lemma` refers to (3).

```
(pick-any n
  (!prove-subsets "({n}.next*).data" "elms({n})"
                  [elms-def star-pow-lemma scalar-lemma ind-lemma]))
```

Here `prove-subsets` is a defined method, which we will now explain. Although Otter is helpful in skipping tedious steps, its autonomous mode is not powerful as a completely automatic theorem prover. More powerful theorem-proving algorithms that guide the proof search by exploiting heuristics for a particular problem domain can be encoded in Athena as *methods*, which are similar to the tactics and tacticals of HOL-like systems. Athena's semantics guarantee soundness: the result of any method call is always a logical consequence of the assumption base in which the call takes place.

A simple example of a method is `prove-subsets`, which captures the following "tactic" for arbitrary sets S_1 and S_2: to prove $S_1 \subseteq S_2$ from a set of assumptions Δ, consider an arbitrary x, suppose that $x \in S_1$, and then try to prove $x \in S_2$ under the assumptions $\Delta \cup \{x \in S_1\}$. The justification for this tactic (i.e., the fact from which the desired goal will be derived once the subgoals have been established) is simply the definition of set containment. Such tactics are readily expressible as Athena methods[2].

[2] Since sets in this problem domain are structured (i.e., elements are usually tuples), these methods employ some additional heuristics to increase efficiency.

Checking both directions (soundness and completeness) of the correctness proof takes about 1 sec in our current implementation. The whole proof for this example (including the lemma library and other auxiliary code) is available online: `http://mulsaw.lcs.mit.edu/prioni/relmics03`

7 Conclusions

`Prioni` is a tool that integrates model checking and theorem proving for relational reasoning. Several other tools combine model checking and theorem proving but focus on reactive systems and modal logics [19, 18] or general first-order logic [13], whereas `Prioni` focuses on structural system properties. Recently, Frias et al. [4] have given an alternative semantics to Alloy in terms of fork algebras [3] and extended it with features from dynamic logic [6]. Further, Lopez Pombo et al. [16] have used the PVS theorem prover [14] to prove specifications in the extended Alloy. This approach has been used for proving properties of execution traces, whereas `Prioni` has been used for structurally complex data.

A key issue in the usability of a theorem prover tool is the difficulty of finding proofs. We have addressed this issue by lightening the formalization burden through our automatic translation and by providing a lemma library that captures commonly used patterns in relational reasoning. Athena makes it easy to guide the proof, focusing on its interesting parts, while Otter automatically fills in the gaps.

References

[1] TPTP problem library for automated theorem proving. `www.cs.miami.edu/~tptp`.
[2] K. Arkoudas. Type-ω DPLs. MIT AI Memo 2001-27, 2001.
[3] M. F. Frias. *Fork Algebras in Algebra, Logic and Computer Science.* World Scientific Publishing Co., 2002.
[4] M. F. Frias, C. G. L. Pombo, G. A. Baum, N. M. Aguirre, and T. Maibaum. Taking alloy to the movies. In *Proc. Formal Methods Europe (FME)*, Sept. 2003.
[5] M. J. C. Gordon and T. F. Melham. *Introduction to HOL, a theorem proving environment for higher-order logic.* Cambridge University Press, Cambridge, England, 1993.
[6] D. Harel, D. Kozen, and J. Tiuryn. *Dynamic Logic.* The MIT Press, Cambridge, MA, 2000.
[7] D. Jackson. Micromodels of software: Modelling and analysis with Alloy, 2001. `http://sdg.lcs.mit.edu/alloy/book.pdf`.
[8] D. Jackson, I. Schechter, and I. Shlyakhter. ALCOA: The Alloy constraint analyzer. In *Proc. 22nd International Conference on Software Engineering (ICSE)*, Limerick, Ireland, June 2000.
[9] D. Jackson and K. Sullivan. COM revisited: Tool-assisted modeling of an architectural framework. In *Proc. 8th ACM SIGSOFT Symposium on the Foundations of Software Engineering (FSE)*, San Diego, CA, 2000.

[10] S. Khurshid and D. Jackson. Exploring the design of an intentional naming scheme with an automatic constraint analyzer. In *Proc. 15th IEEE International Conference on Automated Software Engineering (ASE)*, Grenoble, France, Sep 2000.

[11] S. Khurshid, D. Marinov, and D. Jackson. An analyzable annotation language. In *Proc. ACM SIGPLAN 2002 Conference on Object-Oriented Programming, Systems, Languages, and Applications (OOPSLA)*, Seattle, WA, Nov 2002.

[12] D. Marinov and S. Khurshid. TestEra: A novel framework for automated testing of Java programs. In *Proc. 16th IEEE International Conference on Automated Software Engineering (ASE)*, San Diego, CA, Nov. 2001.

[13] W. McCune. Mace: Models and counter-examples. `http://www-unix.mcs.anl.gov/AR/mace/`, 2001.

[14] S. Owre, J. Rushby, and N. Shankar. PVS: A prototype verification system. In *Proc. 11th International Conference on Automated Deduction (CADE)*, volume 607 of *Lecture Notes in Artificial Intelligence*, pages 748–752, Saratoga, NY, June 1992.

[15] L. Paulson. *Isabelle, A Generic Theorem Prover.* Lecture Notes in Computer Science. Springer-Verlag, 1994.

[16] C. L. Pombo, S. Owre, and N. Shankar. A semantic embedding of the a_g dynamic logic in PVS. Technical Report SRI-CSL-02-04, SRI International, Menlo Park, CA, May 2003.

[17] A. Riazanov and A. Voronkov. The design and implementation of VAMPIRE. *AI Communications*, 15(2–3), 2002.

[18] S. Rajan, N. Shankar, and M. K. Srivas. An integration of model checking with automated proof checking. In *Proceedings of the 7th International Conference On Computer Aided Verification*, volume 939, Liege, Belgium, 1995. Springer Verlag.

[19] N. Shankar. Combining theorem proving and model checking through symbolic analysis. *Lecture Notes in Computer Science*, 1877, 2000.

[20] M. Vaziri and D. Jackson. Checking properties of heap-manipulating procedures with a constraint solver. In *Proc. 9th International Conference on Tools and Algorithms for Construction and Analysis of Systems (TACAS)*, Warsaw, Poland, April 2003.

[21] L. Wos, R. Overbeek, E. Lusk, and J. Boyle. *Automated Reasoning, Introduction and Applications.* McGraw-Hill, Inc., 1992.

Fixed-Point Characterisation of Winning Strategies in Impartial Games

Roland Backhouse[1] and Diethard Michaelis[2]

[1] School of Computer Science and Information Technology
University of Nottingham, Nottingham NG8 1BB, England
rcb@cs.nott.ac.uk
[2] Beethovenstr. 55, 90513 Zirndorf, Germany
diethard.michaelis@t-online.de

Abstract. We use fixed-point calculus to characterise winning strategies in impartial, two-person games. A byproduct is the fixed-point characterisation of winning, losing and stalemate positions. We expect the results to be most useful in teaching calculational reasoning about least and greatest fixed points.

Game theory [BCG82] is an active area of research for computing scientists. For example, it is a frutiful source of examples illustating complexity theory, and it is also used as the basis for the semantics of model checking. Our interest in the area is as a test case for the use of formalisms for the constructive derivation of algorithms. Game theory is well-suited to our goals because it is about constructing winning strategies. Moreover, examples of games are easy to explain to students, they carry no theoretical overhead, and motivation is for free.

In the study of games, as in the book "Winning Ways" [BCG82], a basic assumption is that all games are terminating. Only limited attention has been paid to games where non-termination is possible (so-called "loopy" games [BCG82, chapter 12]). In this paper, we study impartial two-person games, in which non-termination is a possibility. We show how to characterise winning, losing and stalemate positions in terms of least and greatest fixed points of conjugate predicate transformers.

The division of *positions* into winning, losing and stalemate positions is well-known (see, for example, [SS93] or [BCG82]) . The contribution of this paper is to focus on winning *strategies*; we formalise their construction in point-free relation algebra. A byproduct is the fixed-point characterisation of the different types of position.

In order to satisfy length restrictions, several proofs are omitted. A full version is available at the first author's website.

1 Impartial Two-Person Games

An *impartial, two-person game* is defined by a binary relation, denoted here by M. Elements of the domain of M are called *positions*; pairs of positions related by M are called *moves*.

R. Berghammer et al. (Eds.): RelMiCS/Kleene-Algebra Ws 2003, LNCS 3051, pp. 34–47, 2004.

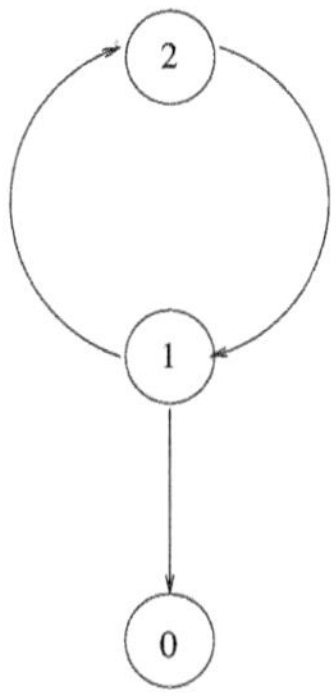

Fig. 1. The lollipop game

Figure 1 is an example of a (non-well-founded) move relation. The positions are the nodes in the figure, and moves are indicated by arrows. We call it the *lollipop game.*

A game is started in a given position. Each player takes it in turn to move; in position s, a move is to a position t such that $s\,\mathsf{M}\,t$. The game ends when a player cannot move; the player whose turn it is to move then loses.

In fig. 1, the game ends in position 0. In position 1, there is a choice of two moves, either to position 0 or to position 2. In position 2, there is no choice of move; the only move that can be made is to position 1. The move relation is not well-founded because it is possible for a game to continue indefinitely, by cycling between positions 1 and 2.

Allowing the move relation to be non-well-founded introduces additional difficulties in the development of the theory. For example, in traditional game theory, a fundamental element is the definition of an equivalence relation on games; that this relation is reflexive is established by a "tit-for-tat" winning strategy (the "Tweedledum and Tweedledee Argument" in [BCG82]). But, tit-for-tat is invalid in the case of non-well-founded game relations.

Throughout this paper, we use the Dijkstra-Scholten [DS90] notation for predicates and predicate transformers. In particular, we use square brackets to indicate that a predicate is true at all positions. For a given relation R, $dom.R$ and $rng.R$ are predicates characterising the domain and range of R, respectively. Formally, for all positions s, $dom.R.s \equiv \langle \exists t :: sRt \rangle$ and, for all positions t, $rng.R.t \equiv \langle \exists s :: sRt \rangle$. The composition of relations R and S is denoted by $R \bullet S$. Specifically, for all s and u, $s\,(R \bullet S)\,u \equiv \langle \exists t :: sRt \wedge tSu \rangle$.

2 Strategies and Position Predicates

2.1 Winning Strategies

A *winning* position in a game is one from which the first player has a strategy to choose moves that guarantee that the game ends, within a finite number of moves, on the second player's turn to move. A *losing* position is one from which there are only moves into winning positions. All other positions are *stalemate* positions.

Formally, a *winning strategy* for game M is a relation W on positions, with the following three properties:

$$W \subseteq \mathsf{M} \quad , \tag{1}$$

$$W \bullet \mathsf{M} \;\text{ is well-founded} \quad , \tag{2}$$

$$\langle \forall t : rng.W.t : \langle \forall u : t\,\mathsf{M}\,u : dom.W.u \rangle \rangle \quad . \tag{3}$$

In words, a winning strategy is a subset of the move relation —(1)— , such that repeatedly invoking the strategy, followed by making an arbitrary move, cannot continue indefinitely —(2)— , and, from every position in the range of the strategy, every move is to a position in the domain of the strategy —(3)— .

A position s is a *winning position* if s is in the domain of a winning strategy. A position t is a *losing position* iff every move from t is to a winning position. A position that is not a winning position or a losing position is a *stalemate position*.

In the lollipop game (fig. 1), node 0 is a losing position, since it is vacuously true that every move from this position is to a winning position. Node 1 is a winning position — a winning strategy is to move from position 1 to position 0. (Formally, the winning strategy is the relation $\{(1,0)\}$.) Finally, node 2 is a losing position, since every move from this position is to node 1, which we have determined to be a winning position.

There are no stalemate positions in the lollipop game. For several entertaining, non-trivial examples of games with stalemate positions, see [BCG82, chapter 12].

2.2 Winning and Losing

From the definition of winning and losing positions, we can identify two properties of positions that they must satisfy. First, losing equivales every move is into a winning position: for all positions t,

$$lose.t \;\equiv\; \langle \forall u : t\,\mathsf{M}\,u : win.u \rangle \quad . \tag{4}$$

(This is by definition.) Second, from a winning position there is always a move into a losing position: for all positions s,

$$win.s \;\Rightarrow\; \langle \exists t : s\,\mathsf{M}\,t : lose.t \rangle \quad . \tag{5}$$

The proof is straightforward:

$$\begin{array}{ll} & win.s \\ = & \{ \quad \text{definition of } win \quad \} \\ & \langle \exists W : WinningStrategy.W : dom.W.s \rangle \end{array}$$

$$
\begin{array}{ll}
= & \{ \quad \text{definitions of } dom.W \text{ and } rng.W,\ (3) \quad \} \\
 & \langle \exists W : WinningStrategy.W : \langle \exists t :: s\,W\,t \wedge \langle \forall u : t\,\mathsf{M}\,u : dom.W.u \rangle\rangle\rangle \\
\Rightarrow & \{ \quad \text{calculus and (1)} \quad \} \\
 & \langle \exists t :: s\,\mathsf{M}\,t \wedge \langle \exists W : WinningStrategy.W : \langle \forall u : t\,\mathsf{M}\,u : dom.W.u \rangle\rangle\rangle \\
\Rightarrow & \{ \quad \text{(4) and calculus} \quad \} \\
 & \langle \exists t : s\,\mathsf{M}\,t : lose.t \rangle \quad .
\end{array}
$$

Note that (5) is an implication, not an equivalence. (See the last two steps in the calculation.) Knowing that there is a move from position s to a losing position (i.e. $\langle \exists t : s\,\mathsf{M}\,t : lose.t \rangle$) is not sufficient to construct a winning strategy with domain containing s. This is illustrated by the lollipop game (fig. 1). An ignorant player might repeatedly choose to move from node 1 (a winning position) to node 2 (a losing position) in the —mistaken— belief that a winning strategy is simply to always leave the opponent in a losing position.

The converse implication is nevertheless true. Demonstrating, by formal calculation, that this is the case is the driving force behind several of our calculations.

2.3 The Predicate Transformers *Some* and *All*

From (4), we abstract the predicate transformer $All.R$, defined by

$$All.R.p.s \;\equiv\; \langle \forall t : sRt : p.t \rangle \quad , \tag{6}$$

and, from (5), we abstract the predicate transformer $Some.R$, defined by

$$Some.R.p.s \;\equiv\; \langle \exists t : sRt : p.t \rangle \quad . \tag{7}$$

In both definitions, R is a relation on positions, p is a predicate on positions, and s and t are positions.

The properties (3), (4) and (5) can be reformulated in a point-free form using these predicate transformers. Effective calculation is considerably enhanced by the convention of regarding a predicate on positions as a partial identity relation; the relation obtained by restricting the domain of a relation, R say, to positions satisfying a predicate, p say, is then simply the relation $p{\bullet}R$. Similarly, $R{\bullet}p$ is the relation obtained by restricting the range of relation R to positions satisfying predicate p. In this way, properties (3), (4) and (5) become:

$$W \;=\; W \bullet All.\mathsf{M}.(dom.W) \quad , \tag{8}$$

$$[lose \;\equiv\; All.\mathsf{M}.win] \quad , \text{ and} \tag{9}$$

$$[win \;\Rightarrow\; Some.\mathsf{M}.lose] \quad . \tag{10}$$

We record some simple properties of *Some* and *All* for later use. (The rules given here are used more than once. Other rules that are used once only are stated at the appropriate point in a calculation.) A property of *All* and *rng* is that, for all relations R and all predicates p,

$$[All.R.p \equiv All.R.(rng.R \Rightarrow p)] \quad . \tag{11}$$

The function *Some* is monotonic in its first argument. That is, for all relations R and S, and all predicates p,

$$[Some.R.p \Rightarrow Some.S.p] \quad \Leftarrow \quad R \subseteq S \quad . \tag{12}$$

(The function *All* is *anti*monotonic in its first argument. But, we don't use this rule.) The predicate transformers *Some.R* and *All.R* are monotonic. That is, for all relations R, and predicates p and q,

$$[Some.R.p \Rightarrow Some.R.q] \Leftarrow [p \Rightarrow q] \quad , \quad \text{and} \tag{13}$$

$$[All.R.p \Rightarrow All.R.q] \Leftarrow [p \Rightarrow q] \quad . \tag{14}$$

Consequently, for all relations R, $All.R \circ Some.R$ and $Some.R \circ All.R$ are also monotonic. (We use "$\circ$" for the composition of functions.)

A crucial observation is that, for all relations R, *Some.R* and *All.R* are *conjugate* predicate transformers. That is,

$$All.R \quad = \quad \neg \circ Some.R \circ \neg \quad . \tag{15}$$

(This is just De Morgan's rule.) A simple consequence is that the predicate transformers $All.R \circ Some.R$ and $Some.R \circ All.R$ are also conjugate.

3 Fixed Points

The main significance of the monotonicity properties (13) and (14) is the guaranteed existence of the least and greatest fixed points of compositions of these predicate transformers (where predicates are ordered as usual by "only-if" — i.e. implication everywhere—).

In this section, we first give a very brief summary of fixed-point calculus (subsection 3.1) before motivating a possible relationship between the winning and losing positions in a game, and fixed points of the predicate transformers $Some.R \circ All.R$ and $All.R \circ Some.R$ (subsection 3.2). That these predicate transformers are conjugates leads us to give a brief summary of the properties of fixed points of conjugate (monotonic) predicate transformers (subsection 3.3). The section is concluded by an analysis of moves of different type (subsection 3.4). Taken as a whole, the section establishes strong evidence for the claim that the winning and losing positions are characterised as least fixed points, but does not prove that this is indeed the case.

3.1 Basic Fixed-Point Calculus

We assume that the reader is familiar with fixed-point calculus. The fixed points we consider are of monotonic functions from relations to relations, and from predicates to predicates (so-called predicate transformers). The ordering on relations is the subset ordering —a relation is a set of pairs— , and the ordering on predicates is "only-if" (i.e. implication everywhere).

We use μ to denote the function that maps a monotonic endofunction to its *least* fixed point, and ν to denote the function that maps a monotonic endofunction to its *greatest* fixed point. So, for example, $\mu(All.R \circ Some.R)$ denotes the least fixed point of the predicate transformer $All.R \circ Some.R$, and $\nu(All.R \circ Some.R)$ denotes its greatest fixed point. Sometimes, we need to be explicit about the ordering relation (for example, in the statement of the rolling rule below). If so, we write it as a subscript to μ or ν.

For predicate transformers, the basic rules of the fixed-point calculus are as follows. The least fixed point μf of the monotonic predicate transformer f is a fixed point of f:

$$[\mu f \equiv f.\mu f] \quad , \tag{16}$$

that is "least" (i.e. "strongest") among all prefix points of f: for all predicates p,

$$[\mu f \Rightarrow p] \ \Leftarrow \ [f.p \Rightarrow p] \quad . \tag{17}$$

The dual rules for the "greatest" (i.e. "weakest") fixed point are obtained by replacing "μ" by "ν", and "$\Rightarrow$" by "$\Leftarrow$".

Rules (16) and its dual are called the *computation rules*, and rules (17) and its dual are called the *induction rules.*

The *rolling rule* is used several times. Suppose f is a monotonic function to A, ordered by $\leq$, from B, ordered by $\sqsubseteq$, and suppose g is a monotonic function to B from A. Then, $f{\circ}g$ is a monotonic endofunction on A, and $g{\circ}f$ is a monotonic endofunction on B. Moreover,

$$\mu_{\leq}(f{\circ}g) \quad = \quad f \,.\, \mu_{\sqsubseteq}(g{\circ}f) \quad . \tag{18}$$

3.2 Winning and Least Fixed Points

Eliminating *lose* from (9) and (10), we get:

$$[win \ \Rightarrow \ (Some.\mathsf{M} \circ All.\mathsf{M}) \,.\, win] \quad .$$

Consequently, by fixed point induction,

$$[win \ \Rightarrow \ \nu(Some.\mathsf{M} \circ All.\mathsf{M})] \quad .$$

Note that nowhere does this calculation exploit property (2) —the relation $W \bullet \mathsf{M}$ is well-founded— of a winning strategy W. By doing so, we can strengthen the property, replacing "greatest" by "least". The key is to use a fixed-point characterisation of well-foundedness: a relation R is well-founded exactly when the least fixed point of the predicate transformer $All.R$ is everywhere true (see eg. [DBvdW97]). Formally, suppose W is a winning strategy, and suppose p is an arbitrary predicate. We calculate a condition on p, in terms of the predicate transformers $Some.\mathsf{M}$ and $All.\mathsf{M}$, that guarantees $[dom.W \Rightarrow p]$ as follows.

$$
\begin{array}{cl}
 & [dom.W \;\Rightarrow\; p] \\
\Leftarrow & \{ \quad (2) \text{ — } W \bullet \mathsf{M} \text{ is well-founded, fixed point induction (17)} \quad \} \\
 & [All.(W \bullet \mathsf{M}).(dom.W \Rightarrow p) \;\Rightarrow\; (dom.W \;\Rightarrow\; p)] \\
= & \{ \quad \text{aiming to remove leftmost “}(dom.W \Rightarrow)\text{” , apply (11)} \quad \} \\
 & [All.(W \bullet \mathsf{M}).(rng.(W \bullet \mathsf{M}) \wedge dom.W \Rightarrow p) \;\Rightarrow\; (dom.W \;\Rightarrow\; p)] \\
= & \{ \quad (3) \text{ — in particular,} \\
 & \qquad [rng.(W \bullet \mathsf{M}) \wedge dom.W \equiv rng.(W \bullet \mathsf{M})] \quad \} \\
 & [All.(W \bullet \mathsf{M}).(rng.(W \bullet \mathsf{M}) \Rightarrow p) \;\Rightarrow\; (dom.W \;\Rightarrow\; p)] \\
= & \{ \quad (11) \quad \} \\
 & [All.(W \bullet \mathsf{M}).p \;\Rightarrow\; (dom.W \;\Rightarrow\; p)] \\
= & \{ \quad \text{Leftmost “}(dom.W \Rightarrow)\text{” has now been removed;} \\
 & \qquad \text{now introduce } Some\text{:} \\
 & \qquad \text{for all } R,\ [dom.R \equiv Some.R.\mathsf{true}], \text{ calculus} \quad \} \\
 & [All.(W \bullet \mathsf{M}).p \;\wedge\; Some.W.\mathsf{true} \;\;\Rightarrow\;\; p] \\
= & \{ \quad All \text{ distributes through composition} \quad \} \\
 & [All.W.(All.\mathsf{M}.p) \;\wedge\; Some.W.\mathsf{true} \;\;\Rightarrow\;\; p] \\
\Leftarrow & \{ \quad \text{for all } R,\ p \text{ and } q, \\
 & \qquad [All.R.p \wedge Some.R.q \;\;\Rightarrow\;\; Some.R.(p \wedge q)] \quad \} \\
 & [Some.W.(All.\mathsf{M}.p) \;\;\Rightarrow\;\; p] \\
\Leftarrow & \{ \quad W \subseteq \mathsf{M} \text{ — (1), } Some \text{ is monotonic — (12)} \quad \} \\
 & [(Some.\mathsf{M} \circ All.\mathsf{M}).p \Rightarrow p] \quad .
\end{array}
$$

We conclude that $[dom.W \Rightarrow p]$ if p is any prefix point of $Some.\mathsf{M} \circ All.\mathsf{M}$. Since, $\mu(Some.\mathsf{M} \circ All.\mathsf{M})$ is the least prefix point of $Some.\mathsf{M} \circ All.\mathsf{M}$, we conclude that

$$
[win \;\;\Rightarrow\;\; \mu(Some.\mathsf{M} \circ All.\mathsf{M})] \quad . \tag{19}
$$

A simple calculation gives the corresponding property of $lose$:

$$
\begin{array}{cl}
 & \mu(All.\mathsf{M} \circ Some.\mathsf{M}) \\
= & \{ \quad \text{rolling rule: (18)} \quad \} \\
 & All\,.\,\mathsf{M}\,.\,\mu(Some.\mathsf{M} \circ All.\mathsf{M})
\end{array}
$$

$\Leftarrow$ { (14) and (19) }

$All.\mathsf{M}.win$

$=$ { (4), definition of All }

$lose$.

That is,

$$[lose \;\Rightarrow\; \mu(All.\mathsf{M} \circ Some.\mathsf{M})] \quad . \tag{20}$$

3.3 Conjugate Monotonic Predicate Transformers

We remarked earlier that $Some.\mathsf{M} \circ All.\mathsf{M}$ is conjugate to $All.\mathsf{M} \circ Some.\mathsf{M}$. In this section, we give a brief summary of fixed-point theory applied to conjugate, monotonic predicate transformers.

Suppose f and g are conjugate, monotonic predicate transformers. Then,

$$\neg \circ g = f \circ \neg \quad \wedge \quad \neg \circ f = g \circ \neg \quad . \tag{21}$$

Negation is a monotonic function from predicates ordered by "only-if" to predicates ordered by "if". So, by the rolling rule (18) for fixed points,

$$[\neg \mu f \;\equiv\; \nu g] \quad . \tag{22}$$

An easy consequence of (22) is the following lemma.

Lemma 23 If f and g are conjugate, monotonic predicate transformers, the predicates μf, μg and $\nu f \wedge \nu g$ are mutually distinct:

$$[\neg(\mu f \wedge \mu g) \;\wedge\; \neg(\mu f \wedge (\nu f \wedge \nu g)) \;\wedge\; \neg(\mu g \wedge (\nu f \wedge \nu g))]$$

and together cover all positions:

$$[\mu f \vee \mu g \vee (\nu f \wedge \nu g)] \quad .$$

□

3.4 Application to Win-Lose Equations

Since $Some.\mathsf{M} \circ All.\mathsf{M}$ and $All.\mathsf{M} \circ Some.\mathsf{M}$ are conjugate, monotonic predicate transformers, (19) and lemma 23 suggest that the winning positions are given by the predicate

$$\mu(Some.\mathsf{M} \circ All.\mathsf{M}) \quad ,$$

the losing positions are given by

$$\mu(All.\mathsf{M} \circ Some.\mathsf{M}) \quad ,$$

and the stalemate positions are given by

$$\nu(Some.\mathsf{M} \circ All.\mathsf{M}) \wedge \nu(All.\mathsf{M} \circ Some.\mathsf{M}) \quad .$$

We abbreviate $\mu(Some.\mathsf{M} \circ All.\mathsf{M})$ to $\mathcal{N}$ and $\mu(All.\mathsf{M} \circ Some.\mathsf{M})$ to $\mathcal{P}$. We also abbreviate $\neg\mathcal{N} \wedge \neg\mathcal{P}$ to $\mathcal{O}$.

In the standard game-theory nomenclature [BCG82], "$\mathcal{O}$" is short for "open", "$\mathcal{N}$" is short for "next player wins" and "$\mathcal{P}$" is short for "previous player wins". We use the nomenclature here *without interpretation* in order to emphasise that we have yet to establish that $\mathcal{N}$ describes the winning positions, $\mathcal{P}$ describes the losing positions, and $\mathcal{O}$ describes the stalemate positions.

The following lemma characterises the moves that are possible and not possible between the different types of position. (We use "$\perp\!\!\!\perp$" to denote the empty relation.)

Lemma 24

(a) From $\mathcal{O}$, there is always a move to $\mathcal{O}$: $[\mathcal{O} \Rightarrow Some.\mathsf{M}.\mathcal{O}]$.
(b) Every move from $\mathcal{O}$ to $\neg\mathcal{O}$ is to $\mathcal{N}$: $\mathcal{O} \bullet \mathsf{M} \bullet \neg\mathcal{O} \;=\; \mathcal{O} \bullet \mathsf{M} \bullet \mathcal{N}$.
(c) Moves from $\neg\mathcal{O}$ to $\mathcal{O}$ must start at $\mathcal{N}$: $\neg\mathcal{O} \bullet \mathsf{M} \bullet \mathcal{O} \;=\; \mathcal{N} \bullet \mathsf{M} \bullet \mathcal{O}$.
(d) From $\mathcal{N}$, there is always a move to $\mathcal{P}$: $[\mathcal{N} \Rightarrow Some.\mathsf{M}.\mathcal{P}]$.
(e) Every move from $\mathcal{P}$ is to $\mathcal{N}$: $\mathcal{P} \bullet \mathsf{M} \;=\; \mathcal{P} \bullet \mathsf{M} \bullet \mathcal{N}$.
(f) Consequently, there are no moves from $\mathcal{P}$ to $\mathcal{O}$: $\mathcal{P} \bullet \mathsf{M} \bullet \mathcal{O} \;=\; \perp\!\!\!\perp$,
(g) and there are no moves between $\mathcal{P}$ and itself: $\mathcal{P} \bullet \mathsf{M} \bullet \mathcal{P} \;=\; \perp\!\!\!\perp$.
(h) Every move to $\mathcal{P}$ is from $\mathcal{N}$: $\mathsf{M} \bullet \mathcal{P} \;=\; \mathcal{N} \bullet \mathsf{M} \bullet \mathcal{P}$.
(i) Consequently, there are no moves to $\mathcal{P}$ from $\mathcal{O}$: $\mathcal{O} \bullet \mathsf{M} \bullet \mathcal{P} \;=\; \perp\!\!\!\perp$.

□

Figure 2 summarises the discussion so far. The three disjoint predicates on positions, $\mathcal{P}$, $\mathcal{N}$ and $\mathcal{O}$, are shown as "clouds". A solid arrow indicates the (definite) existence of a move of a certain type. A dotted arrow indicates the possible existence of a move of a certain type. So, the two solid arrows indicate that, for every $\mathcal{O}$-position, there is a move to an $\mathcal{O}$-position, and, for every $\mathcal{N}$-position, there is a move to a $\mathcal{P}$-position. The dotted arrows indicate that there may be moves from some $\mathcal{O}$-positions to $\mathcal{N}$-positions, from some $\mathcal{N}$-positions to $\mathcal{O}$-positions or $\mathcal{N}$-positions, and from some $\mathcal{P}$-positions to $\mathcal{N}$-positions. Just as important, the absence of arrows indicates the impossibility of certain moves. No moves are possible *from* $\mathcal{P}$-positions to $\mathcal{O}$-positions or $\mathcal{P}$-positions, and no moves are possible *to* $\mathcal{P}$-positions from $\mathcal{O}$-positions or $\mathcal{P}$-positions.

Suppose that $\mathcal{N}$ does indeed characterise the winning positions. Then, lemma 24 establishes that $\mathcal{O}$ characterises the stalemate positions. After all, from $\mathcal{O}$ it is always possible to remain in $\mathcal{O}$. Moving out of it would place the opponent in an $\mathcal{N}$-position, which we have assumed is a winning position. So, from $\mathcal{O}$, the best strategy for both players is to remain in $\mathcal{O}$, waiting for the opponent to make a mistake, and thus continuing the game indefinitely. This is what is meant by "stalemate".

4 Constructing a Winning Strategy

We now turn to the converse of (19). The proof is constructive: we exhibit an algorithm that constructs a winning strategy. Our algorithm is motivated by the

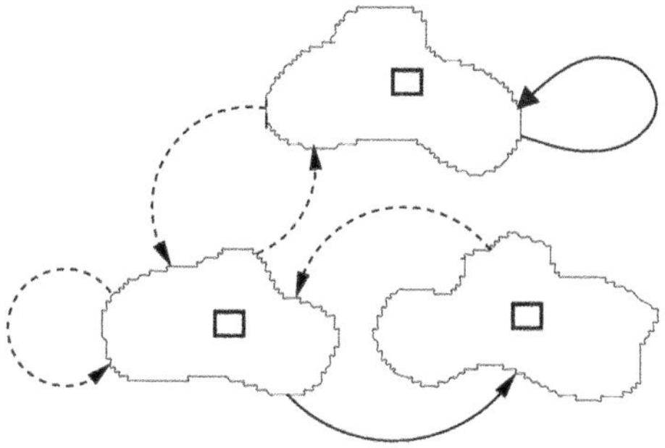

Fig. 2. Moves between $\mathcal{P}$-, $\mathcal{N}$- and $\mathcal{O}$-positions. Solid lines mean there *is* a move, dotted lines mean there *may be* a move

following property of the moves from $\mathcal{N}$ to $\mathcal{P}$. (Note that $\mathcal{N}\bullet\mathsf{M}\bullet\mathcal{P}$ is the set of moves from an $\mathcal{N}$-position to a $\mathcal{P}$-position.)

$$[\mathcal{N}\bullet\mathsf{M}\bullet\mathcal{P} \ \equiv\ \langle\mu R :: \mathsf{M}\bullet All.\mathsf{M}.(dom.R)\rangle] \ . \tag{25}$$

The fixed-point characterisation suggests how to proceed. We assume that the number of positions is finite and consider an algorithm that incrementally computes a winning strategy W. Initially, W is set to the empty relation (which is easily seen to be a winning strategy). Subsequently, W is augmented by exploiting the fact that positions satisfying $All.\mathsf{M}.(dom.W)$ are losing positions. Specifically, moves from positions not in the domain of W to positions satisfying $All.\mathsf{M}.(dom.W)$ are added to W. Formally, we introduce the function f from relations to relations defined by, for all relations R,

$$f.R \ =\ \neg(dom.R)\bullet\mathsf{M}\bullet All.\mathsf{M}.(dom.R) \ , \tag{26}$$

and the function g defined by, for all relations R,

$$g.R \ =\ R\cup f.R \ . \tag{27}$$

Then, the algorithm is as follows:

$\{$ number of positions is finite $\}$

  $W := \perp\!\!\perp$ $\{$ $\perp\!\!\perp$ denotes the empty relation $\}$

; $\{$ **Invariant:** $WinningStrategy.W$ $\}$

  do $\neg(W = g.W) \ \rightarrow\ W := g.W$

  od

$\{$ $WinningStrategy.W \ \wedge\ [dom.W = win = \mathcal{N}]$ $\}$.

Note that the function g is not monotonic. (Take, for example, M to be the relation $\{(a,b),(a,c)\}$, R to be the empty relation and S to be $\{(a,b)\}$. Then,

$R \subseteq S$. But, $g.R = \mathsf{M}$, whereas $g.S = S$. That is, $\neg(g.R \subseteq g.S)$.) So, although the algorithm computes a fixed point of g, standard fixed-point theory does not predict its existence. Nor can the rules of standard fixed-point calculus be used to reason about a fixed point of g.

The assumption that the number of positions is finite guarantees termination because each iteration either adds at least one position to the domain of W or truthifies the termination condition, and there is an upper bound on the number of positions.

In the case that the number of positions is infinite, the algorithm can sometimes be used to determine whether a given position is a winning or losing position, by repeatedly executing the loop body until the position is added to the domain of W (in which case it is a winning position) or to the range of W (in which case it is a losing position). However, examples of games are easily constructed for which some winning positions are not eventually added to the domain of W, so the procedure does *not* constitute a semi-decision procedure for enumerating the winning positions. (The claim made by the authors in the draft paper is thus *wrong*.)

The invariant property is simply that W is a winning strategy. Crucial to establishing the converse of (19) is the claim that, on termination,

$$[dom.W = win = \mathcal{N}] \quad .$$

That is, a maximal winning strategy has been constructed (i.e. $[dom.W = win]$) and the winning positions are precisely characterised by the least fixed point of the predicate transformer $Some.\mathsf{M} \circ All.\mathsf{M}$ (i.e. $[win = \mathcal{N}]$).

To establish the conditional correctness of the algorithm, we must verify the invariant property. The invariant is obviously truthified by the assignment $W := \perp\!\!\perp$. To establish that it is maintained by the loop body, it suffices to verify three properties:

$$g.W \subseteq \mathsf{M} \equiv W \subseteq \mathsf{M} \quad , \tag{28}$$

$$WellFounded.(g.W \bullet \mathsf{M}) \Leftarrow WinningStrategy.W \quad , \tag{29}$$

$$g.W = g.W \bullet All.\mathsf{M}.(dom.(g.W)) \Leftarrow WinningStrategy.W \quad . \tag{30}$$

(Compare (28) with (1), (29) with (2), and (30) with (8), the point-free form of (3).)

Property (28) is obviously true. To prove (29), we use the theorem that, for all relations R and S, $R \cup S$ is well-founded if R is well-founded, S is well-founded, and $S \bullet R = \perp\!\!\perp$. (This is a special case of a general theorem on the well-foundedness of the union of well-founded relations [DBvdW97].) So,

$$WellFounded.(g.W \bullet \mathsf{M})$$

$$\Leftarrow \quad \{ \quad \text{above mentioned theorem, } g.W = W \cup f.W \quad \}$$

$$WellFounded.(W \bullet \mathsf{M}) \wedge WellFounded.(f.W \bullet \mathsf{M}) \wedge$$

$$W \bullet \mathsf{M} \bullet f.W \bullet \mathsf{M} = \perp\!\!\perp$$

$$= \quad \{ \quad W \bullet \mathsf{M} \text{ is well-founded, since } WinningStrategy.W \quad \}$$

$$WellFounded.(f.W \bullet \mathsf{M}) \;\wedge\; W \bullet \mathsf{M} \bullet f.W \bullet \mathsf{M} \;=\; \perp\!\!\!\perp \;.$$

Continuing with each conjunct individually,

$$
\begin{array}{ll}
 & WellFounded.(f.W \bullet \mathsf{M}) \\
= & \quad \{ \quad \text{fixed-point characterisation of well-founded [DBvdW97]} \quad \} \\
 & \langle \nu X :: f.W \bullet \mathsf{M} \bullet X \rangle \;=\; \perp\!\!\!\perp \\
\Leftarrow & \quad \{ \quad \text{for all } R,\ \langle \nu X :: R \bullet X \rangle = R \bullet R \bullet \langle \nu X :: R \bullet X \rangle \\
 & \qquad\quad \perp\!\!\!\perp \text{ is zero of composition} \quad \} \\
 & f.W \bullet \mathsf{M} \bullet f.W \;=\; \perp\!\!\!\perp \\
\Leftarrow & \quad \{ \quad \text{definition of } f,\ \perp\!\!\!\perp \text{ is zero of composition} \quad \} \\
 & All.\mathsf{M}.(dom.W) \bullet \mathsf{M} \bullet \neg(dom.W) \;=\; \perp\!\!\!\perp \\
\Leftarrow & \quad \{ \quad All.\mathsf{M}.(dom.W) \bullet \mathsf{M} \;=\; All.\mathsf{M}.(dom.W) \bullet \mathsf{M} \bullet dom.W \\
 & \qquad\quad \perp\!\!\!\perp \text{ is zero of composition} \quad \} \\
 & dom.W \bullet \neg(dom.W) \;=\; \perp\!\!\!\perp \\
= & \quad \{ \quad \text{composition of predicates is their conjunction} \quad \} \\
 & \mathsf{true} \;.
\end{array}
$$

The proof that $W \bullet \mathsf{M} \bullet f.W \bullet \mathsf{M} \;=\; \perp\!\!\!\perp$ is similar.

Property (30) is proved as follows.

$$
\begin{array}{ll}
 & g.W \bullet All.\mathsf{M}.(dom.(g.W)) \\
\supseteq & \quad \{ \quad g.W \supseteq W;\ dom \text{ and } All.\mathsf{M} \text{ are monotonic} \quad \} \\
 & g.W \bullet All.\mathsf{M}.(dom.W) \\
= & \quad \{ \quad g.W = W \cup f.W,\ \text{distributivity} \quad \} \\
 & W \bullet All.\mathsf{M}.(dom.W) \;\cup\; f.W \bullet All.\mathsf{M}.(dom.W) \\
= & \quad \{ \quad \text{by assumption, } WinningStrategy.W. \\
 & \qquad\quad \text{So, by (8), } [W \;\equiv\; W \bullet All.\mathsf{M}.(dom.W)]. \\
 & \qquad\quad \text{Also, } f.W = f.W \bullet All.\mathsf{M}.(dom.W) \text{ as composition} \\
 & \qquad\quad \text{of partial identity relations is idempotent} \quad \} \\
 & W \cup f.W \\
= & \quad \{ \quad \text{definition of } g \quad \} \\
 & g.W \;.
\end{array}
$$

We have thus proved that

$$g.W \bullet All.\mathsf{M}.(dom.(g.W)) \supseteq g.W$$

(under the assumption that W is a winning strategy). Equality of the left and right sides (that is (30)) follows from the fact that $All.\mathsf{M}.(dom.(g.W))$ is a partial identity relation.

We now turn to the postcondition. We know that $WinningStrategy.W$ is an invariant. So, at all stages, including on termination, $[dom.W \Rightarrow win]$. We show that the converse implication follows from the fact that, on termination, W is a fixed point of g. (Note that g does not have a unique fixed point: $\mathsf{M}\bullet\mathcal{P}$ is also a fixed point of g. We leave the proof to the reader.)

We have, for all relations R,

$$
\begin{array}{ll}
 & R = g.R \\
= & \{ \quad \text{set calculus, definition of } g \quad \} \\
 & \neg(dom.R) \bullet \mathsf{M} \bullet All.\mathsf{M}.(dom.R) \subseteq R \\
\Rightarrow & \{ \quad dom \text{ is monotonic} \quad \} \\
 & [dom.(\neg(dom.R) \bullet \mathsf{M} \bullet All.\mathsf{M}.(dom.R)) \Rightarrow dom.R] \\
= & \{ \quad \text{for all predicates } p \text{ and } q, \\
 & \quad\quad [dom.(p \bullet \mathsf{M} \bullet All.\mathsf{M}.q) \equiv p \wedge (Some.\mathsf{M} \circ All.\mathsf{M}).q] \quad \} \\
 & [\neg(dom.R) \wedge (Some.\mathsf{M} \circ All.\mathsf{M}).(dom.R) \Rightarrow dom.R] \\
= & \{ \quad \text{predicate calculus} \quad \} \\
 & [(Some.\mathsf{M} \circ All.\mathsf{M}).(dom.R) \Rightarrow dom.R] \\
\Rightarrow & \{ \quad \text{fixed-point induction, definition of } \mathcal{N} \quad \} \\
 & [\mathcal{N} \Rightarrow dom.R] \quad .
\end{array}
$$

That is, for all relations R,

$$(R = g.R) \Rightarrow [\mathcal{N} \Rightarrow dom.R] \quad . \tag{31}$$

Combining (31) with (19), we get that, on termination,

$$[(win \Rightarrow \mathcal{N}) \wedge (\mathcal{N} \Rightarrow dom.W)] \quad .$$

But, since W is a winning strategy, by definition of win, $[dom.W \Rightarrow win]$. Thus, by antisymmetry of implication, on termination,

$$[dom.W = \mathcal{N} = win] \quad .$$

5 Conclusion

There is a very large and growing amount of literature on game theory —too much for us to try to summarise here— and fixed-point characterisations of winning, losing and stalemate positions are likely to be well-known. Nevertheless, the properties we have proved are often justified informally, or simply assumed.

For example, Schmidt and Ströhlein [SS93] state lemma 24, but they provide no formal justification. Berlekamp, Conway and Guy [BCG82, chapter 12] assert that every position is in $\mathcal{N}$, $\mathcal{P}$ or $\mathcal{O}$, and describe the algorithm for calculating $\mathcal{P}$-positions. But, their account is entirely informal (in the spirit of the rest of the book, it has to be said).

Surprisingly, exploitation of the simple fact that the predicate transformers $All.\mathsf{M}$ and $Some.\mathsf{M}$ are conjugate seems to be new; if it is not, then it appears to be not as well-known as it should be. Also, our focus on winning strategies rather than winning positions appears to be novel; we have yet to encounter any publication that formalises the notion of a winning strategy.

However, it would be wrong to claim that this paper presents novel results. But that is not our goal in writing the paper. Our goal is to use simple games to explain, in a calculational framework, the all-important concepts of least and greatest fixed points to students of computing science. Of course, we would not present this paper, as written here, to our students —concrete examples like those discussed by Berlekamp, Conway and Guy are vital— but it is on the extent to which our paper succeeds in providing a basis for teaching material that we would wish the paper to be judged.

Acknowledgements

We are grateful to the referees for their careful and thorough reading of the paper, and for their suggestions for improvement.

References

[BCG82] Elwyn R. Berlekamp, John H. Conway, and Richard K. Guy. *Winning Ways*, volume I and II. Academic Press, 1982. 34, 35, 36, 42, 47

[DBvdW97] Henk Doornbos, Roland Backhouse, and Jaap van der Woude. A calculational approach to mathematical induction. *Theoretical Computer Science*, 179(1–2):103–135, 1 June 1997. 40, 44, 45

[DS90] E. W. Dijkstra and C. S. Scholten. *Predicate Calculus and Program Semantics*. Springer-Verlag, Berlin, 1990. 35

[SS93] G. Schmidt and T. Ströhlein. *Relations and Graphs, Discrete Mathematics for Computer Scientists*. EATCS Monographs on Theoretical Computer Science. Springer-Verlag, Berlin Heidelberg, 1993. 34, 47

Checking the Shape Safety of Pointer Manipulations

Adam Bakewell, Detlef Plump, and Colin Runciman

Department of Computer Science
University of York, UK
{ajb,det,colin}@cs.york.ac.uk

Abstract. We present a new algorithm for checking the shape-safety of pointer manipulation programs. In our model, an abstract, data-less pointer structure is a *graph*. A *shape* is a language of graphs. A pointer manipulation program is modelled abstractly as a set of graph rewrite rules over such graphs where each rule corresponds to a pointer manipulation step. Each rule is annotated with the intended shape of its domain and range and our algorithm checks these annotations.
We formally define the algorithm and apply it to a binary search tree insertion program. Shape-safety is undecidable in general, but our method is more widely applicable than previous checkers, in particular, it can check programs that temporarily violate a shape by the introduction of intermediate shape definitions.

1 Introduction

In imperative programming languages, pointers are key to the efficiency of many algorithms. But pointer programming is an error-prone weak point in software development. The type systems of most current programming languages cannot detect non-trivial pointer errors which violate the intended shapes of pointer data structures. From a programming languages viewpoint, programmers need means by which to specify the shapes of pointer data structures, together with safety checkers to guarantee statically that a pointer program always preserves these shapes.

For example, Figure 1 defines a simple program for insertion in binary search trees, written in a pseudo-C notation. Ideally the type system of this language should allow a definition of `BT` to specify exactly the class of binary trees, and the type checker would verify that whenever the argument `t` is a pointer to a binary tree and `insert` returns, the result is a pointer to a binary tree. Such a system would guarantee that the program does not create any dangling pointers or shape errors such as creating sharing or cycles within the tree and that there are no null pointer dereferences. It would not guarantee the stronger property that `insert` does insert `d` properly at the appropriate place in the tree because that is not a pointer safety issue.

The method developed in our *Safe Pointers by Graph Transformation* project [SPG] is to specify the shape of a pointer data-structure by graph reduction rules, see Section 2 and [BPR03b]. Section 3 models the operations upon the

R. Berghammer et al. (Eds.): RelMiCS/Kleene-Algebra Ws 2003, LNCS 3051, pp. 48–61, 2004.

```
BT *insert(datum d, BT *t) = {
  a˜:= t;
  while branch(a) && a->data != d do
    if a->data > d
    then a˜:= a->left
    else a˜:= a->right;
  if leaf(a)
  then *a := branch{data=d, left=leaf, right=leaf};
  return(t)
}
```

Fig. 1. Binary search tree insertion program

data structure by more general graph transformation rules. Section 5 describes our language-independent algorithm for checking shape preservation, which is based on Fradet and Le Metayer's algorithm [FM97, FM98]. It automatically proves the shape safety of operations such as search, insertion and deletion in cyclic lists, linked lists and binary search trees. It can also handle operations that temporarily violate shapes if the intermediate shapes are specified, see Section 4. Section 6 considers related work and concludes.

This paper formalises the overview we gave in [BPR03a], a much more detailed explanation including the proofs omitted from this paper and a number of alternative checking algorithms is provided by the technical report [Bak03].

2 Specifying Shapes by Graph Reduction

A *shape* is a language of labelled, directed graphs. This sections summarises our method of specifying shapes (see [BPR03b]) and presents an example specification of binary trees.

A *graph* $G = \langle V_G, E_G, s_G, t_G, l_G, m_G \rangle$ consists of: a finite set of nodes V_G; a finite set of arcs E_G; total functions $s_G, t_G : E_G \to V_G$ assigning a source and target vertex to each arc; a partial node labelling function $l_G : V_G \to \mathcal{C}_V$; and a total arc labelling function $m_G : E_G \to \mathcal{C}_E$. Graph G is an abstract model of a pointer data structure which retains only the pointer fields. Each node models a record of pointers. Nodes are labelled from the *node alphabet* $\mathcal{C}_V$ to indicate their tag. Graph arcs model a pointer field of their source node; their label, drawn from the *arc alphabet* $\mathcal{C}_E$ indicates which pointer field. The *label type* function $type : \mathcal{C}_V \to \wp(\mathcal{C}_E)$ specifies that if node v is labelled l and the source of arc e is v then the label of e must be in $type(l)$ and e must be the only such arc. Together, $\langle \mathcal{C}_V, \mathcal{C}_E, type \rangle$ form a *signature* Σ and G is a Σ*-graph*.

Graphs may occur in rewrite rules or as language (shape) members. Language members are always Σ*-total* meaning that every node v is labelled with some l and the labels of the arcs whose source is v together equal $type(l)$; so they model closed pointer structures with no missing or dangling pointers.

A *graph morphism* $g : G \to H$ consists of a node mapping $g_V : V_G \to V_H$ and an arc mapping $g_E : E_G \to E_H$ that preserve sources, targets and labels: $s_H \circ g_E = g_V \circ s_G$, $t_H \circ g_E = g_V \circ t_G$, $m_H \circ g_E = m_G$ and $l_H(g_V(v)) = l_G(v)$ for all nodes v where $l_G(v) \neq \perp$. An *isomorphism* is a morphism that is injective and surjective in both components and maps unlabelled nodes to unlabelled nodes. If there is an isomorphism from G to H they are *isomorphic*, denoted by $G \cong H$. Applying morphism $g : G \to H$ to graph G yields a graph gG where: $V_{gG} = g_V V_G$ (i.e. apply g_V to each node in V_G); $E_{gG} = g_E E_G$; $s_G(e) = n \Leftrightarrow s_{gG}(g_E(e))$ and similarly for targets; $m_G(e) = m \Leftrightarrow m_{gG}(g_E(e)) = m$; $l_G(n) = l \Leftrightarrow l_{gG}(g_V(n)) = l$.

A *graph inclusion* $H \supseteq G$ is a graph morphism $g : G \to H$ such that $g(x) = x$ for all nodes and arcs x in G.

A *rule* $r = \langle L \supseteq K \subseteq R \rangle$ consists of three graphs: the *interface* graph K and the *left* and *right* graphs L and R which both include K. Intuitively, a rule deletes nodes and arcs in $L - K$, preserves those in K and allocates those in $R - K$. Our pictures of rules show the left and right graphs; the interface is always just their common nodes which are indicated by numbers.

Graph G *directly derives* graph H through rule $r = \langle L \supseteq K \subseteq R \rangle$, injective morphism g and isomorphism i, written $G \Rightarrow H$ or $G \Rightarrow_{r,g,i} H$, if the diagram below consists of pushouts (1) and (2) and an i arrow (see [HMP01] for a full definition of pushouts).

$$\begin{array}{ccccccc} L & \supseteq & K & \subseteq & R & & \\ g\downarrow & (1) & \downarrow & (2) & \downarrow & & \\ G & \supseteq & D & \subseteq & H' & \xrightarrow{i} & H \end{array}$$

Injectivity means that distinct nodes in L must be distinct in gL; the pushout construction means that deleted nodes, those in $gL - gK$, cannot be adjacent to any arcs in D if the derivation exists (the *dangling condition*).

If $H \cong G$ or H is derived from G by a sequence of direct derivations using rules in set $\mathcal{R}$ we write $G \Rightarrow^*_{\mathcal{R}} H$ or $G \Rightarrow^* H$. If no graph can be directly derived from G through a rule in $\mathcal{R}$ we say G is *$\mathcal{R}$-irreducible*.

A *GRS* (graph reduction specification) $S = \langle \Sigma, \mathcal{R}, Acc \rangle$ consists of a signature Σ, a set of Σ-total rules $\mathcal{R}$ and a Σ-total $\mathcal{R}$-irreducible *accepting graph Acc*. It defines a language $\mathcal{L}(S) = \{G \mid G \Rightarrow^*_{\mathcal{R}} Acc\}$.

So a GRS is a reversed graph grammar: *Acc* corresponds to the start graph and $\mathcal{R}$ corresponds to reversed production rules. The rules are *Σ-total* meaning that if $G \Rightarrow_{\mathcal{R}} H$ then G is a Σ-total graph iff H is a Σ-total graph. So GRSs are guaranteed to define languages of pointer structure models.

A GRS S is *polynomially terminating* if there is a polynomial p such that for every derivation $G \Rightarrow_{\mathcal{R}} G_1 \Rightarrow_{\mathcal{R}} \cdots \Rightarrow_{\mathcal{R}} G_n$, $n \leq p(\#V_G + \#E_G)$. It is *closed* if $G \in \mathcal{L}(S)$ and $G \Rightarrow_{\mathcal{R}} H$ implies $H \in \mathcal{L}(S)$. A *PGRS* is a polynomially terminating and closed GRS. Membership of PGRS languages is decidable in polynomial time; this and sufficient conditions for closedness, polynomial termination and Σ-totality are discussed in [BPR03b].

$$\Sigma_{BT} = \langle\, \{R, B, L\}, \{o, l, r\},\ \{R \mapsto \{o\}, B \mapsto \{l, r\}, L \mapsto \{\}\}\rangle$$

$Acc_{BT} =$ 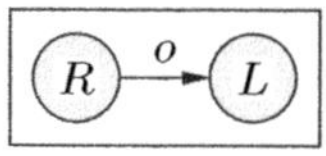

BtoL : 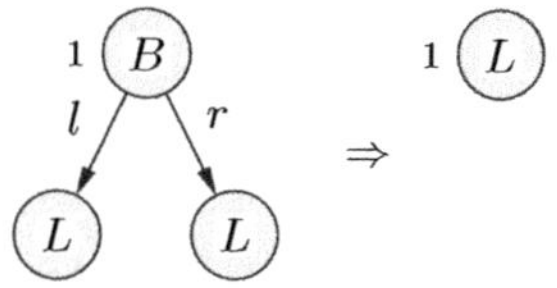

Fig. 2. A PGRS of rooted binary trees, *BT*.

For example, a *rooted binary tree* is a graph containing one root node labelled R and a number of other nodes labelled B(ranch) or L(eaf). Branch nodes have l(eft) and r(ight) outgoing arcs, the root has one o(rigin) outgoing arc and leaf nodes have no outgoing arcs. Branches and leaves all have one incoming arc, the root has no incoming arcs and every node is reachable from the root. If every branch contains a data item, the leaves contain no data and the data is ordered it is a *binary search tree.*

The data-less shape is specified by the PGRS *BT* in Fig. 2. The *BT* signature allows nodes to be labelled R, B or L, where R-nodes have an o-labelled outgoing arc, B-nodes have two outgoing arcs labelled l and r, and L-nodes have no outgoing arcs. The *BT* accepting graph Acc_{BT} is the smallest possible tree and every other tree reduces to it by repeatedly applying the reduction rule `BtoL` to its branches. No non-tree reduces to Acc_{BT} because `BtoL` is matched injectively and cannot be applied if deleted nodes are adjacent to arcs outside the left-hand side of the rule, see [BPR03b] for an example. *BT* is polynomially terminating and closed because `BtoL` is size reducing and *non-overlapping.* See [BPR03b] for full details.

3 Graph Transformation Models of Pointer Programs

Textbooks on data structures often present pointer programs pictorially and then formalise them as imperative programs. In our approach a pictorial presentation is a formal graph-transformation model of a program.

A *model pointer program* in the sense of this paper is a set of rules with a strategy for their application (see [HP01] for more on the the syntax and semantics of such programs). Programs may temporarily violate the shape of a graph so the rule construction is not as restricted as the Σ-total reduction rules.

The rules in Fig. 3 model all the pointer manipulation steps in a *binary search tree insertion* program such as that in Figure 1. They manipulate graphs over two signatures, *AT* is an extension of *BT* which allows A-labelled nodes with two outgoing arcs labelled a and o. The idea is to model insertion by replacing the R-labelled tree root with an A-labelled *auxiliary root*, moving the a-arc to the insertion position and changing the tree structure at that point appropriately. So the control strategy is to apply the `Begin` rule once, then apply `GoLeft` and `GoRight` any number of times, then apply either `Insert` or `Found`. The *seman-*

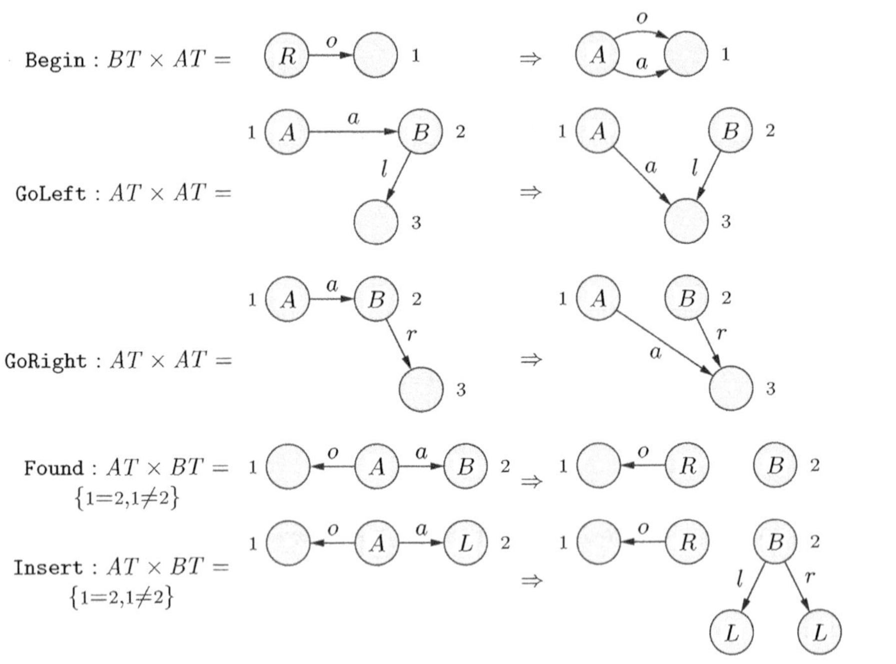

Fig. 3. Transformation rules modelling binary search tree insertion. In `Found` and `Insert`, nodes 1 and 2 may either be distinct or identical

tics of our program is the following binary relation $\rightarrow_{ins}$ on BT graphs, which represents every possible insertion of every possible element in every possible tree.

$$\rightarrow_{ins} = (\Rightarrow_{\{\texttt{Begin}\}}) \circ (\Rightarrow^*_{\{\texttt{GoLeft},\texttt{GoRight}\}}) \circ (\Rightarrow_{\{\texttt{Insert},\texttt{Found}\}})$$

A simple type-checker verifies that the declared range shape of `Begin` matches the domain shape of `GoLeft` and `GoRight` and that the declared range shape of `GoLeft` and `GoRight` matches their domain shape and the domain shape of `Insert` and `Found`. Shape checking aims to prove the individual rule shape annotations.

The `Begin` rule relabels the root A and introduces an auxiliary pointer a to the origin; this is a simple model of procedure call. Then, if the branch pointed to by a contains the datum to insert, the procedure should just return, removing a and relabelling the root back to R, which is done by `Found`. If a points to a leaf the datum is not present in the tree so a new branch should be allocated to hold it and the procedure should return, this is done by `Insert`. If a points to a branch and the datum to insert is less than the branch datum, a should move to insert in its left child, this is done by `GoLeft`; `GoRight` is similar.

$\Sigma_{AT} = \Sigma_{BT} \cup \langle \{A\}, \{a\}, \{A \mapsto \{a, o\}\} \rangle$

$\mathcal{R}_{AT} = \{\texttt{BtoL}, \texttt{BtoLl}, \texttt{BtoLr}\}$

$Acc_{AT} =$ 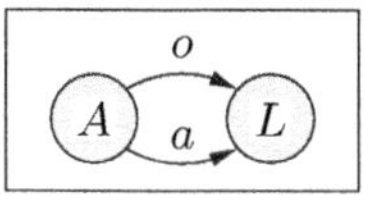

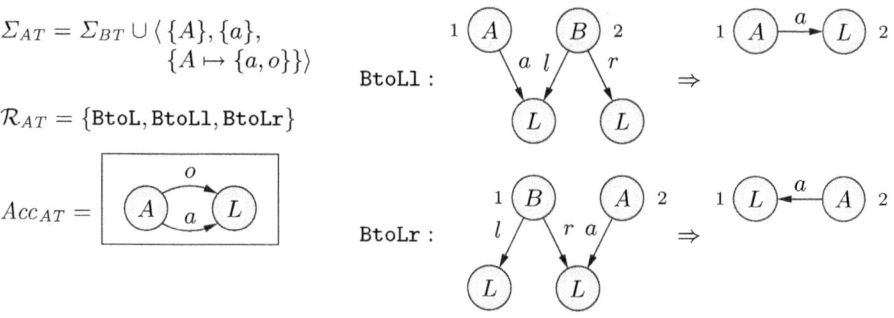

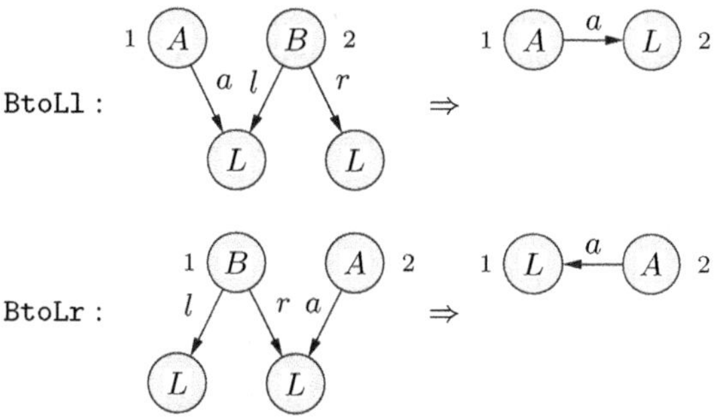

Fig. 4. PGRS of rooted binary trees with an auxiliary pointer, AT

A slightly simpler model of tree insertion is possible by giving the auxiliary pointer a separate parent from the origin (see [BPR03b]) but this version illustrates the ability of our method to check shape-changing rules.

4 Specifying Intermediate Shapes

The insertion rules of Fig. 3 temporarily violate the BT shape by relabelling the root and introducing an auxiliary pointer a. Temporary shape violation is essential in many pointer algorithms such as red-black tree insertion or in-place list reversal. Our approach to such algorithms is to define their intermediate shapes by PGRSs, annotate each of their rules with the intended shapes of their domain and range, and check these shape-changing rules.

Another approach would be to separate the heap-resident branch and leaf nodes from the stack-resident root nodes in our graphs. This way the tree insertion rules can be treated as non heap-shape changing but this approach would not allow the list reversal or red-black insertion examples to be treated as non-shape changing; in general, specifying intermediate shapes seems a better solution.

The rules in Fig. 3 are annotated with the intended shape of their domain and range: during the search phase the shape should be rooted binary trees where the root node is labelled A and has an auxiliary arc a pointing somewhere in the tree, in addition to the origin arc. This shape AT is specified formally by the PGRS in Fig. 4. The new rules `BtoLl` and `BtoLr` reduce branches if one of their children is the target of the auxiliary pointer a, which they move up the tree. Thus the smallest AT-graph, which is the accepting graph, has one leaf pointed to by both arcs of the root A.

5 Checking Shape Safety

An annotated rule $t : S \times T$ is *shape-safe* if for every derivation $G \Rightarrow_t H$, $G \in \mathcal{L}(S)$ implies $H \in \mathcal{L}(T)$.

Our checking algorithm builds an ARG (abstract reduction graph), which is a finite representation of graphs in $\mathcal{L}(S)$ and the domain of $\Rightarrow_t$ and rewrites this ARG to a *normal form* (Section 5.2). Then it builds another ARG to represent graphs in $\mathcal{L}(T)$ and the range of $\Rightarrow_t$ and tests whether this *includes* the left ARG (Section 5.3). To help represent infinite languages finitely, ARGs only include *basic contexts* (Section 5.1); the ARG inclusion test proves shape-safety for basic contexts; the *congeniality* test extends the safety result to all of $\mathcal{L}(S)$ in the domain of $\Rightarrow_t$ (Section 5.4).

The algorithm is necessarily incomplete because shape safety is undecidable in general. This follows from [FM97] which reduces the inclusion problem for context-free graph languages to a variant of shape safety. In practice, ARG construction may not terminate and if it does then some safe rules may fail the tests. The closedness property of PGRSs helps to improve ARG construction and the success rate of the algorithm.

5.1 Graph Contexts

An ARG represents all the contexts for the left (or right) graph of a rule. Intuitively, if $G \in \mathcal{L}(S) \cap dom(\Rightarrow_t)$ then G is the left graph of t glued into some graph context C. We denote this by $C(\!|L|\!)$. The ARG represents the set of all such contexts C as a kind of graph automaton: the derivation $C(\!|L|\!) \Rightarrow^* Acc_S$ can be broken down into a sequence of derivations $C_1(\!|L_1|\!) \Rightarrow L_2, \ldots, C_n(\!|L_n|\!) \Rightarrow Acc_S$ where C_i is the smallest context needed for the ith derivation to take place and C may be obtained by gluing all the C_i together. So the ARG is an automaton whose nodes are the L_i's of all such possible derivation sequences and whose arcs are labelled with the C_i necessary for the derivation from source to target.

There are two issues addressed in this section before the formal definition of ARGs: 1. All the represented contexts must be valid graphs; our definition of graph contexts ensures this. 2. The ARG must be finite; this is not always possible but it is often achievable by restricting the ARG to represent basic contexts only.

A *context* is a graph in which some nodes are internal. *Internal* nodes must be labelled and have a full set of outarcs, their inarcs cannot be extended when the context is extended. This restriction prevents ARGs representing invalid graphs, which otherwise would arise, for example if nodes allocated during a derivation are then glued into a context which must exist before those nodes.

Formally, a *Σ-context* is a pair $C = (G, I)$ where G is a Σ-graph and $I \subseteq V_G$ is a set of *internal* nodes such that $\forall v \in I. l_G(v) \neq \bot \wedge \{m_g(e) \mid s_G(e) = v\} = type\ (l_G(v))$. The *boundary* of C, $boundary(C) = V_G - I$, is all the external nodes of C. Equality, intersection, union and inclusion extend to contexts from graphs in the obvious way. A reduction rule is converted to a pair of contexts by the following function:
$ruletocxts(\langle L \supseteq K \subseteq R \rangle) = ((L, V_L - V_K), (R, V_R - V_K))$.

A *context morphism* $g : (G, I) \to (H, J)$ is a graph morphism $g : G \to H$ which preserves internal nodes and the indegree of internal nodes: $g_V I \subseteq$

J and $v \in I$ implies $indegree_G(v) = indegree_H(g(v))$ where $indegree_A(v) = \#\{e|t_A(e) = v\}$.

A *direct derivation* of context D from C through rule r and morphisms g, h is given by $C \Rightarrow_{r,g,h} D$ where $r = \langle L \supseteq K \subseteq R \rangle$ and the following diagram is two pushouts and h is a context isomorphism.

$$\begin{array}{ccccccc} (L, V_L - V_K) & \supseteq & (K, \emptyset) & \subseteq & (R, V_R - V_K) & & \\ g\downarrow & & \downarrow & & \downarrow & & \\ C & \supseteq & B & \subseteq & D' & \xrightarrow{h} & D \end{array}$$

A direct derivation preserves all the external nodes of C: only internal nodes can be deleted and allocated nodes are internal. $C(\!|D|\!)$ means *glue* D into C. It is defined if the following diagram is a pushout: the arrows are context inclusion morphisms; $C \cap D$ is discrete, unlabelled and all its nodes external; and D and $C(\!|D|\!)$ are contexts.

$$\begin{array}{ccc} C \cap D & \rightarrow & D \\ \downarrow & & \downarrow \\ C & \subseteq & C(\!|D|\!) \end{array}$$

The pushout construction means that $C(\!|D|\!)$ includes everything in C and D and nothing else (and only nodes internal in C or D are internal in $C(\!|D|\!)$); the context inclusions guarantee that every internal node of D has exactly the same inarcs and outarcs in $C(\!|D|\!)$; the restrictions on $C \cap D$ guarantee that C is the smallest context needed to form $C(\!|D|\!)$. Note that internal nodes of D cannot occur in C but external nodes of D can be made internal in $C(\!|D|\!)$ by being internal in C; and C does not have to be a proper context as its internal nodes can lack some of the inarcs or outarcs they have in $C(\!|D|\!)$. So $(\!||\!)$ is associative but not commutative.

A useful property is that $(\!||\!)$ cannot prevent reducibility by breaking the dangling condition, so if $C \Rightarrow D$ then $X(\!|C|\!) \Rightarrow X(\!|D|\!)$.

In a *basic direct derivation* of context C glued in context X, there must be a non-trivial overlap between C and X. Formally, $basic(X(\!|C|\!) \Rightarrow_{r,g,id} D)$ if $r = \langle L \supseteq K \subseteq R \rangle$; context X is minimal, $X(\!|C|\!) = gL \cup C$; the derivation exists, $X(\!|C|\!) \Rightarrow_{r,g,id} D$; and the overlap is non-trivial, $gL \cap C \neq gK \cap C$.

A non-basic derivation leaves C unchanged (but the reduction rule left graph may overlap some of C). Every derivation of the form $X(\!|C|\!) \Rightarrow^* Acc$ can be reordered and split into two consecutive derivations $X(\!|C|\!) \Rightarrow^* Y(\!|C|\!) \Rightarrow^* Acc$ where the direct derivations in the first sequence are all non-basic and those in the second sequence are all basic. The left ARG represents all such Y; the inclusion test checks that the transformation is safe for all graph contexts of the form $Y(\!|C|\!)$ and the shape congeniality test extends the result to all graphs of the form $X(\!|C|\!)$.

$ARG(C, S) = Build_{CXT}(D, \boxed{D}, S)$ *where* $D = reduce(C, \Rightarrow_S)$
$Build_{CXT}(C, A, S = \langle \Sigma, \mathcal{R}, Acc \rangle) =$
for $each(D, X) \in (\{(reduce(D, \Rightarrow_{\mathcal{R}}), X) \mid r \in \mathcal{R}, basic(X(\!|C|\!)) \Rightarrow_{r,g,id} D)\}$
$\cup\{(X(\!|C|\!), X) \mid X(\!|C|\!) \cong Acc, C \ncong Acc\})/\cong$
do if $\exists C' \in V_A$, context isomorphism $i.D = iC'$ **then** $A := A \cup \boxed{C \xrightarrow{X} C'}$
else $Build_{CXT}(D, A \cup \boxed{C \xrightarrow{X} D}, S)$
return A
$reduce(C, \Rightarrow) =$ **if** $C \Rightarrow C'$ **then** $reduce(C', \Rightarrow)$ **else** C

Fig. 5. Context-based ARG construction algorithm

5.2 Abstract Reduction Graphs

An *ARG* $A = \langle V, E, m, s, t \rangle$ is a directed graph comprising a set of contexts V (the nodes); a set of arcs E; an arc labelling function $m : E \to Context$ and arc source and target functions $s, t : E \to V$.

For transformation rule $t : S \times T$, where $ruletocxts(t) = (C, D)$, the left ARG, $ARG(C, S)$ is produced by the algorithm in Fig. 5.

Intuitively, an ARG is built by starting with node C. For every context X (up to isomorphism, denoted $/\cong$ in Fig. 5) such that $X(\!|C|\!)$ has a basic derivation to D or $X(\!|C|\!)$ is Acc, we add an arc from C to node D, or node Acc, labelled X (the boxed expressions in Fig. 5 denote graphs pictorially). The process repeats on these new nodes. In general, $Build_{CXT}$ is non-terminating so safety checking often fails with certain GRSs. The report [Bak03] considers conditions for termination — a terminating, or size-reducing, GRS is not sufficient.

The closedness property of PGRSs allows us to *reduce* the contexts D — and the initial context C — before adding it to the ARG (and makes *reduce* deterministic). This reduces ARG size and improves the likelihood of $Build_{CXT}$ terminating.

Raw ARGs can be very large and they can include garbage paths that do not lead to the accepting graph. Therefore we use a system of *ARG normalisation* rules to eliminate excess nodes by merging and deleting arcs where possible. Fig. 6 shows an ARG: the left node is the context C; the right node is the accepting graphs of AT; positively numbered graph nodes are external; Unnumbered or negatively numbered graph nodes in ARG labels are internal. So for example, the top loop arc in this ARG means that gluing the central node into the loop label forms a context from which we can derive the central node (after renaming the boundary node 2 to 3).

Every path from C to Acc_S in $A = ARG(C, S)$ represents a context of C. The *context-paths* of A are $cxtpaths(A) = \{p \in paths(A) | s_p = C \wedge t_p = Acc_S\}$ where the *paths* in graph A are all sequences of arcs such that the target of each arc is the source of its successor in the sequence, $paths(G) = \{\langle e_i \rangle_{i=1}^{n} | e_i \in E_A \wedge 1 \leq$

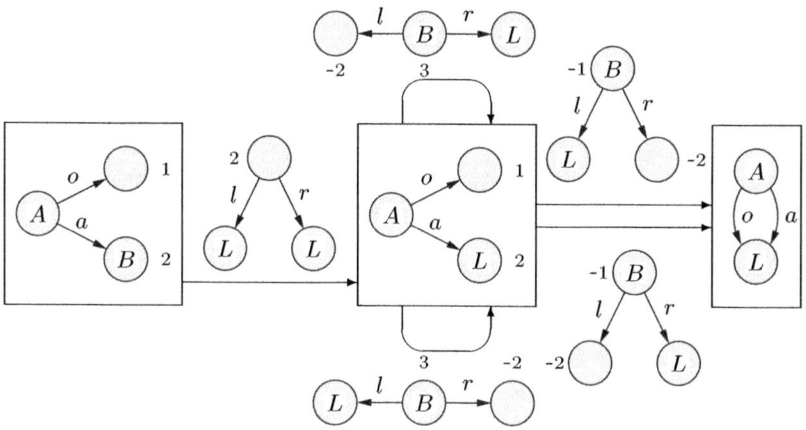

Fig. 6. Normalised ARG for the left hand side of `Found` (nodes 1 and 2 distinct)

$i < n \Rightarrow t_A(e_i) = s_A(e_{i+1})\}$; the *source* of a path $p = \langle e_i \rangle_{i=1}^{n}$ is $s_p = s_A(e_1)$ and the *target* of p is $t_p = t_A(e_n)$.

The *context represented* by a context-path p may be extracted by the following function which glues the arc labels together and uses a renaming morphism α to ensure that the right nodes are identified at each gluing.

$$\begin{aligned}
&\mathit{cof}\,\langle\rangle = \emptyset \\
&\mathit{cof}\,\langle C \xrightarrow{X,g} D\rangle + p = g\alpha_{C,X,g,D,P}P(\!|X|\!) \\
&\text{where } D = \mathit{reduce}(X(\!|C|\!), \Rightarrow) \\
&\qquad P = \mathit{cof}\,p \\
&\qquad \alpha_{C,X,\sigma,D,Y} : (\mathit{Rng}(\sigma) - \mathit{Dom}(\sigma)) \cup (VE_Y - VE_D) \rightarrow \\
&\qquad\qquad (VE - VE_{C \cup X \cup D}) - \mathit{Rng}(\sigma)
\end{aligned}$$

The example ARG in Fig. 6 represents all the graphs depicted in Fig. 7: starting from an instance of the `Found` rule left graph we can reduce the branch pointed to by a to a leaf with one `BtoL` derivation then any number of `BtoLl` or `BtoLr` derivations (following the cycle in the centre of the ARG) move the a-arc up until it points to a child of the root branch; finally a `BtoLl` or `BtoLr` derivation at the root results in the accepting graph. So the basic contexts include the path from the a-arc up to the origin; the reduction of the other sub-trees to leaves are always part of the non-basic derivations.

The ARGs generated by our algorithm have the following properties. *ARG context-path completeness* says every basic context is represented by some context-path: if $G \cong X(\!|C|\!)$ and $G \Rightarrow_S^* \mathit{Acc}_S$ then there is a path $p \in \mathit{cxtpaths}(\mathit{ARG}(C,S)).G \Rightarrow_S^* \mathit{cof}\,p(\!|C|\!) \Rightarrow_S^* \mathit{Acc}_S$. *ARG context-path soundness* says every context-path represents some basic context: if $\mathit{ruletocxts}(r) = (C, D)$ and $G \Rightarrow^* \mathit{cof}\,p(\!|C|\!)$ and $p \in \mathit{cxtpaths}(\mathit{ARG}(C,S))$ then $G \Rightarrow_S^* \mathit{Acc}_S$ and $G \in \mathit{dom}(\Rightarrow_r)$.

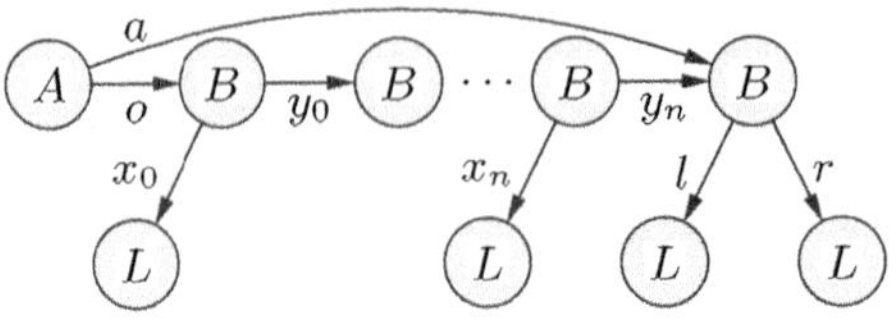

$n \geq 0$
$(x_i, y_i) \in \{(l, r), (r, l)\}$

Fig. 7. Graphs represented by Fig. 6

5.3 ARG Inclusion

To check the shape-safety of a transformation for all basic contexts we can construct the ARG of its right graph then check that every context represented by the left ARG is represented by the right ARG. In practice this is undecidable so we ask whether the right ARG *includes* the left ARG. A more powerful algorithm uses the left ARG as a guide for the construction of the right ARG such that the construction succeeds if the inclusion exists.

Let $A = ARG(L, S)$ and $B = ARG(R, T)$. Then B *includes* A, $B \gtrsim A$, if there is:

1. A total injective morphism $m : VE_A \to VE_B$ such that (i). $m(L) = R$ and (ii). $m(Acc_S) = Acc_T$,
2. A total function $\beta : V_A \to (V \to V)$ assigning a node morphism to each node of A such that (i). $\beta(v) : boundary(v) \to boundary(m(v))$ is a total bijection and (ii). $\beta(L)$ is an identity,
3. A total function $\sigma : E_A \to (VE \to VE)$ assigning a graph isomorphism to each arc $e = C \xrightarrow{X} C'$ of A, where $m(e) = D \xrightarrow{Y} D'$, such that (i). $\sigma(e)X = Y$, (ii). $\sigma(e)|_C = \beta(C)$ and (iii). $\sigma(e)|_{C''} = h\alpha_h\beta(C')\alpha_g^{-1}g^{-1}|_{C''}$ where $reduce(X(\!|C|\!), \Rightarrow_S) = gC'$ and $reduce(Y(\!|D|\!), \Rightarrow_S) = hD'$ and $C'' = g\alpha_g C'$, $\alpha_g = \alpha_{C,X,g,C',C'}$ and $\alpha_h = \alpha_{D,Y,h,D',D'}$.

Fig. 8 shows a right ARG of the Found rule. This includes the left ARG in Fig. 6: there is an obvious isomorphism between these ARGs; the boundary nodes of corresponding contexts are the same; the context labels of corresponding arcs are the same (though their internal node and arc ids may differ). So all the contexts represented by the left ARG are represented by the right ARG too.

In general, we have the result that if $A = ARG(L, S)$ and $B = ARG(R, T)$ and $B \gtrsim A$ and $p \in cxtpaths(A)$ then there is a $q \in cxtpaths(B)$ such that $cof\,p(\!|L|\!) \cong cof\,q(\!|L|\!)$.

5.4 GRS Congeniality

To complete the shape-safety proof we need to extend the result from the basic contexts to all contexts. Formally we say that GRSs S and T are *congenial* for transformation t if whenever $X(\!|C|\!) \Rightarrow_S^* Y(\!|C|\!)$ then $X(\!|D|\!) \Rightarrow_T^* Y(\!|D|\!)$ where $(C, D) = ruletocxts(t)$ and no reduction in the $\Rightarrow_S^*$ sequence is basic.

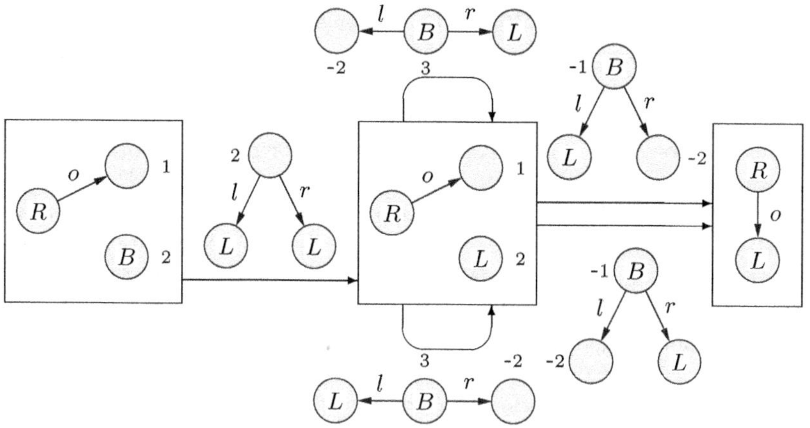

Fig. 8. Normalised ARG for the right hand side of `Found` (nodes 1 and 2 distinct)

The following principles are used to check congeniality. For every rule $r = \langle L \supseteq K \subseteq R \rangle \in \mathcal{R}_S$:
1. $r \in \mathcal{R}_T$ and K is discrete and unlabelled and every node in V_K is a source or target in L.
2. $r \notin \mathcal{R}_T$ and r can only be used in basic reductions of C.

These principles suffice for our example: in `Begin`, `GoLeft` and `GoRight` the rule range shape includes all the reduction rules of the domain shape, and they satisfy principle 1 above. So every reduction of any context to a basic context is still possible after application of these rules. The `Found` and `Insert` rules are more difficult because the `BtoLl` and `BtoLr` reduction rules are not part of BT so we need to use principle 2: as the `Found` and `Insert` left graphs both include the a-arc of the root node, every overlap with these reduction rules must be basic, otherwise the graph would have two roots and could not be in $\mathcal{L}(AT)$.

Thus all the insertion rules are shape safe and so insertion is shape safe.

6 Related Work and Conclusion

Our checking algorithm automatically proves the shape safety of operations such as search, insertion and deletion in cyclic lists, linked lists and binary search trees. It is more widely applicable than Fradet and Le Metayer's original algorithm [FM97, FM98] because our checking method is strictly more powerful and our ARGs are more precise through the use of graph contexts. In Fradet and Le Metayer's framework shape specifications are restricted to context-free graph grammars, this means that ARG construction is guaranteed to terminate and they do not need the congeniality condition (but checking is still undecidable). However their checking method is not applicable to shape changing rules such as the tree insertion algorithm here.

Context-exploiting shapes are generated by hyperedge-replacement rules that are extended with context [DHM03]. This paper shows that there is a reasonable (and decidable) class of *shaped transformation rules* that preserve context-exploiting shapes. Hence there is no need for type checking if shapes and transformations are restricted to these classes. It is unclear at present how severe this restriction is and how it compares to PGRSs with arbitrary transformations.

Other approaches to shape safety are more distant from our work as they are based on logics or types. The *logic of reachability expressions*, or *shape analysis*, [BRS99, SRW02] can be used for deciding properties of linked data structures expressed in a 3-valued logic. The PALE [MS01] tool can check specifications expressed in a *pointer assertion logic*, an extension of the earlier *graph types* specification method [KS93]. The logic of *bunched implications* [IO01] can be used as a language for describing and proving properties of pointer structures. *Alias Types* [WM01] are a pseudo-linear type system which accept shape safe programs on structures such as lists, trees and cycles. *Role analysis* [KLR02] checks programs annotated with role specifications which restrict the incoming and outgoing pointers of each record type and specify properties such as which pointer sequences form identities. Generally speaking, these approaches cannot or do not deal with context-sensitive shapes.

We can specify shapes beyond the reach of context-free graph grammars, such as red-black trees [BPR03b]. However, our current checking method is often non-terminating on such shapes. Essentially the problem occurs when a GRS has rules that can cause context D to be larger than context C in some basic derivation $X(\!|C|\!) \Rightarrow D$, and this growth is repeatable without limit, causing ARG construction to non-terminate. This situation often occurs with non-context-free GRSs such as balanced trees, preventing us from checking operations on such shapes.

The main areas for further work are to develop languages for safe pointer programming based on our existing specification and checking methods, and to overcome some of the limitations of the current approach to enable the automatic checking of operations on non-context-free shapes.

References

[Bak03] A Bakewell. Algorithms for checking the shape safety of graph transformation rules. Technical report, 2003. Available from [SPG]. 49, 56

[BPR03a] A Bakewell, D Plump, and C Runciman. Checking the shape safety of pointer manipulations — extended abstract. In *Participant's Proceedings of the 7th International Seminar on Relational Methods in Computer Science (RelMiCS 7), Malente, Germany*, pages 144–151. University of Kiel, 2003. Available from [SPG]. 49

[BPR03b] A Bakewell, D Plump, and C Runciman. Specifying pointer structures by graph reduction. Technical Report YCS-2003-367, Department of Computer Science, University of York, 2003. Available from [SPG]. 48, 49, 50, 51, 53, 60

[BRS99] M Benedikt, T Reps, and M Sagiv. A decidable logic for describing linked data structures. In *Proc. European Symposium on Programming Languages and Systems (ESOP '99)*, volume 1576 of *LNCS*, pages 2–19. Springer-Verlag, 1999. 60

[DHM03] F Drewes, B Hoffmann, and M Minas. Context-exploiting shapes for diagram transformation. *Machine Graphics and Vision*, 12(1):117–132, 2003. 60

[FM97] P Fradet and D Le Métayer. Shape types. In *Proc. Principles of Programming Languages (POPL '97)*, pages 27–39. ACM Press, 1997. 49, 54, 59

[FM98] P Fradet and D Le Métayer. Structured Gamma. *Science of Computer Programming*, 31(2–3):263–289, 1998. 49, 59

[HMP01] A Habel, J Müller, and D Plump. Double-pushout graph transformation revisited. *Math. Struct. in Comp. Science*, 11:637–688, 2001. 50

[HP01] A Habel and D Plump. Computational completeness of programming languages based on graph transformation. In *Proc. Foundations of Software Science and Computation Structures (FOSSACS 2001)*, volume 2030 of *LNCS*, pages 230–245. Springer-Verlag, 2001. 51

[IO01] S Ishtiaq and P W O'Hearn. BI as an assertion language for mutable data structures. In *Proc. Principles of Programming Languages (POPL '01)*, pages 14–26. ACM Press, 2001. 60

[KLR02] V Kuncak, P Lam, and M Rinard. Role analysis. In *Proc. Principles of Programming Languages (POPL '02)*, pages 17–32. ACM Press, 2002. 60

[KS93] N Klarlund and M I Schwartzbach. Graph types. In *Proc. Principles of Programming Languages (POPL '93)*, pages 196–205. ACM Press, 1993. 60

[MS01] A Møller and M I Schwartzbach. The pointer assertion logic engine. In *Proc. ACM SIGPLAN '01 Conference on Programming Langauge Design and Implementation (PLDI '01)*. ACM Press, 2001. 60

[SPG] Safe Pointers by Graph Transformation, project webpage. `http://www-users.cs.york.ac.uk/~ajb/spgt/`. 48, 60

[SRW02] M Sagiv, T Reps, and R Wilhelm. Parametric shape analysis via 3-valued logic. *ACM Transactions on Programming Languages and Systems*, 24(3):217–298, 2002. 60

[WM01] D Walker and G Morrisett. Alias types for recursive data structures. In *Types in Compilation (TIC '00), Selected Papers*, volume 2071 of *LNCS*, pages 177–206. Springer-Verlag, 2001. 60

Applying Relational Algebra in 3D Graphical Software Design

Rudolf Berghammer[1] and Alexander Fronk[2]

[1] Institute of Computer Science and Applied Mathematics
University of Kiel, D-24098 Kiel, Germany
[2] Software Technology
University of Dortmund, D-44221 Dortmund, Germany

Abstract. Programming in the large is generally supported by tools allowing to design software in a graphical manner. 2D representations of object-oriented classes and their relations, such as e.g. UML diagrams, serve for automatically generating code frames and hence allow to manipulate code by manipulating these diagrams. We focus on 3D representations of various relations between classes implemented in JAVA. Editing these diagrams requires not only to retain correct JAVA code. Moreover, layout is given a meaning, and thus graphical constraints need to be considered as well to describe both the syntax and semantics of our graphical language underlying 3D representations. We model these constraints with relational algebra, and contribute to the practical employment of relational methods in computer science. The RELVIEW system is used to efficiently check these constraints during editing operations to determine syntactically valid diagrams.

1 Introduction

Visualizing software design is most commonly done by means of UML. Classes and their relations, for example, are visualized through UML class diagrams. Other engineering sciences often use 3D CAD systems, but three dimensions are rarely used in the visualization of large software systems. In contrast to the products constructed in classical engineering, the software artifacts are abstract as they do not have a substantial physical corpus. No natural 3D representation of software exists, thus making it difficult to find suitable visual concepts. Additionally, the visualization of large systems raises many problems due to the limitations of the display in space and resolution.

We, however, aim at exploiting three dimensions to support comprehension and overview of the design of large software systems. Thereby, we do not merely convert UML class diagrams into 3D. As the main emphasis, we concentrate on the differences between two and three dimensions. The benefits of a 3D layout for comprehending the static structure of large object-oriented software systems more quickly and easily are discussed in [1]. In [3], we report on this experience based on a first prototype implemented in the course of a Master's thesis [8] for visualizing relations between JAVA classes as 3D diagrams. In [2], finally, we

R. Berghammer et al. (Eds.): RelMiCS/Kleene-Algebra Ws 2003, LNCS 3051, pp. 62–74, 2004.

report on the next step, viz the editing of diagrams, which shifts 3D visualizations to a 3D software modelling language.

We declaratively define our visual language by means of predicates such that a 3D diagram is a term of the language – we say: is valid – if all predicates are satisfied. Editing is seen as term replacement, i.e., editing may transfer valid diagrams into other valid diagrams. In constructive approaches to language definition using, for example, structural induction or production rules, a word or term is contained in a language if it can be constructed in a specific way. Our approach simulates this process by editing operations obeying certain constraints. This means that the validity of 3D diagrams is proved by checking the constraints imposed on the visual language. As a consequence, speaking about editing operations is a crucial and vital aspect for a predicate-based approach for describing 3D visual languages.

To keep both the set of constraints formulated by predicates and editing thereupon manageable, we adopt the approach proposed in [13]. This approach is technically based on a formal structure that separates the abstract syntax graph (ASG) of the visual language from the spatial relations graph (SRG) containing visual elements and their relations displayed graphically. For the latter a physical layout is used which contains all technical information needed to display diagrams on the computer screen. This formal ASG/SRG structure reflects the differences between abstract and concrete syntax. The underlying JAVA code is needed to generate a 3D diagram. In turn, JAVA code has to be generated from a valid diagram. Hence, the source code represents some data visualized three-dimensionally. We obtain an abstract syntax graph of the JAVA code as usual. The ASG simultaneously represents the abstract inner logical structure of a 3D diagram by means of a graph. It interprets a diagram using only its syntactical structure; no information about its layout is needed. The physical layout assembles what geometric elements are displayed on a screen together with their individual location, size, color, and so on. The spatial relations graph abstracts from physical layout by introducing graphical items related to geometric elements in the physical layout. The SRG serves as an intermediate data structure describing how graphical items are related to each other. With these relations, a concrete syntax is introduced. We understand an SRG of such a visualization as a graphical interpretation of the underlying JAVA source code.

In the present paper, we focus on a relation-algebraic formalization of constraints concerning the 3D arrangements and their relations to JAVA. The Kiel RELVIEW system allows to efficiently check them after editing operations. Hence, we obtain a decision basis for distinguishing valid from invalid manipulations. The paper is organized as follows. Section 2 introduces some JAVA class relations of interest and their 3D representations by means of the ASG; Section 3 discusses editing constraints of our 3D diagrams by means of the SRG, and relates them to the ASG; Section 4 gives a relation-algebraic formalization of constraints and discusses the employment of RELVIEW, thereby focusing on sample constraints and their efficient testing. The paper concludes in Section 5 and sketches some related work.

2 Java Class Relations and Their 3D Representations

Based on abstract syntax, we introduce some JAVA class relations of interest in the following. Then we show how to represent them three-dimensionally. At the end of the section we also sketch the underlying physical layout and its purpose.

2.1 The Abstract Syntax

JAVA software is statically structured by packages, interfaces and classes. Packages assemble classes, interfaces and (sub-)packages within a strict hierarchy. Interfaces define types and consist only of method signatures and constant attributes. As they do not have any implementations, multiple inheritance between interfaces is allowed. Only simple inheritance is allowed between classes, but a class can implement several interfaces at once, and implementation can be deferred to sub-classes. Further static relationships are, for instance, *locality*, *association* and *aggregation*, the *uses* relation, and the *implements* relation. These latter relations and others are of no concern for the present paper.

We write $R : X \leftrightarrow Y$ if R is a relation between the two sets X and Y, and $R_{x,y}$ if R contains the pair (x, y). In summary, we investigate three different relations of interest, viz:

$$\begin{array}{ll} \mathsf{cH} : C \leftrightarrow C & \text{class hierarchy} \\ \mathsf{iH} : I \leftrightarrow I & \text{interface hierarchy} \\ \mathsf{pC} : C \cup I \cup P \leftrightarrow P & \text{package containment} \end{array}$$

Here C, I, and P are the sets of all classes, interfaces, and packages, resp. of the JAVA code under consideration. We have $\mathsf{cH}_{x,y}$ if x is a sub-class of y. The relation iH is defined in the same way. Finally, $\mathsf{pC}_{x,p}$ holds if the class (or interface, or package) x is contained in the package p.

Notice that we have to deal with two kinds of information stored in the abstract syntax graph of the underlying JAVA code: information about classes, interfaces, and packages, i.e., about elements of the abstract syntax graph, and derived information about hierarchies or package containment. The latter are not elements of the abstract syntax graph but must also be displayed graphically. To keep track of these relations in our 3D diagrams, we use different geometric arrangements for each relation, thereby visually clearly separating them from each other.

2.2 The Concrete Syntax

In this section, we introduce the concrete syntax of our 3D diagrams by means of pictorials. We visualize JAVA classes and interfaces by boxes and spheres, respectively. Pipes connect them and model different kinds of associations which are not of importance in the present paper. Fig. 1 shows some connected entities arranged as so-called *webs*.

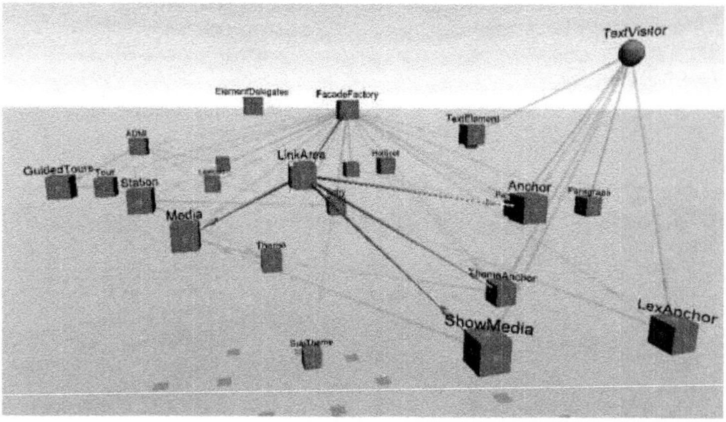

Fig. 1. Some boxes, spheres, and pipes arranged as a web

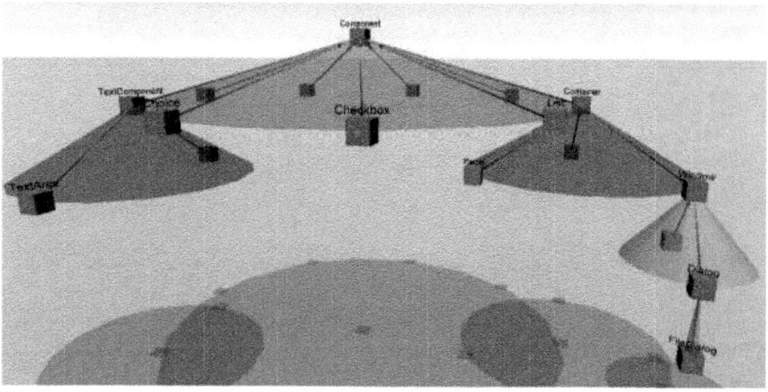

Fig. 2. Some cone trees

To represent class hierarchies, we use *cone trees* as presented in [15]. Cone trees relate graphical objects in a tree-like manner (see the example in Fig. 2). The root of each sub-tree forms the top of a semi-transparent cone and the root's direct children are arranged around the cone's base. A superclass is always displayed in a cone's root. This structure is applied recursively. Individual rotation of each cone allows to bring certain information into the foreground without losing context.

We use *information cubes* as introduced in [14] to display package hierarchies, i.e. to model package containment. Information cubes glue together related information and may contain arbitrary arrangements, in particular sub-cubes, webs, or cone trees. Cubes are semi-transparent and allow visual access to the information within. An example is depicted in Fig. 3.

Interface hierarchies are displayed using the *walls* [11]. Walls form 2D planes as shown in Fig. 4, where we present walls of interfaces and classes, resp. Us-

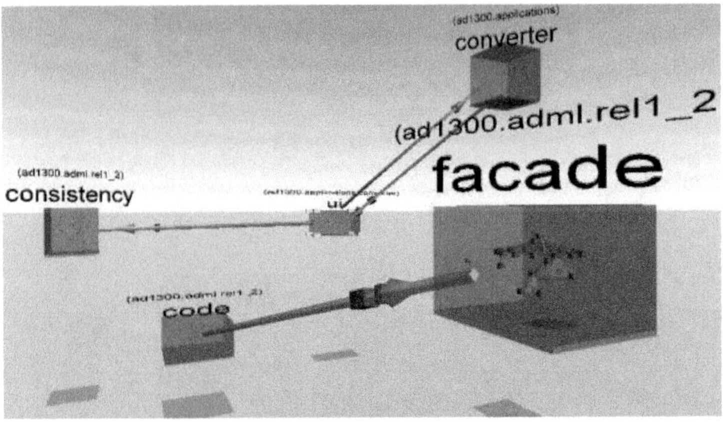

Fig. 3. Some information cubes

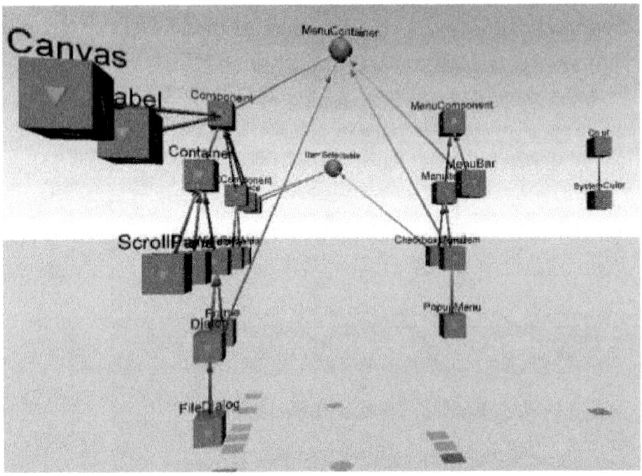

Fig. 4. Walls of interfaces and classes

ing the effect of depth, perspective distortion offers a broad view on entities assembled within the wall.

2.3 The Physical Layout

We assume a 3D coordinate system underlying our diagrams. The coordinate system is seen as independent from coordinates given by a computer screen to technically display the diagrams. This allows an observer to fly through a diagram or, for example, to zoom into information cubes without changing the physical properties of the diagram or its elements. We refer to this 3D coordinate system, which allows us to unambiguously determine the position of each geometric entity, by means of the following relation:

$$\mathsf{point} : POINT \leftrightarrow ENTITY \qquad \text{points of an entity}$$

Here $POINT$ is the set of all 3D coordinates (points), and $ENTITY$ is the set of all geometric entities. We assume that different geometric entities always have different point sets and can thereby unambiguously be identified.

With the help of point we can easily define relations such as inside, outside, or overlaps on $ENTITY$. We will use inside in Sect. 3:

$$\mathsf{inside}_{e,e'} :\Longleftrightarrow \forall p : \mathsf{point}_{p,e} \rightarrow \mathsf{point}_{p,e'}$$
$$\mathsf{outside}_{e,e'} :\Longleftrightarrow \forall p : \mathsf{point}_{p,e} \rightarrow \neg\mathsf{point}_{p,e'}$$
$$\mathsf{overlaps}_{e,e'} :\Longleftrightarrow \exists p : \mathsf{point}_{p,e} \wedge \mathsf{point}_{p,e'}$$

These element-wise definitions say (cf. [16]) that inside equals the right residual point $\backslash$ point. Using inside, also the above mentioned demand that different geometric entities always have different point sets can easily be described by relation-algebraic means. It is equivalent to inside being anti-symmetric, i.e., to the inclusion $\mathsf{inside} \cap \mathsf{inside}^\mathsf{T} \subseteq \mathbb{I}$ with $\mathbb{I}$ as identity relation on the set $ENTITY$ and $\mathsf{inside}^\mathsf{T}$ as transpose of inside. Similarly, outside equals point $\backslash$ $\overline{\mathsf{point}}$, the right residual of point with its negation, and overlaps is given as the composition $\mathsf{point}^\mathsf{T}; \mathsf{point}$.

Further relations like leftOf and rightOf can be determined by the x-axis, the y-axis yields e.g., above and below, whereas the z-axis leads to inFrontOf, behind etc. Notice that these relations are independent from the observer's viewpoint and are only associated with the 3D coordinate system.

3 Editing under Relation-Based Invariants

Having introduced the language of 3D diagrams for representing JAVA class diagrams, we now concentrate on the manipulation of diagrams. First, we introduce some relations which relate the bricks of our visual language to JAVA code. Based on them, we then formulate a list of properties which have to be kept invariant to yield valid diagrams.

3.1 Relating Geometric Entities with Java Code

The meaning of a 3D diagram is the JAVA code it represents. We must distinguish 3D editing operations changing this JAVA code from manipulations leaving the semantic background untouched. In [1], we identified several levels of such manipulations and distinguished 1) operations changing the model neither syntactically nor semantically (such as changing the user's viewpoint, zooming in and out, flying through the model, fading in and out some details, and rearranging the 3D model for better layout), 2) operations changing the diagram but not its underlying JAVA code (such as changing some color, changing an entity's size), and 3) operations changing the meaning (such as extracting an entity from an information cube and moving it into another cube, or changing the label of a box, i.e., changing the name of the class represented by this box).

Level independently, both properties of JAVA code, such as acyclic inheritance hierarchies, as well as certain geometrical invariants describing JAVA independent properties of 3D diagrams have to hold as soon as code is generated from a 3D diagram. In the remainder of this section, we discuss some of them (cf. also [2]). To be technically precise, we call cones, cubes and walls the *displayable entities* of a 3D diagram, and boxes and spheres the *atomic displayable entities*. Let C, I, and P be as in Section 2, and $TREE$, $CUBE$, $WALL$, BOX, and $SPHERE$ be the pairwise disjoint sets of cone trees, cubes, walls, boxes, and spheres, resp. Following Section 2 again, we denote the set of all displayable entities, i.e., the union of $TREE$, $CUBE$, $WALL$, BOX, and $SPHERE$, as $ENTITY$. To relate geometric entities to JAVA code, i.e., to relate the SRG and the ASG, we consider the following relations:

$$\mathsf{C} : BOX \leftrightarrow C \qquad \text{class as box}$$

$$\mathsf{I} : SPHERE \leftrightarrow I \qquad \text{interface as sphere}$$

$$\mathsf{P} : CUBE \leftrightarrow P \qquad \text{package as cube}$$

$$\mathsf{tM} : TREE \leftrightarrow C \qquad \text{tree memberships}$$

$$\mathsf{wM} : WALL \leftrightarrow I \qquad \text{wall membership}$$

$$\mathsf{cM} : CUBE \leftrightarrow C \cup I \cup P \qquad \text{cube membership}$$

The relations C, I, and P are univalent and total. Using the usual notation for functions, $\mathsf{C}(b)$, $\mathsf{I}(s)$, and $\mathsf{P}(c)$ is the JAVA code graphically represented by a box b, a sphere s, or a cube c, resp. The relation tM relates a cone tree t to a class x if the box representing x is a vertex of t. The relations wM and cM are defined analogously.

3.2 The Relation-Based Invariants

The following properties serve as constraints when manipulating 3D visualizations. Their maintenance needs to be checked during editing to determine valid 3D diagrams from which JAVA code can be generated. The proper set of such invariants describing the syntax of our 3D-diagrams is established by translating JAVA code issues into graphical properties concerning the SRG. We focus on a small selection of them to explain our approach.

1. Whenever a class represented by a box b is a member of a package p represented by a cube c, the box must be inside the cube or, recursively, of a cube representing a sub-package of p. Using the relations introduced so far and function notation for C and P, this property reads as

$$\forall\, b, c : \mathsf{cM}_{c,\mathsf{C}(b)} \leftrightarrow \mathsf{inside}_{b,c} \vee \exists\, c' : \mathsf{cM}_{c,\mathsf{P}(c')} \wedge \mathsf{inside}_{b,c'} \,. \tag{1}$$

2. Each box b is contained in at most one cube c, i.e., each JAVA class is a member of at most one package. An obvious formalization of this property using the cardinality operator on sets is

$$\forall b : |\{c : \mathsf{cM}_{c,\mathsf{C}(b)}\}| \leq 1\,. \tag{2}$$

3. Each geometric entity e displayed inside a wall w must be a sphere representing an interface (this invariant holds for cone trees and boxes analogously). Hence, we demand that

$$\forall e, w : \mathsf{inside}_{e,w} \rightarrow e \in \mathit{SPHERE} \wedge \exists i : i = \mathsf{I}(e)\,. \tag{3}$$

4. Each wall w must at least contain two spheres to establish a hierarchy (this invariant holds for cones and cubes analogously). The obvious formalization of this fact uses again the cardinality operator. We get

$$\forall w : |\{i : \mathsf{wM}_{w,i}\}| \geq 2\,. \tag{4}$$

5. Each sphere s displayed within a wall w must have at least one connection to any other sphere s' inside this wall. This implies that s' is also displayed within w, and the relation wM associates w to the interfaces represented by s and s'. This demand is formalized by

$$\forall s, w : \mathsf{inside}_{s,w} \rightarrow \mathsf{wM}_{w,\mathsf{I}(s)} \wedge \exists s' : \mathsf{inside}_{s',w} \wedge \mathsf{wM}_{w,\mathsf{I}(s')}\,. \tag{5}$$

6. For layout purposes, we require that each cone tree t has to be depicted completely within a cube c, i.e., the entire class hierarchy described by t is part of the package c stands for. This demand leads to the formula

$$\forall t, c : (\exists b : \mathsf{tM}_{t,\mathsf{C}(b)} \wedge \mathsf{inside}_{b,c}) \rightarrow \mathsf{inside}_{t,c}\,. \tag{6}$$

4 Mechanizing the Checking of Invariants

In this section, we translate the invariants of the last section into relation-algebraic versions, i.e., into formulae over relational expressions. Then we introduce the tool RELVIEW and show how it can be used for checking the invariants.

4.1 Translating Invariants into Relation-Algebraic Formulae

For constructing relational expressions, we use the empty relation $\perp\!\!\!\perp$, the universal relation $\top\!\!\!\top$, the identity relation $\mathbb{I}$, and the basic operations of relational algebra, viz union $R \cup S$, intersection $R \cap S$, negation $\overline{R}$, transposition R^{T}, and composition $R; S$.

We start the translation with formula (1) of Section 3. Using pure relational notation, $\mathsf{cM}_{c,\mathsf{C}(b)}$ is equivalent to $\exists x : \mathsf{C}_{b,x} \wedge \mathsf{cM}_{c,x}$, and $\mathsf{cM}_{c,\mathsf{P}(c')}$ is equivalent to $\exists p : \mathsf{P}_{c',p} \wedge \mathsf{cM}_{c,p}$. Hence, formula (1) rewrites to

$$\forall b, c : (\exists x : \mathsf{C}_{b,x} \wedge \mathsf{cM}_{c,x}) \leftrightarrow \mathsf{inside}_{b,c} \vee (\exists c' : \exists p : \mathsf{P}_{c',p} \wedge \mathsf{cM}_{c,p} \wedge \mathsf{inside}_{b,c'})\,.$$

Next, we apply transposition, followed by composition. This yields

$$\forall\, b,c : (\mathsf{C};\mathsf{cM}^{\mathsf{T}})_{b,c} \;\leftrightarrow\; \mathsf{inside}_{b,c} \vee (\mathsf{inside};\mathsf{P};\mathsf{cM}^{\mathsf{T}})_{b,c}\,.$$

Now we exploit union and equality of relations obtaining as a relation-algebraic version of (1) the equation

$$\mathsf{C};\mathsf{cM}^{\mathsf{T}} = \mathsf{inside} \cup \mathsf{inside};\mathsf{P};\mathsf{cM}^{\mathsf{T}}\,. \tag{1'}$$

To transform (2) into a relation-algebraic version, we again replace $\mathsf{cM}_{c,\mathsf{C}(b)}$ by $\exists\, x : \mathsf{C}_{b,x} \wedge \mathsf{cM}_{c,x}$ and use composition, yielding

$$\forall\, b : |\{c : (\mathsf{C};\mathsf{cM}^{\mathsf{T}})_{b,c}\}| \leq 1\,.$$

Hence, (2) holds, if and only if $\mathsf{C};\mathsf{cM}^{\mathsf{T}}$ is univalent. It is well known how to express univalence as a relation-algebraic formula (see [16]). In our case we get

$$\mathsf{cM};\mathsf{C}^{\mathsf{T}};\mathsf{C};\mathsf{cM}^{\mathsf{T}} \subseteq \mathbb{I}\,. \tag{2'}$$

For formula (3), we exploit that *SPHERE* is a subset of *ENTITY*, and, hence, there exists an injective function $\mathsf{inj} : \mathit{SPHERE} \leftrightarrow \mathit{ENTITY}$, viz the identity. Using inj, formula (3) becomes

$$\forall\, e,w : \mathsf{inside}_{e,w} \;\rightarrow\; \exists\, s : (\mathsf{inj}_{s,e} \wedge \exists\, i : i = \mathsf{I}(s))\,.$$

A little reflection shows that this formula is equivalent to

$$\forall\, e,w : \mathsf{inside}_{e,w} \;\rightarrow\; \exists\, s : (\mathsf{inj}^{\mathsf{T}}_{e,s} \wedge \exists\, i : \mathsf{I}_{s,i} \wedge \mathbb{T}_{i,w})\,.$$

Now we can use composition followed by the definition of relational inclusion and obtain the desired relation-algebraic version of (3) as

$$\mathsf{inside} \subseteq \mathsf{inj}^{\mathsf{T}};\mathsf{I};\mathbb{T}\,. \tag{3'}$$

The univalent part $R \cap \overline{R;\overline{\mathbb{I}}}$ of a relation R contains a pair (x,y) if and only if R associates with x no further element (see [16]). Using this fact, it is obvious how to translate formula (4) into a relation-algebraic form. We get

$$\mathsf{wM} \cap \overline{\mathsf{wM};\overline{\mathbb{I}}} = \mathbb{O}\,. \tag{4'}$$

Replacing $\mathsf{wM}_{w,\mathsf{I}(s)}$ by $\exists\, i : \mathsf{I}_{s,i} \wedge \mathsf{wM}_{w,i}$, and $\mathsf{wM}_{w,\mathsf{I}(s')}$ by $\exists\, i : \mathsf{I}_{s',i} \wedge \mathsf{wM}_{w,i}$, formula (5) becomes

$$\forall\, s,w : \mathsf{inside}_{s,w} \;\rightarrow\; (\exists\, i : \mathsf{I}_{s,i} \wedge \mathsf{wM}_{w,i}) \wedge \exists\, s' : (\mathsf{inside}_{s',w} \wedge \exists\, i : \mathsf{I}_{s',i} \wedge \mathsf{wM}_{w,i})\,.$$

Now we delete the existential quantifications $\exists i$ by introducing composition and transposition, and replace logical conjunction by relational intersection. We get

$$\forall\, s,w : \mathsf{inside}_{s,w} \;\rightarrow\; (\mathsf{I};\mathsf{wM}^{\mathsf{T}})_{s,w} \wedge \exists\, s' : (\mathsf{inside} \cap \mathsf{I};\mathsf{wM}^{\mathsf{T}})_{s',w}\,.$$

To remove the remaining existential quantification of s', we add the conjunct $\mathsf{T}_{s,s'}$ before $(\mathsf{inside} \cap \mathsf{I};\mathsf{wM}^\mathsf{T})_{s',w}$ and then use composition. The result is

$$\forall s, w : \mathsf{inside}_{s,w} \rightarrow (\mathsf{I};\mathsf{wM}^\mathsf{T})_{s,w} \wedge (\mathsf{T};(\mathsf{inside} \cap \mathsf{I};\mathsf{wM}^\mathsf{T}))_{s,w} .$$

Finally, the definition of intersection followed by that of inclusion transforms formula (5) into the relation-algebraic form

$$\mathsf{inside} \subseteq \mathsf{I};\mathsf{wM}^\mathsf{T} \cap \mathsf{T};(\mathsf{inside} \cap \mathsf{I};\mathsf{wM}^\mathsf{T}) . \tag{5'}$$

Transforming formula (6) into a relation-algebraic version works as follows. We replace $\mathsf{tM}_{t,\mathsf{C}(b)}$ by $\exists x : \mathsf{C}_{b,x} \wedge \mathsf{tM}_{t,x}$, yielding

$$\forall t, c : (\exists b, x : \mathsf{C}_{b,x} \wedge \mathsf{tM}_{t,x} \wedge \mathsf{inside}_{b,c}) \rightarrow \mathsf{inside}_{t,c} ,$$

which in turn is equivalent to

$$\forall t, c : (\exists x : \mathsf{tM}_{t,x} \wedge \exists b : \mathsf{C}^\mathsf{T}_{x,b} \wedge \mathsf{inside}_{b,c}) \rightarrow \mathsf{inside}_{t,c} .$$

From this formula we obtain the relation-algebraic form

$$\mathsf{tM};\mathsf{C}^\mathsf{T};\mathsf{inside} \subseteq \mathsf{inside} , \tag{6'}$$

if we replace the two existential quantifications by two compositions, and thereafter the universal quantification by relational inclusion.

4.2 Using RelView for Checking Invariants

RelView (see [5, 4] for more details) is a computer system for calculating with relations and relational programming representing all data as relations. Because RelView computations frequently use very large relations, for instance, membership, inclusion, and size comparison on powersets, the system uses a sophisticated and very efficient implementation of relations [6] via reduced ordered binary decision diagrams (ROBDDs).

The RelView system can manage as many relations simultaneously as memory allows. The user can manipulate and analyze them by pre-defined operations and tests, user-defined relation-algebraic expressions, functions, and programs as well. Many relational constants and operations on relations are available in the language of RelView, especially the basic operations of relational algebra, viz `^` (transposition), `-` (negation), `|` (union), `&` (intersection), and `*` (composition). The pre-defined tests include `incl`, `eq`, and `empty` for testing inclusion, equality, and emptiness of relations, respectively. This functionality can be accessed through command buttons and simple mouse-clicks. But the usual way is to combine them into expressions, functions, or programs. Currently, we are wrapping the entire RelView functionality into a Java library to make it available for use within other tools and in domains open to relational modelling. One such domain is checking properties of three-dimensional graphical languages as presented in this paper. The tool we built is realized as a plug-in of the development environment Eclipse [7] and exploits the functional core of RelView [12] for testing invariants.

Due to the expressiveness of its language and the efficient ROBDD implementation of relations, the RelView system is an excellent tool for checking invariants as defined in Section 3 by evaluating their relation-algebraic versions. For example, equation (1′) looks in the syntax of RelView as follows:

```
eq(C*cM^, inside | inside * P * cM^).                                    (1'')
```

Checking invariant (1), hence, means to evaluate (1″) and to look whether its value is *true*, which in RelView is represented by an universal relation on a singleton set. (See [4] for more details on the relational modelling of truth values and their operations). Due to its equivalence to the relation-algebraic formula (4′), invariant (4) holds if and only if the value of

```
empty(up(wM))                                                            (4'')
```

is *true*, where the user-defined RelView-function `up(R) = R & -(R*-I(R))` computes the univalent part of a relation.

Even large Java systems mean no problem for RelView. For instance, in [3] the Java API of JDK 1.1.8 is analyzed. Its 679 source files contain 150,289 lines of code which lead to 767 elements (609 classes, 134 interfaces, 24 packages). All relationships between these elements summarize to 5,287 pairs. Relations of this size are efficiently checkable in RelView.

5 Conclusion and Related Work

We have shown how relational algebra can be put to use for checking the syntax of three-dimensional representations of relations between Java classes. The syntax is constituted by language-based properties, such as *cyclic inheritance is not allowed*, and geometrical constraints, such as *cone trees must be displayed either totally inside a cube or totally outside.* We defined them in a predicate logical style and transformed them into an efficiently testable relation-algebraic version. The relations together with their language based properties establish both the syntax and semantics of our 3D diagrams. For example, moving a wall is safe as long as two walls do not intersect. Hence, dragging a wall has to be verified after dropping it at some place. This is done by checking constraints. By relating the SRG to the ASG we determine the Java code underlying the verified diagram. Relational algebra is thus used to both efficiently check syntax and semantics of manipulations carried out on Java code graphically displayed in 3D.

There are some other approaches which apply relational algebra for analyzing and modifying software. In this context, Grok [10] should be mentioned. In contrast to RelView, it has specifically been tailored for the use of relational algebra in software architecture and maintenance. Hence, it allows to solve many tasks in this application domain more easily than RelView. However, compared to the ROBDD-representation of relations in RelView, the Grok-system uses a very simple data structure for relations. If non-relational constructs like loops are frequently needed within a GROK-program, manipulations are executed slowly in the case of very large input relations (see [9]).

References

[1] K. Alfert and A. Fronk. 3-dimensional visualization of Java class relations. In *Proc. 5th World Conference on Integrated Design Process Technology*. Society for Design and Process Science, 2000. On CD-ROM.

[2] K. Alfert and A. Fronk. Manipulation of 3-dimensional visualization of Java class relations. Society for Design and Process Science, 2002.

[3] K. Alfert, A. Fronk, and F. Engelen. Experiences in 3-dimensional visualization of Java class relations. *Transactions of the SDPS: Journal of Integrated Design and Process Science*, 5(3):91–106, September 2001.

[4] R. Behnke, R. Berghammer, T. Hoffmann, B. Leoniuk, and P. Schneider. Applications of the RELVIEW system. In *Proc. 3rd International Workshop on Tool Support for System Specification, Development and Verification*, Advances in Computer Science, pages 33–47. Springer, 1998.

[5] R. Behnke, R. Berghammer, E. Meyer, and P. Schneider. RELVIEW– A system for calculating with relations and relational programming. In *Proc. 1st International Conference on Fundamental Approaches to Software Engineering*, volume 1382 of *Lecture Notes in Computer Science (LNCS)*, pages 318–321. Springer, 1998.

[6] R. Berghammer, B. Leoniuk B., and U. Milanese. Implementation of relational algebra using binary decision diagrams. In *Proc. 6th International Workshop on Relational Methods in Computer Science*, volume 2561 of *Lecture Notes in Computer Science (LNCS)*, pages 241–357. Springer, 2002.

[7] Eclipse.org Consortium. `http://www.eclipse.org`. Last visited June 23, 2003.

[8] F. Engelen. 3-Dimensionale Darstellung von Java-Klassenstrukturen. Master's thesis, Lehrstuhl für Software-Technologie, Universität Dortmund, 2000.

[9] H. M. Fahmy, R. C. Holt, and J. R. Cordy. Wins and losses of algebraic transformations of software architecture. In *Proc. IEEE 16th International Conference on Automated Software Engineering*, November 2001.

[10] R. C. Holt. Binary relational algebra applied to software architecture. Technical Report 345, Computer Systems Research Institute, University of Toronto, 1996.

[11] H. Koike and H.-C. Chu. How does 3-D visualization work in Software Engineering? Empirical study of a 3-D version/module visualization system. In *Proc. 20th International Conference on Software Engineering*, pages 516–519, Kyoto, Japan, April 1998.

[12] Kiel University. `http://www.informatik.uni-kiel.de/~progsys/relview/kure`. Last visited June 23, 2003.

[13] J. Rekers and A. Schürr. Defining and parsing visual languages with layered graph grammars. *Journal of Visual Languages and Computing*, 8(1):27–55, 1997.

[14] J. Rekimoto and M. Green. The information cube: Using transparency in 3D information visualization. In *Proc. 3rd Annual Workshop on Information Technologies & Systems*, pages 125–132, 1993.

[15] G. G. Robertson, S. K. Card, and J. D. Mackinlay. Information visualizing using 3D interactive animations. *Communications of the ACM*, 36(4):57–71, 1993.

[16] G. Schmidt and Th. Ströhlein. *Relations and graphs*. EATCS Monographs on Theoretical Computer Science. Spinger, 1993.

Investigating Discrete Controllability with Kleene Algebra*

Hans Bherer[1], Jules Desharnais[1], Marc Frappier[2], and Richard St-Denis[2]

[1] Département d'informatique et de génie logiciel, Université Laval
Québec, QC, G1K 7P4 Canada
{Hans.Bherer,Jules.Desharnais}@ift.ulaval.ca
[2] Département de mathématiques et d'informatique, Université de Sherbrooke
Sherbrooke, QC, J1K 2R1 Canada
{Marc.Frappier,Richard.St-Denis}@dmi.usherb.ca

1 Introduction

A Discrete Event System (DES) is a dynamic system whose evolution is governed by the instantaneous occurrence of physical events. DES arise in many areas such as robotics, manufacturing, communication networks, and transportation. They are often modelled by languages or automata over an alphabet of symbols denoting the events. In 1987, Ramadge and Wonham initiated a very successful approach to the control of DES [10, 13], which was subsequently extended by themselves and others. Textbooks or course notes on the subject include [1, 7, 12].

In this paper, we present a new view of the concept of controllability used in discrete control theory by adopting a formalization based on Kleene Algebra (KA). This allows us to formulate new problems (and solve them) by incorporating additional parameters in the formalization. Moreover, the new results hold for models other than languages, namely path algebras and algebras of relations. Finally, because the results are proved in a calculational style based on an axiomatization of KAs, they can more easily be checked than those of standard presentations.

In Sect. 2, we present the basic notions of control theory and a sample of results drawn from [11] and Chap. 3 of [7]. This material concerns control theory under complete observation. Sect. 3 presents the necessary background on KA. Sect. 4 contains the KA formalization of the concept of controllability and related results. And there is a short conclusion.

2 Standard DES Control Theory under Complete Observation

Assume a finite set Σ of *events* that occur in a DES to be controlled, called the *plant*. The plant is described by two languages P and P_{m} over Σ, such that $P_{\mathsf{m}} \subseteq$

* This research is supported by FQRNT (Fonds québécois de la recherche sur la nature et les technologies).

R. Berghammer et al. (Eds.): RelMiCS/Kleene-Algebra Ws 2003, LNCS 3051, pp. 74–85, 2004.

$P = \mathsf{PR}(P) \neq \emptyset$, where $\mathsf{PR}(P)$ denotes the *prefix closure* of P. The language P_{m} is called the *marked* language of the plant, while P is called the *generated* language. By definition, any generated language P is *prefix-closed*, i.e., $\mathsf{PR}(P) \subseteq P$ (equivalently, $\mathsf{PR}(P) = P$). One may think of P_{m} as the language accepted by a (not necessarily finite) automaton according to the standard definition, while P is the language accepted by the same automaton, except that all its states are accepting; note that the latter language is prefix-closed.

There is a subset $\Sigma_{\mathsf{u}} \subseteq \Sigma$ of *uncontrollable* events, whose complement $\Sigma_{\mathsf{c}} \stackrel{\mathsf{def}}{=} \Sigma - \Sigma_{\mathsf{u}}$ contains the *controllable* events. A *controller* (or *supervisor*) is a function $f{:}P \to 2^{\Sigma_{\mathsf{c}}}$. For a sequence $s \in P$, $f(s) \subseteq \Sigma_{\mathsf{c}}$ is the set of controllable events that are disabled by the controller after the occurrence of s. The generated language P^f and the marked language P^f_{m} of the plant controlled by supervisor f are defined as follows:[1]

$$\begin{aligned} &\epsilon \in P^f \ , \\ &\forall(s,\sigma \mid s \in \Sigma^* \wedge \sigma \in \Sigma : s \in P^f \ \wedge\ s\sigma \in P \wedge \sigma \notin f(s) \ \Rightarrow\ s\sigma \in P^f) \ , \\ &P^f_{\mathsf{m}} \stackrel{\mathsf{def}}{=} P^f \cap P_{\mathsf{m}} \ . \end{aligned} \qquad (1)$$

In other words, if a sequence s belongs to the supervised behaviour P^f, and the plant generates the sequence $s\sigma$, and the event σ is not prohibited by the supervisor after s, then $s\sigma$ belongs to the supervised behaviour. The equation $P^f_{\mathsf{m}} = P^f \cap P_{\mathsf{m}}$ means that the supervisor is *nonmarking*, that is, it does not contribute to marking sequences of the supervised marked language.

A supervisor is said to be *nonblocking* iff $\mathsf{PR}(P^f_{\mathsf{m}}) = P^f$.

Example 1. Consider the following automaton with marked language P_{m} and generated language P as indicated (regular expressions are used to simplify the description of languages). The automaton models a simple (unrealistic) machine. The marked language consists of sequences of events that bring the machine in the idle state after completion of a part or after being repaired. Starting and repairing the machine can be controlled, but production of a part once the machine has started cannot be controlled and, of course, breaks cannot either. Hence, $\Sigma_{\mathsf{u}} = \{\mathsf{p}, \mathsf{b}\}$.

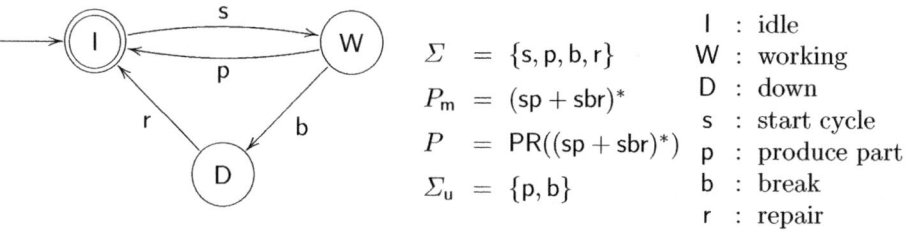

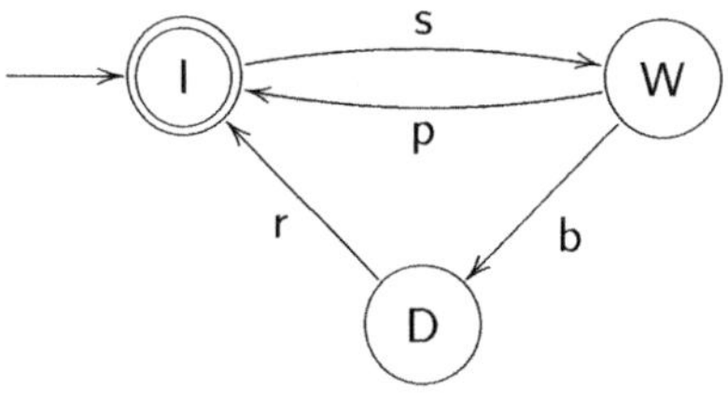

$$\begin{aligned} \Sigma &= \{\mathsf{s},\mathsf{p},\mathsf{b},\mathsf{r}\} \\ P_{\mathsf{m}} &= (\mathsf{sp}+\mathsf{sbr})^* \\ P &= \mathsf{PR}((\mathsf{sp}+\mathsf{sbr})^*) \\ \Sigma_{\mathsf{u}} &= \{\mathsf{p},\mathsf{b}\} \end{aligned}$$

I : idle
W : working
D : down
s : start cycle
p : produce part
b : break
r : repair

[1] Quantifiers have three arguments: a list of variables, the domain over which the quantification applies, and the quantified expression; for instance, $\forall(x \mid P : Q)$ is read "for all x satisfying P, Q holds", or "for all x, $P \Rightarrow Q$", while $\exists(x \mid P : Q)$ is read "there exists an x satisfying P and Q". When the second argument is true, it is omitted.

Suppose that the specification of the controller f is

$$f(u) = \begin{cases} \{\mathsf{r}\} & \text{if } u = v\mathsf{sb} \text{ for some } v \in P \\ \emptyset & \text{otherwise} \end{cases} .$$

The role of the controller is to prevent repairs (the owner of the machine goes bankrupt and cannot afford a repair). Then the generated and marked language of the controlled system are the following.

$$\begin{aligned} P^f &= (\mathsf{sp})^* + (\mathsf{sp})^*\mathsf{s} + (\mathsf{sp})^*\mathsf{sb} \\ P^f_{\mathsf{m}} &= (\mathsf{sp})^* \end{aligned}$$

Since $\mathsf{PR}(P^f_{\mathsf{m}}) = (\mathsf{sp})^* + (\mathsf{sp})^*\mathsf{s} \neq P^f$, the supervisor is blocking.

Now, let S be a language intended as the specification of a DES. We want to implement specification S by adding a controller to the plant described by P and P_{m}. The question is then whether there exists a controller such that the controlled system has behaviour S. The following definitions will help characterize situations where such a supervisor exists.

Definition 2. *Given Σ_{u} and a generated language P, a language S is said to be* controllable[2] *(with respect to P and Σ_{u}) iff* $\mathsf{PR}(S)\Sigma_{\mathsf{u}} \cap P \subseteq \mathsf{PR}(S)$.

This means that if a prefix s of a controllable language is extended by an uncontrollable event u, and the sequence su is generated by the plant, then su is a prefix of the controllable language. Note that S is controllable iff $\mathsf{PR}(S)$ is controllable.

Definition 3. *Given the marked language P_{m}, a language $S \subseteq \Sigma^*$ is said to be* P_{m}-closed *iff* $\mathsf{PR}(S) \cap P_{\mathsf{m}} \subseteq S$ *(equivalently,* $\mathsf{PR}(S) \cap P_{\mathsf{m}} = S \cap P_{\mathsf{m}}$*).*

This is a form of *relative closure* (closure relative to P_{m}).

Theorem 4. *Let a DES with generated language P and marked language P_{m} be given. Let $S \subseteq \Sigma^*$ be a specification.*

1. *There exists a supervisor f such that $P^f = S$ iff $\emptyset \neq S = \mathsf{PR}(S) \subseteq P$ and S is controllable.*
2. *There exists a nonblocking supervisor f such that $P^f_{\mathsf{m}} = S$ iff $\emptyset \neq S = \mathsf{PR}(S) \cap P_{\mathsf{m}}$ and S is controllable.*

When the conditions of Theorem 4 do not hold and a supervisor cannot be found, one can still look for approximations to the desired language (i.e., the specification) for which a supervisor can be found. Theorem 5 presents results in this respect. The following convention is used in the statement of the theorem:

– $\mathsf{SUP}X(S)$ denotes the supremal sublanguage of S (i.e., the greatest language included in S) satisfying property X.

[2] Even though it is the plant that we want to control, it is the specification that is deemed to be controllable or not; this terminology is traditional [7, 11].

- $\mathsf{INF}X(S)$ denotes the infimal superlanguage of S (i.e., the least language containing S) satisfying property X.
- Property X can be C (controllability with respect to P), P (prefix closure), R (P_{m} relative closure), or a combination of these.
- The notation $\overline{A}$ denotes the complement of A with respect to Σ^*, while $A - B \stackrel{\text{def}}{=} A \cap \overline{B}$. We also use the *right-detachment operator* $\lfloor$ [9], defined as follows: for two languages A and B, $A \lfloor B \stackrel{\text{def}}{=} \{s \mid \exists(t \mid t \in B : st \in A)\}$.

Theorem 5. *Let Σ_{u}, a generated language P, a marked language P_{m} and a specification $S \subseteq P$ be given.*

1. $\mathsf{SUPC}(S)$ *exists. If S is prefix-closed*[3], *then* $\mathsf{SUPC}(S)$ *can be computed as the limit of the following iteration:*
   - $S_0 \stackrel{\text{def}}{=} S$,
   - $S_{i+1} \stackrel{\text{def}}{=} S_i - ((P - S_i)\lfloor \Sigma_{\mathsf{u}})\Sigma^* = S_i \cap \overline{((P \cap \overline{S_i})\lfloor \Sigma_{\mathsf{u}})\Sigma^*}$.
2. $\mathsf{INFC}(S)$ *need not exist.*
3. $\mathsf{SUPPC}(S) = S - ((P - S)\lfloor \Sigma_{\mathsf{u}}^*)\Sigma^* = S \cap \overline{((P \cap \overline{S})\lfloor \Sigma_{\mathsf{u}}^*)\Sigma^*}$.
4. $\mathsf{INFPC}(S) = \mathsf{PR}(S)\Sigma_{\mathsf{u}}^* \cap P$.
5. SUPRC *exists. If S is prefix-closed*[4], *then* SUPRC *can be computed as the limit of the following iteration:*
   - $S_0 \stackrel{\text{def}}{=} S - (P_{\mathsf{m}} - S)\Sigma^* = S \cap \overline{(P_{\mathsf{m}} \cap \overline{S})\Sigma^*}$,
   - $S_{i+1} \stackrel{\text{def}}{=} S_i - ((P - S_i)\lfloor \Sigma_{\mathsf{u}})\Sigma^* = S_i \cap \overline{((P \cap \overline{S_i})\lfloor \Sigma_{\mathsf{u}})\Sigma^*}$.
6. $\mathsf{INFRC}(S)$ *need not exist.*

Example 6. Consider the following automaton.

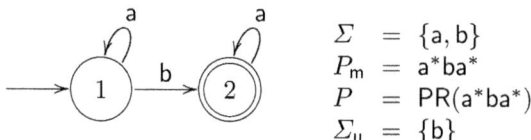

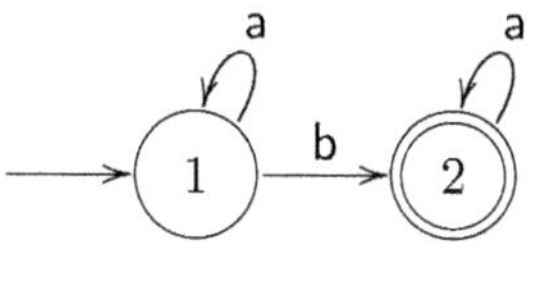

$$\begin{aligned} \Sigma &= \{\mathsf{a}, \mathsf{b}\} \\ P_{\mathsf{m}} &= \mathsf{a}^*\mathsf{b}\mathsf{a}^* \\ P &= \mathsf{PR}(\mathsf{a}^*\mathsf{b}\mathsf{a}^*) \\ \Sigma_{\mathsf{u}} &= \{\mathsf{b}\} \end{aligned}$$

1. $S = \mathsf{a}^*$ is not controllable, since
$$\mathsf{PR}(S)\Sigma_{\mathsf{u}} \cap P = \mathsf{a}^*b \cap \mathsf{PR}(\mathsf{a}^*\mathsf{b}\mathsf{a}^*) = \mathsf{a}^*b \not\subseteq a^* = \mathsf{PR}(S)$$
To enforce S, the controller would have to disable b, but b is not controllable. Using Theorem 5(1), one finds that $\mathsf{SUPC}(S) = \emptyset$. But since by definition $P^f \neq \emptyset$ (see (1)), there is no controller for $\mathsf{SUPC}(S)$. However, $\mathsf{INFC}(S)$ exists; it is easy to see that $\mathsf{INFC}(S) = \mathsf{INFPC}(S) = \mathsf{a}^*(\epsilon + \mathsf{b})$.
2. $S = \mathsf{ba} + \mathsf{aba}$ is controllable, since $\mathsf{PR}(S)\Sigma_{\mathsf{u}} \cap P = (\epsilon + \mathsf{b} + \mathsf{ba} + \mathsf{a} + \mathsf{ab} + \mathsf{aba})\mathsf{b} \cap \mathsf{PR}(\mathsf{a}^*\mathsf{b}\mathsf{a}^*) = \mathsf{b} + \mathsf{ab} \subseteq \epsilon + \mathsf{b} + \mathsf{ba} + \mathsf{a} + \mathsf{ab} + \mathsf{aba} = \mathsf{PR}(S)$. By Theorem 4, there exists a supervisor f such that $P^f = \mathsf{PR}(S)$, since $\mathsf{PR}(S)$ is controllable and $\emptyset \neq \mathsf{PR}(S) \subseteq P$. However, there is no supervisor f such that $P_{\mathsf{m}}^f = S$, since $\mathsf{PR}(S) \cap P_{\mathsf{m}} = \mathsf{b} + \mathsf{ba} + \mathsf{ab} + \mathsf{aba} \neq S$.

---

[3] This restriction is not explicitly given in [7] and may be implicit from the context; a non-prefix-closed S can also be used if $P - S_i$ is replaced by $P - \mathsf{PR}(S_i)$ in the definition of S_{i+1}.

[4] Same comment as for SUPC.

3 Kleene Algebra Basics

We use the following definition from [4], which follows Conway [2], since we want to admit general recursive definitions, not just the Kleene star. There are different definitions, such as that in [6], for instance.

Definition 7. *A* (standard) Kleene algebra (KA) *is a sextuple* $(K, \leq, 0, \top, \cdot, 1)$ *satisfying the following properties:*

1. $(K, \leq)$ *is a complete lattice with least element* 0 *and greatest element* $\top$*. The supremum of a subset* $L \subseteq K$ *is denoted by* $\sqcup L$*. The supremum of two elements* $x, y \in K$ *is denoted by* $x + y$*.*
2. $(K, \cdot, 1)$ *is a monoid.*
3. *The operation* $\cdot$ *is universally disjunctive (i.e., it distributes through arbitrary suprema) in both arguments.*

Because the lattice of a KA is complete, and because $+$ and $\cdot$ are monotonic, we can recover the Kleene star by defining $a^* \stackrel{\mathsf{def}}{=} \mu(x \mid: a \cdot x + 1)$. Since $+$ and $\cdot$ are continuous, we also have

$$a^* = \bigsqcup(i \mid i \geq 0 : a^i) \ ,$$

where $a^0 = 1$ and $a^{i+1} = a \cdot a^i$.

In the sequel, we assume KAs to be Boolean (i.e., the underlying lattice is a Boolean algebra), with meet and complement denoted by $\sqcap$ and $^{-}$, respectively[5]. This allows the definition of the *left* and *right detachment* operators [9] (*relative converses* in [5]) by the following axiom of *exchange*:

$$a \rfloor c \sqcap b = 0 \Leftrightarrow a \cdot b \sqcap c = 0 \Leftrightarrow c \lfloor b \sqcap a = 0 \ . \tag{2}$$

Models of Boolean KAs include algebras of languages over Σ^* for an alphabet Σ (this is the setting of Sect. 2), algebras of path sets in a directed graph [8], algebras of concrete relations and abstract relation algebras. For languages,

$$\begin{aligned} A \lfloor B &= \{s \mid \exists(t \mid t \in B : st \in A)\} \ , \\ A \rfloor B &= \{t \mid \exists(s \mid s \in A : st \in B)\} \ . \end{aligned} \tag{3}$$

For relations, $A \lfloor B = A \cdot B^{\smile}$ and $A \rfloor B = A^{\smile} \cdot B$, where $A^{\smile}$ is the converse of A.

Here is a sample of relevant laws.

Theorem 8. *Let* $a, b, c \in K$ *and* $X \subseteq K$*.*

1. $0 \cdot a = a \cdot 0 = 0$ *(zero of* $\cdot$*).*
2. $1 \cdot a = a \cdot 1 = a$ *(identity of* $\cdot$*).*
3. $\top \cdot \top = \top$*.*
4. $a \leq b \Rightarrow a \cdot c \leq b \cdot c$ *and* $a \leq b \Rightarrow c \cdot a \leq c \cdot b$ *(monotonicity of* $\cdot$*).*
5. $a \leq b \Rightarrow a^* \leq b^*$ *(monotonicity of* **).*
6. $a \lfloor 0 = 0 \lfloor a = 0$ *(zero of* $\lfloor$*).*
7. $a \lfloor 1 = a$ *(right unit of* $\lfloor$*).*
8. $a \leq b \Rightarrow a \lfloor c \leq b \lfloor c$ *and* $a \leq b \Rightarrow c \lfloor a \leq c \lfloor b$ *(monotonicity of* $\lfloor$*).*
9. $(\sqcup X) \lfloor a = \sqcup (X \lfloor a)$ *and* $a \lfloor (\sqcup X) = \sqcup (a \lfloor X)$ *(disjunctivity).*

[5] The precedence of the operators, from lowest to highest, is $(+, \sqcap), (\cdot, \rfloor, \lfloor), ^{-}$.

4 Formalizing Controllability Properties with Kleene Algebra

We now formalize the notion of controllability with Kleene algebra. This is done in the following definition. We then explain how to understand the definition and how it is linked to Definition 2.

Definition 9. *Let $p, r, s, u \in K$. We say that s is r-*controllable *with respect to p and the* uncontrollable *element u, notated* $\mathsf{CTR}(s, p, r, u)$, *iff*

$$(s\lfloor r)\cdot u \sqcap p\lfloor r \leq s\lfloor r. \tag{4}$$

We now make a few remarks on this definition. Unless otherwise indicated, we assume a KA of languages, to facilitate the connection with Definition 2.

1. In a KA of languages, $a\lfloor\top$ is the set of prefixes of a (i.e., $a\lfloor\top = \mathsf{PR}(a)$), as one can check from (3). To make the correspondence with Definition 2 more exact, we would define controllability by $(s\lfloor\top)\cdot u \sqcap p\lfloor\top \leq s\lfloor\top$ (i.e., $\top$-controllability). The drawback is that in an algebra of relations, any relation s is $\top$-controllable for any p, u. Indeed, for any relation s, $s\lfloor\top = s\cdot\top^{\smile} = s\cdot\top$, whence $(s\lfloor\top)\cdot u = s\cdot\top\cdot u \leq s\cdot\top = s\lfloor\top$. Choosing an arbitrary r provides greater generality. The expression $s\lfloor r$ denotes the set of prefixes of sequences of s that have sequences of r as suffixes. Having $r \neq \top$ means that there is a limited capability to look at prefixes (to look in the past, so to speak).
2. The elements p and s correspond to the language P of the plant and specification S of Sect. 2, except that p is not assumed to be prefix-closed; instead of writing p in (4), we write $p\lfloor r$. The symbol "r" reminds of the "removing" action of $\lfloor r$. In a faithful translation of Definition 2, we would have $u = \Sigma_{\mathsf{u}}$. But we can choose anything else.

The following theorem provides some basic properties of relation CTR.

Theorem 10. *Let I be an arbitrary index set.*

1. $\forall(i \mid i \in I : \mathsf{CTR}(s_i, p, r, u)) \Rightarrow \mathsf{CTR}(\bigsqcup(i \mid i \in I : s_i), p, r, u)$ *and the implication is strict.*
2. $\forall(i \mid i \in I : \mathsf{CTR}(s, p_i, r, u)) \Leftrightarrow \mathsf{CTR}(s, \bigsqcup(i \mid i \in I : p_i), r, u)$.
3. $\forall(i \mid i \in I : \mathsf{CTR}(s, p, r, u_i)) \Leftrightarrow \mathsf{CTR}(s, p, r, \bigsqcup(i \mid i \in I : u_i))$.
4. $\mathsf{CTR}(0, p, r, u)$, $\mathsf{CTR}(s, 0, r, u)$, $\mathsf{CTR}(s, p, 0, u)$, $\mathsf{CTR}(s, p, r, 0)$.
5. $\mathsf{CTR}(p, p, r, u)$, $\mathsf{CTR}(s, p, r, 1)$, $\mathsf{CTR}(\top, p, 1, u)$, $\mathsf{CTR}(\top, p, r^*, u)$.
6. $p_1 \leq p_2 \Rightarrow (\mathsf{CTR}(s, p_2, r, u) \Rightarrow \mathsf{CTR}(s, p_1, r, u))$ *(anti-monotonicity of* CTR *with respect to its second argument).*
7. $u_1 \leq u_2 \Rightarrow (\mathsf{CTR}(s, p, r, u_2) \Rightarrow \mathsf{CTR}(s, p, r, u_1))$ *(anti-monotonicity of* CTR *with respect to its fourth argument).*

Proof.

1. $\forall(i \mid i \in I : \mathsf{CTR}(s_i, p, r, u))$
$\Leftrightarrow$ $\langle$ *Definition of* CTR *(Definition 9)* $\rangle$
$\forall(i \mid i \in I : (s_i\lfloor r)\cdot u \sqcap p\lfloor r \leq s_i\lfloor r)$
$\Rightarrow$ $\langle$ *Monotonicity of* $\sqcup$ $\rangle$
$\bigsqcup(i \mid i \in I : (s_i\lfloor r)\cdot u \sqcap p\lfloor r) \leq \bigsqcup(i \mid i \in I : s_i\lfloor r)$
$\Leftrightarrow$ $\langle$ *Distributivity of* $\sqcap$ *over* $\sqcup$ $\rangle$
$\bigsqcup(i \mid i \in I : (s_i\lfloor r)\cdot u) \sqcap p\lfloor r \leq \bigsqcup(i \mid i \in I : s_i\lfloor r)$
$\Leftrightarrow$ $\langle$ *Distributivity of* $\cdot$ *over* $\sqcup$ *(Definition 7(3))* $\rangle$
$\bigsqcup(i \mid i \in I : s_i\lfloor r)\cdot u \sqcap p\lfloor r \leq \bigsqcup(i \mid i \in I : s_i\lfloor r)$
$\Leftrightarrow$ $\langle$ *Distributivity of* $\lfloor$ *over* $\sqcup$ *(Theorem 8(9))* $\rangle$
$(\bigsqcup(i \mid i \in I : s_i)\lfloor r)\cdot u \sqcap p\lfloor r \leq \bigsqcup(i \mid i \in I : s_i)\lfloor r$
$\Leftrightarrow$ $\langle$ *Definition of* CTR *(Definition 9)* $\rangle$
$\mathsf{CTR}(\bigsqcup(i \mid i \in I : s_i), p, r, u)$.
The following counterexample shows that the implication is strict: $\mathsf{CTR}(1 + \top, \top, 1, \top)$ *holds, but* $\mathsf{CTR}(1, \top, 1, \top)$ *does not.*
2. $\forall(i \mid i \in I : \mathsf{CTR}(s, p_i, r, u))$
$\Leftrightarrow$ $\langle$ *Definition of* CTR *(Definition 9)* $\rangle$
$\forall(i \mid i \in I : (s\lfloor r)\cdot u \sqcap p_i\lfloor r \leq s\lfloor r)$
$\Leftrightarrow$ $\langle$ *Property of* $\sqcup$ $\rangle$
$\bigsqcup(i \mid i \in I : (s\lfloor r)\cdot u \sqcap p_i\lfloor r) \leq s\lfloor r$
$\Leftrightarrow$ $\langle$ *Distributivity of* $\sqcap$ *over* $\sqcup$ $\rangle$
$(s\lfloor r)\cdot u \sqcap \bigsqcup(i \mid i \in I : p_i\lfloor r) \leq s\lfloor r$
$\Leftrightarrow$ $\langle$ *Distributivity of* $\lfloor$ *over* $\sqcup$ *(Theorem 8(9))* $\rangle$
$(s\lfloor r)\cdot u \sqcap \bigsqcup(i \mid i \in I : p_i)\lfloor r \leq s\lfloor r$
$\Leftrightarrow$ $\langle$ *Definition of* CTR *(Definition 9)* $\rangle$
$\mathsf{CTR}(s, \bigsqcup(i \mid i \in I : p_i), r, u)$.
3. *The proof is similar to the previous one.*
4. *All four properties follow directly from Definition 9 and Theorem 8(1,6).*
5. $\mathsf{CTR}(p, p, r, u)$ *and* $\mathsf{CTR}(s, p, r, 1)$ *are trivial.* $\mathsf{CTR}(\top, p, 1, u)$ *follows from Theorem 8(7), because* $s\lfloor r = \top\lfloor 1 = \top$. *Due to* $1 \leq r^*$ *and monotonicity of* $\lfloor$ *(Theorem 8(8)), the case of* $\mathsf{CTR}(\top, p, r^*, u)$ *is similar.*
6. *Assume* $p_1 \leq p_2$ *and set* $I \stackrel{\text{def}}{=} \{1, 2\}$ *in part 2 of this theorem.*
7. *Assume* $u_1 \leq u_2$ *and set* $I \stackrel{\text{def}}{=} \{1, 2\}$ *in part 3 of this theorem.* □

Next comes a generalization of the notion of P_{m}-closure (Definition 3).

Definition 11. *Given* $r, t \in K$, *an element* $s \in K$ *is said to be r-t-*closed, *notated* $\mathsf{CLR}(s, r, t)$, *iff*

$$s\lfloor r \sqcap t \leq s \ .$$

Remark 12. There are two special cases of Definition 11 worth mentioning. The first one is that $\mathsf{CLR}(s, r, 0)$ holds, which means that no closure constraint is imposed. The second one is that $\mathsf{CLR}(s, \top, \top)$ is equivalent to prefix-closure of s.

Lemma 13. *Let I be an arbitrary index set. Then*

$$\forall(i \mid i \in I : \mathsf{CLR}(s_i, r, t)) \Rightarrow \mathsf{CLR}(\bigsqcup(i \mid i \in I : s_i), r, t)$$

and the implication is strict.

The proof of the lemma is omitted, since it is similar to that of Theorem 10(1).

In the next lemma, we give equivalent expressions for $\mathsf{CTR}(s, p, r, u)$ and $\mathsf{CLR}(s, r, t)$. These will be used to describe maximal elements satisfying certain properties in Corollary 15.

Lemma 14. *1.* $\mathsf{CTR}(s, p, r, u) \Leftrightarrow s \leq \overline{((p\lfloor r \sqcap \overline{s\lfloor r})\lfloor u)\cdot r}$.

2. $\mathsf{CTR}(s, p, r, u) \Leftrightarrow p \leq \overline{((s\lfloor r)\cdot u \sqcap \overline{s\lfloor r})\cdot r}$.

3. $\mathsf{CTR}(s, p, r, u) \Leftrightarrow u \leq \overline{(s\lfloor r)\rfloor(p\lfloor r \sqcap \overline{s\lfloor r})}$.

4. $\mathsf{CLR}(s, r, t) \Leftrightarrow s \leq \overline{(t \sqcap \overline{s})\cdot r}$.

Proof.

1. $\mathsf{CTR}(s, p, r, u)$
$\Leftrightarrow$ ⟨ Definition of CTR (Definition 9) ⟩
$(s\lfloor r)\cdot u \sqcap p\lfloor r \leq s\lfloor r$
$\Leftrightarrow$ ⟨ Shunting ⟩
$(s\lfloor r)\cdot u \sqcap p\lfloor r \sqcap \overline{s\lfloor r} \leq 0$
$\Leftrightarrow$ ⟨ Exchange (2) ⟩
$s\lfloor r \sqcap (p\lfloor r \sqcap \overline{s\lfloor r})\lfloor u \leq 0$
$\Leftrightarrow$ ⟨ Exchange (2) ⟩
$s \sqcap ((p\lfloor r \sqcap \overline{s\lfloor r})\lfloor u)\cdot r \leq 0$
$\Leftrightarrow$ ⟨ Shunting ⟩
$s \leq \overline{((p\lfloor r \sqcap \overline{s\lfloor r})\lfloor u)\cdot r}$

2. $\mathsf{CTR}(s, p, r, u)$
$\Leftrightarrow$ ⟨ Definition of CTR (Definition 9) ⟩
$(s\lfloor r)\cdot u \sqcap p\lfloor r \leq s\lfloor r$
$\Leftrightarrow$ ⟨ Shunting ⟩
$(s\lfloor r)\cdot u \sqcap p\lfloor r \sqcap \overline{s\lfloor r} \leq 0$
$\Leftrightarrow$ ⟨ Exchange (2) ⟩
$p \sqcap ((s\lfloor r)\cdot u \sqcap \overline{s\lfloor r})\cdot r \leq 0$
$\Leftrightarrow$ ⟨ Shunting ⟩
$p \leq \overline{((s\lfloor r)\cdot u \sqcap \overline{s\lfloor r})\cdot r}$

3. $\mathsf{CTR}(s, p, r, u)$
$\Leftrightarrow$ ⟨ Definition of CTR (Definition 9) ⟩
$(s\lfloor r)\cdot u \sqcap p\lfloor r \leq s\lfloor r$
$\Leftrightarrow$ ⟨ Shunting ⟩
$(s\lfloor r)\cdot u \sqcap p\lfloor r \sqcap \overline{s\lfloor r} \leq 0$
$\Leftrightarrow$ ⟨ Exchange (2) ⟩

$$
\begin{array}{l}
u \sqcap (s\lfloor r)\rfloor(p\lfloor r \sqcap \overline{s\lfloor r}) \leq 0 \\
\Leftrightarrow \quad \langle \text{ Shunting } \rangle \\
u \leq \overline{(s\lfloor r)\rfloor(p\lfloor r \sqcap \overline{s\lfloor r})}
\end{array}
$$

4. The proof is similar to the previous ones. □

The following corollary shows how to find maximal elements below given elements s', p', u'. It also describes lattice structures below these maximal elements. The notation $\downarrow a$, for $a \in K$, is defined by $\downarrow a \stackrel{\text{def}}{=} \{x \mid x \leq a\}$.

Corollary 15. *Let $s', p', u' \in K$.*

1. *The set of elements below s' that are r-t-closed and r-controllable with respect to p and u,*

$$S' \stackrel{\text{def}}{=} \{s \mid s \leq s' \ \wedge \ \mathsf{CLR}(s, r, t) \ \wedge \ \mathsf{CTR}(s, p, r, u)\} \ ,$$

is a complete lattice with least element 0 and greatest element

$$\mathsf{SUP}_{s'}(p, r, u) \stackrel{\text{def}}{=} \nu(s \mid: s' \sqcap \overline{(t \sqcap \overline{s})\cdot r} \sqcap \overline{((p\lfloor r \sqcap \overline{s\lfloor r})\lfloor u)\cdot r}) \ . \quad (5)$$

The join $\sqcup_{s'}$ in this lattice is that of the underlying KA (i.e., $\sqcup$). For $S \subseteq S'$, the meet $\sqcap_{s'} S$ is

$$\sqcap_{s'} S = \bigsqcup(s \mid s \in S' \wedge s \leq S : s) \ ,$$

where the abbreviation $s \leq S$ means $\forall(x \mid x \in S : s \leq x)$.

2. *Given s, r, u, the largest p such that $\mathsf{CTR}(s, p, r, u)$ is*

$$p_{\mathsf{max}}(s, r, u) \stackrel{\text{def}}{=} \overline{((s\lfloor r)\cdot u \sqcap \overline{s\lfloor r})\cdot r} \ .$$

The largest element p below p' such that $\mathsf{CTR}(s, p, r, u)$ is

$$\mathsf{SUP}_{p'}(s, r, u) \stackrel{\text{def}}{=} p' \sqcap p_{\mathsf{max}}(s, r, u) \ . \quad (6)$$

The set $\downarrow\mathsf{SUP}_{p'}(s, r, u)$ is a complete lattice whose join is the same as that of the underlying KA and whose meet is also the same when applied to nonempty sets. Every $p \in \downarrow\mathsf{SUP}_{p'}(s, r, u)$ satisfies $\mathsf{CTR}(s, p, r, u)$.

3. *Given s, p, r, the largest u such that $\mathsf{CTR}(s, p, r, u)$ is*

$$u_{\mathsf{max}}(s, p, r) \stackrel{\text{def}}{=} \overline{(s\lfloor r)\rfloor(p\lfloor r \sqcap \overline{s\lfloor r})} \ .$$

The largest element u below u' such that $\mathsf{CTR}(s, p, r, u)$ is

$$\mathsf{SUP}_{u'}(s, p, r) \stackrel{\text{def}}{=} u' \sqcap u_{\mathsf{max}}(s, p, r) \ . \quad (7)$$

The set $\downarrow\mathsf{SUP}_{u'}(s, p, r)$ is a complete lattice whose join is the same as that of the underlying KA and whose meet is also the same when applied to nonempty sets. Every $u \in \downarrow\mathsf{SUP}_{u'}(s, p, r)$ satisfies $\mathsf{CTR}(s, p, r, u)$.

Proof.

1. Define $f(s) \overset{\mathsf{def}}{=} s' \sqcap \overline{(t \sqcap \overline{s})\cdot r} \sqcap \overline{((p\lfloor r \sqcap \overline{s\lfloor r})\lfloor u)\cdot r}$. Function f is monotonic and thus has the greatest fixed point given in (5), due to completeness of the KA K. By Lemma 14(1,4), $s \leq f(s) \Leftrightarrow s \in S'$. Thus, $\nu f \in S'$. For $S \subseteq S'$, $\sqcup S \in S'$, by Theorem 10(1) and Lemma 13. Thus $\sqcup_{s'} S \overset{\mathsf{def}}{=} \sqcup S$ is the join on S'. Because S' is a complete join semilattice with a greatest element, it is a complete lattice. The expression for $\sqcap_{s'} S$ is then standard lattice theory [3].
2. The expression for $p_{\mathsf{max}}(s, r, u)$ follows from Lemma 14(2). By Theorem 10(6), every $p \in {\downarrow} p_{\mathsf{max}}(s, r, u)$ satisfies $\mathsf{CTR}(s, p, r, u))$ and thus $\mathsf{SUP}_{p'}(s, r, u)$ is the largest p below p' satisfying $\mathsf{CTR}(s, p, r, u)$. That ${\downarrow}\mathsf{SUP}_{u'}(s, p, r)$ is a complete lattice with stated join and meet is standard lattice theory [3].
3. The proof is the same as that for p_{max} and $\mathsf{SUP}_{p'}$, but for using Lemma 14(3) and Theorem 10(7). □

Since the symbol p represents the plant, $\mathsf{SUP}_{p'}(s, r, u)$ is the least constrained (largest) plant p below p' such that s is r-controllable with respect to p and u. Similarly, $\mathsf{SUP}_{u'}(s, p, r)$ is the largest uncontrollable element u such that s is r-controllable with respect to p and u. These are new results we have not yet seen in the literature, which is rather surprising, given that they seem useful in practice. Indeed, if the specification s is not r-controllable with respect to p and u, one should not be happy with an overrestricting specification $\mathsf{SUP}_{s'}(p, r, u)$ if

- there is a possibility to modify the behaviour of the plant to get $\mathsf{SUP}_{p'}(s, r, u)$ and then compose the modified plant with a controller for $\mathsf{SUP}_{p'}(s, r, u)$, or
- there is a possibility to reduce the uncontrollable events to $\mathsf{SUP}_{u'}(s, p, r)$ (e.g., by adding new actuators) and then compose the plant with a (now derivable) controller.

Modifying the plant as required by these calculations may be costly or even physically impossible. Nevertheless, it may sometimes be possible. In any case, these calculations may be very useful at the design stage to explore various ways to implement a specification.

Using Corollary 15, it is possible to retrieve the classical results presented in Theorem 5(1,3,5) by using the instantiations $s' := S, p := P$ (remembering that for a generated language p, $p = p\lfloor\top$), $r := \top = \Sigma^*$, $u := \Sigma_{\mathsf{u}}$, and $t := 0 = \emptyset$ for calculating $\mathsf{SUPC}(S)$, $t := \top$ for $\mathsf{SUPPC}(S)$ and $t := P_{\mathsf{m}}$ for $\mathsf{SUPRC}(S)$. The advantages of our highly parameterized approach are clearly visible here.

Corollary 15 presents results about the existence and value of largest elements $\mathsf{SUP}_{s'}(p, r, u)$, $\mathsf{SUP}_{p'}(s, r, u)$ and $\mathsf{SUP}_{u'}(s, p, r)$ below given elements s', p' and u', respectively, and such that controllability holds, i.e., $\mathsf{CTR}(\mathsf{SUP}_{s'}(p, r, u), p, r, u)$, $\mathsf{CTR}(s, \mathsf{SUP}_{p'}(s, r, u), r, u)$ and $\mathsf{CTR}(s, p, r, \mathsf{SUP}_{u'}(s, p, r))$ hold.

We now briefly look at the case of least elements above given elements. Let $\mathsf{INF}_{s'}(p, r, u)$, $\mathsf{INF}_{p'}(s, r, u)$ and $\mathsf{INF}_{u'}(s, p, r)$ denote the least elements (if they exist) above given elements s', p' and u', respectively, and such that controllability holds, that is, $\mathsf{CTR}(\mathsf{INF}_{s'}(p,r,u),p,r,u)$, $\mathsf{CTR}(s,\mathsf{INF}_{p'}(s,r,u),r,u)$ and $\mathsf{CTR}(s,p,r,\mathsf{INF}_{u'}(s,p,r))$ hold. The case of $\mathsf{INF}_{p'}(s,r,u)$ is easily solved using Corollary 15(2). If $p' \leq p_{\mathsf{max}}(s,r,u)$, then $\mathsf{INF}_{p'}(s,r,u) = p'$; otherwise, $\mathsf{INF}_{p'}(s,r,u)$ does not exist. The case of $\mathsf{INF}_{u'}(s,p,r)$ is similar. The case of $\mathsf{INF}_{s'}(p,r,u)$ is more complex. The problem is that the meet of two controllable elements is not necessarily controllable, so that there is no analogue of Theorem 10(1) for meets unless additional constraints are imposed[6]. One such constraint is

$$\sqcap(i \mid i \in I : s_i)\lfloor r = \sqcap(i \mid i \in I : s_i\lfloor r) \ . \tag{8}$$

In [7], two languages L_1 and L_2 satisfying $\mathsf{PR}(L_1 \cap L_2) = \mathsf{PR}(L_1) \cap \mathsf{PR}(L_2)$ are said to be *modular*. This is why we say that a family of elements $s_i \in K$, for $i \in I$, is *r-modular* if it satisfies (8).

Theorem 16. *Let I be an arbitrary index set.*

1. *If the family $\{s_i \mid i \in I\}$ is r-modular, then*

$$\forall(i \mid i \in I : \mathsf{CTR}(s_i,p,r,u)) \Rightarrow \mathsf{CTR}(\sqcap(i \mid i \in I : s_i),p,r,u) \ .$$

2. *If $1 \leq r$ and, for all $i \in I$, $s_i\lfloor r = s_i$, then $\{s_i \mid i \in I\}$ is r-modular.*
3. *If, for all $i \in I$, $s_i\lfloor r^* = s_i$, then $\{s_i \mid i \in I\}$ is r^*-modular.*

Assuming prefix-closure of the form $s\lfloor r^* = s$, one can then proceed as in Corollary 15(1) and find an expression for an element $\mathsf{INF}_{s'}(p,r,u)$ which is the least prefix-closed and controllable element above s'. With the proper instantiations, it is the possible to derive Theorem 5(4). If only r-t-closure or no closure is assumed, such infimal elements need not exist (Theorem 5(2,6)).

5 Conclusion

We have presented a sample of results obtained by formalizing with Kleene algebra the concept of controllability that is used in discrete control theory. The formalization has more parameters than the standard approach, where one considers r and u as given fixed parameters (and $r = \top$), so that the focus is on the variability of s. This means that most of the results of this paper where p and u are considered as variables are new. A notable exception is [14], where a real[7] cost is associated to the control of an event, the idea being that actuators have a price; the goal is then to minimize the cost of controlling a given plant.

In addition to making the parameterized approach natural, Kleene algebra is well suited to the presentation of proofs in the calculational style, which makes their verification much easier than classical proofs.

---

[6] For instance (example drawn from [7]), take $s_1 \overset{\mathsf{def}}{=} a + b + ab$, $s_2 \overset{\mathsf{def}}{=} a + ab + ba$, $p \overset{\mathsf{def}}{=} (a^*ba^*)\lfloor\top$ and $u \overset{\mathsf{def}}{=} b$ on the alphabet $\Sigma \overset{\mathsf{def}}{=} \{a,b\}$. One can check that $\mathsf{CTR}(s_1,p,\top,u)$ and $\mathsf{CTR}(s_2,p,\top,u)$ hold, but not $\mathsf{CTR}(s_1 \sqcap s_2,p,\top,u)$.

[7] In the sense that it is a real number.

This is only a beginning and the approach seems promising. We are presently investigating what is the best (most general) way to define a notion of controller as a function $f : K \to K$ (see Sect. 2 for the standard definition). And the generalization from languages to path algebras or relation algebras has yet to be exploited.

References

[1] Cassandras, C. G., Lafortune, S.: Introduction to Discrete Event Systems. Kluwer Academic Publishers, Boston (1999) 74

[2] Conway, J. H.: Regular Algebra and Finite Machines. Chapman and Hall, London (1971) 78

[3] Davey, B. A., Priestley, H. A.: Introduction to Lattices and Order. Cambridge University Press (1990) 83

[4] Desharnais, J., Möller, B.: Characterizing determinacy in Kleene algebras. Information Sciences **139** (2001) 253–273 78

[5] von Karger, B., Hoare, C. A. R.: Sequential calculus. Information Processing Letters **53** (1995) 123–130 78

[6] Kozen, D.: Kleene algebra with tests. ACM Transactions on Programming Languages and Systems **19** (1997) 427–443 78

[7] Kumar, R., Garg, V. K.: Modeling and Control of Logical Discrete Event Systems. Kluwer Academic Publishers, Boston (1995) 74, 76, 77, 84

[8] Möller, B.: Derivation of graph and pointer algorithms. In Möller, B., Partsch, H. A., Schuman, S. A., eds.: Formal Program Development. Volume 755 of Lecture Notes in Computer Science. Springer-Verlag, Berlin (1993) 123–160 78

[9] Möller, B.: Residuals and detachment. Personal communication (2001) 77, 78

[10] Ramadge, P. J. G., Wonham, W. M.: Supervisory control of a class of discrete-event processes. SIAM J. on Control and Optimization **25** (1987) 206–230 74

[11] Ramadge, P. J. G., Wonham, W. M.: The control of discrete event systems. Proceedings of the IEEE **77** (1989) 81–98 74, 76

[12] Wonham, W. M.: Notes on control of discrete event systems. Systems Control Group, Edward S. Rogers Sr. Dept. of Electrical & Computer Engineering, University of Toronto (2002) xiv+356 pages. Available at `http://www.control.utoronto.ca/people/profs/wonham/wonham.html`. 74

[13] Wonham, W. M., Ramadge, P. J. G.: On the supremal controllable sublanguage of a given language. SIAM J. on Control and Optimization **25** (1987) 637–659 74

[14] Young, S. D., Garg, V. K.: Optimal sensor and actuator choices for discrete event systems. In: 31st Allerton Conf. on Communication, Control, and Computing, Allerton, IL (1993) 84

Tracing Relations Probabilistically

Ernst-Erich Doberkat

Chair for Software Technology, University of Dortmund
doberkat@acm.org

Abstract We investigate similarities between non-deterministic and probabilistic ways of describing a system in terms of computation trees. We show that the construction of traces for both kinds of relations follow the same principles of construction. Representations of measurable trees in terms of probabilistic relations are given. This shows that stochastic relations may serve as refinements of their non-deterministic counterparts. A convexity argument formalizes the observation that non-deterministic system descriptions are underspecified when compared to probabilistic ones. The mathematical tools come essentially from the theory of measurable selections.

Keywords: Probabilistic relations, specification techniques (non-deterministic, stochastic), representation theory.

1 Introduction

This paper investigates the relationship between trees and probabilistic relations. Trees arise in a natural way when the behavior of a system is specified through relations. Given a family $(R_n)_{n\in\mathbb{N}}$ of non-deterministic relations $R_n \subseteq X \times X$ over a state space X, R_n describes the behavior of the system at step n, so that the current state x_n may be followed by any state x_{n+1} with $\langle x_n, x_{n+1}\rangle \in R_n$. Rolling out the R_n yields a computation tree, the tree of traces that describe all possible paths through the system. If, on the other hand, a system is described through a sequence $(K_n)_{n\in\mathbb{N}}$ of probabilistic relations K_n, then computing the traces for K_n gives the probabilistic analogue of a computation tree. We are interested in the relationship between the non-deterministic and the probabilistic relations, seeing a probabilistic relation as a refinement of a non-deterministic one: whereas a non-deterministic relation R specifies for a state x through the set $R(x) := \{y | \langle x, y\rangle \in R\}$ all possible subsequent states, a probabilistic relation K attaches a weight $K(x)(dy)$ to each next state y.

The problem discussed in this paper is, then, whether it is possible to find a probabilistic refinement for a given computation tree T. Thus we investigate the problem of finding for T a sequence $(K_n)_{n\in\mathbb{N}}$ of probabilistic relations such that after a computation history $w \in T$ and a given state x at time n the set $\{y | wxy \in T\}$ of all possible next states for this computation is exactly the set of states that are assigned positive probability by $K_n(x)$. This problem is posed in a setting which does not assume that the state space is finite or countable,

R. Berghammer et al. (Eds.): RelMiCS/Kleene-Algebra Ws 2003, LNCS 3051, pp. 86–98, 2004.

rendering tools that work with discrete probabilities useless, and asking for new approaches. If the question above is answered in the positive, then not only single steps in a non-deterministically specified computation can be refined stochastically but also whole traces arising from those specifications have a probabilistic refinement. This sheds further light on the relationship between stochastic and non-deterministic relations (compare [2, 3]).
In fact, it can be shown that under some not too restrictive conditions a computation tree has a probabilistic representation. The restrictions are topological in nature: we first show (Proposition 5) that a probabilistic representation can be established provided the set of all possible offsprings at any given time is compact. This condition is relaxed to the assumption that the state space is σ-compact, using a topological characterization of the body of a tree over the natural numbers (Proposition 6).

Overview We introduce in the next section computation trees and define their probabilistic counterparts. It is shown that a computation tree is spawned by the traces of a sequence of non-deterministic relations. This also works the other way around: each computation tree T exhibits a certain lack of memory in the actions it describes, thus it generates a sequence of relations for which T is just the corresponding tree. The probabilistic analogue is also studied: we show under which conditions the probabilistic counterparts of computation trees are spawned by probabilistic relations; it turns out that memoryless relations between X and the set X^∞ of all X-sequences characterize the situation completely. Section 3 introduces measurable trees as those class of trees for which a characterization is possible. It gives the mentioned representations, first for the compact, then for the σ-compact case. It turns out that the latter case is interesting in its own right .

Acknowledgement The author is grateful to Georgios Lajios for his critical comments. The referees' suggestions and comments are also gratefully acknowledged.

2 Computation Trees

Denote for a set V by V^* as usual the free semigroup on V with ϵ as the empty word; V^∞ is the set of all infinite sequences based on V. If $v \in V^*$, $w \in V^* \cup V^\infty$, then $v \preceq w$ iff v is an initial piece of w, in particular $\sigma \mid_k \preceq \sigma$ for all $\sigma \in V^\infty, k \in \mathbb{N}$, where $(s_n)_{n\in\mathbb{N}} \mid_k := s_0 \dots s_k$ is the prefix of $(s_n)_{n\in\mathbb{N}}$ of length $k+1$. Denote for $M \subseteq V^*$ all words of length n by $\pi_n(M)$.
A *tree T on V* is a subset of V^* which is closed under the prefix operation, thus $w \in T$ and $v \preceq w$ together imply $v \in T$. The *body $[T]$ of T* [7] is the set of all sequences on V each finite prefix is in T, thus

$$[T] := \{\sigma \in V^\infty | \forall k \in \mathbb{N} : \sigma \mid_k \in T\}.$$

Suppose we specify the nth step in a process through relation $R_n \subseteq V \times V$. Execution spawns a tree by rendering explicit the different possibilities opening up for exploitation. Put $\mathcal{R} := (R_n)_{n \in \mathbb{N}}$ and

$$\mathsf{Tree}\,(\mathcal{R}) := \{v \in V^* | v_0 \in \mathsf{dom}(R_0), v_j \in R_{j-1}(v_{j-1}) \text{ for } 1 \leq j \leq |v|\} \cup \{\epsilon\},$$

then $\mathsf{Tree}\,(\mathcal{R})$ is a tree with body

$$[\mathsf{Tree}\,(\mathcal{R})] = \{\alpha \in V^\infty | \alpha_0 \in \mathsf{dom}(R_0) \;\&\; \forall j \geq 1 : \alpha_j \in R_{j-1}(\alpha_{j-1})\}.$$

This is the computation tree associated with $\mathcal{R}$; $\mathsf{Tree}\,(\mathcal{R})$ collects all finite, $[\mathsf{Tree}\,(\mathcal{R})]$ all infinite traces.
In fact, each tree T spawns a sequence of relations: Define

$$R_0^T := V^2 \cap T,$$

and inductively for $k \geq 1$

$$\langle x_k, x_{k+1}\rangle \in R_k^T \Leftrightarrow \exists \langle x_0, x_1\rangle \in R_0^T \exists x_2 \in R_1^T(x_1) \ldots \exists x_{k-1} \in R_{k-2}^T(x_{k-2}) : \\ x_k \in R_{k-1}^T(x_{k-1}) \wedge x_0 x_1 \ldots x_k x_{k+1} \in T.$$

Example 1. Let $T := \{\epsilon\} \cup H_{12}$, where H_k is the tree underlying a heap of size k, the nodes being the binary representations of the corresponding numbers. Then

$$\begin{aligned} R_0^T &= \{\langle 1,0\rangle, \langle 1,1\rangle\}, \\ R_1^T &= \{\langle 0,0\rangle, \langle 0,1\rangle, \langle 1,0\rangle, \langle 1,1\rangle\}, \\ R_2^T &= R_1^T. \end{aligned}$$

We see that

$$\mathsf{Tree}\left(\left(R_n^T\right)_{n \in \mathbb{N}}\right) = \{\epsilon\} \cup H_{15}$$

holds, to that T is not generated from the relations R^T.

The trees T which may be represented through $\mathsf{Tree}\left(\left(R_n^T\right)_{n \in \mathbb{N}}\right)$ are of interest, they turn out to be memoryless in the sense that the behavior described by the tree at time $k+1$ depends directly only on the behavior at time k, once the initial input is provided.
Let for the sets X, Y be A a subset of X, and $f : X \to 2^Y$ be a set-valued map. Then define

$$A \otimes f := \{\langle a, b\rangle | a \in A, b \in f(a)\}$$

as the product of A and f. It is clear that each subset $M \subseteq X \times Y$ can be represented as a product $M = \pi_X[M] \otimes f_M$ with $f_M(a) := \{b \in Y | \langle a, b\rangle \in M\}$, π_X denoting the projection to X.
Using this product, we define memoryless trees through the following observation: $\pi_{n+1}(T)$ can be decomposed as a product

$$\pi_{n+1}(T) = \pi_n(T) \otimes J_n$$

with $J_n : X^n \to 2^X$. Thus the next letter x_n in a word $x_0 \ldots x_n \in T$ is under this decomposition an element of $J_n(x_0, \ldots, x_{n-1})$, and the tree being memoryless means that the latter set depends on x_{n-1} only. Thus J_n is induced by a map $f_n : X \to 2^X$ in the sense that

$$J_n(x_0, \ldots, x_{n-1}) = f_n(x_{n-1})$$

holds for all $\langle x_0, \ldots, x_{n-1} \rangle \in X^n$.

Definition 1. *A tree T over V is called* memoryless *iff for each $n \in \mathbb{N}$ with $n \geq 2$ the set*

$$\pi_{n+1}(T)$$

can be written as

$$\pi_n(T) \otimes J_n,$$

where $J_n : X^n \to 2^X$ is induced by a map $X \to 2^X$.

This means that only the length of the history and the initial input determines the behavior of a memoryless tree. Heaps, for example, are not always memoryless:

Example 2. Let T be the tree according to Example 1, then using the notation of the decomposition above

$$\begin{aligned} J_1(1) &= \{0, 1\} \\ J_2(1, 0) = J_2(11) &= \{0, 1\} \\ J_3(1, 0, 0) = J_3(101) &= \{0, 1\} \\ J_3(1, 1, 0) = J_3(110) &= \{0\} \\ J_3(1, 1, 1) = J_3(111) &= \emptyset. \end{aligned}$$

Clearly, $\{\epsilon\} \cup H_k$ is memoryless iff $k = 2^t - 1$ for some t.

It is not difficult to see that the tree $\mathsf{Tree}\left((R_n)_{n\in\mathbb{N}}\right)$ is memoryless, and that for a memoryless tree T with associated maps $J_n : X \to 2^X$ the equality

$$R^T_{n-1} = \{\langle x, y \rangle | x \in \mathsf{dom}(J_n), y \in J_n(x)\}$$

for all $n \geq 2$ holds.

Proposition 1. *Let T be a tree over V. Then the following conditions are equivalent:*

1. *$T = \mathsf{Tree}(\mathcal{R})$ holds for the sequence $\mathcal{R} = \left(R^T_n\right)_{n\in\mathbb{N}}$ of relations $R^T_n \subseteq V \times V$ defined through T.*
2. *T is memoryless.*

In the remainder of the paper we will not distinguish relations from the associated set valued maps. We will see soon (Prop. 2) that a similar notion will be helpful to characterize the probabilistic analogue of trees.

Turning to the stochastic side of the game, we denote for a measurable space X by $\mathbf{P}(X)$ the set of a probability measures on (the σ-algebra of) X. We usually omit mentioning the σ-algebra underlying a measurable space and talk about its members as *measurable subsets*, or as *Borel subsets*, if X is a metric space, see below.
Define for the measurable map $f : X \to Y$ and for $\mu \in \mathbf{P}(X)$ the *image* of μ under f by

$$\mathbf{P}(f)(\mu)(B) := \mu(f^{-1}[B]),$$

then $\mathbf{P}(f)(\mu) \in \mathbf{P}(Y)$.
A *probabilistic relation* $K : X \rightsquigarrow Y$ *between the measurable spaces* X *and* Y [1, 5, 3, 2] is a measurable map $K : X \to \mathbf{P}(Y)$, consequently it has these properties:

1. for all $x \in X$, $K(x)$ is a probability measure on Y,
2. for all measurable subsets B of X, $x \mapsto K(x)(B)$ is a measurable real-valued map on X, where measurability of real functions always refers to the Borel sets in $\mathbb{R}$.

If $\mu \in \mathbf{P}(X), K : X \rightsquigarrow Y$, define for the measurable subset $A \subseteq X \times Y$

$$(\mu \otimes K)(A) := \int_X K(x)(A_x)\ \mu(dx)$$

with

$$A_x := \{y \in Y | \langle x, y\rangle \in A\}.$$

Consequently, $\mu \otimes K \in \mathbf{P}(X \times Y)$.
A Polish space X is a completely metrizable separable topological space; as usual, we take the Borel sets as the σ-algebra on a Polish space. If X is Polish, so are [6] $\mathbf{P}(X)$ under the topology of weak convergence, X^* under the topological sum of $(X_n)_{n\in\mathbb{N}}$, and X^∞ under the topological product (where the open sets are generated by sets of the form $\prod_{n\in\mathbb{N}} A_n$, with all $A_n \subseteq X$ open, and all but a finite number equal X).
The topology of weak convergence on $\mathbf{P}(X)$ is the smallest topology for which the evaluation maps $\mu \mapsto \int_X f\ d\mu$ are continuous for every bounded and continuous function $f : X \to \mathbb{R}$.
An important example of a Polish space is furnished by the *Baire space* $\mathcal{N} := \mathbb{N}^\infty$, where the natural numbers $\mathbb{N}$ have the discrete topology, so that each subset of $\mathbb{N}$ is open; $\mathcal{N}$ carries the product topology. Closed subsets of $\mathcal{N}$ may be characterized in terms of trees:

Lemma 1. *A set $D \subseteq \mathcal{N}$ is closed iff $D = [T]$ for some tree T over $\mathbb{N}$. The body $[T]$ of a tree T over $\mathbb{N}$ is a Polish space.*

Proof. The first assertion follows from [7, Prop. 2.2.13]. Since closed subsets of Polish spaces are Polish again in their relative topology, the second part is established.

One immediate consequence of working in a Polish space is that *disintegration of measures* is possible: Suppose $\mu \in \mathbf{P}(X_1 \times X_2)$ is a probability measure on the product of the Polish spaces X_1 and X_2. Then there exists a probability μ_1 on X_1 and a probabilistic relation $K : X_1 \rightsquigarrow X_2$ such that $\mu = \mu_1 \otimes K$ holds.
Fix for the rest of the paper X as a Polish space. The product X^n is always equipped with the product topology, the free monoid X^* has always the topological sum, and $\mathbf{P}(X)$ always the topology of weak convergence as the respective topologies.
Now suppose that a sequence $\mathcal{K} := (K_n)_{n\in\mathbb{N}}$ of probabilistic relations $K_n : X \rightsquigarrow X$ is given. Define inductively a sequence $K_0^n : X \rightsquigarrow X^{n+1}$ by setting $K_0^0 := K_0$, and for $x \in X, A \subseteq X^{n+2}$ measurable

$$\begin{aligned} K_0^{n+1}(x)(A) &:= \int_{X^{n+1}} K_{n+1}(x_n)(\{x_{n+1} | \langle x_0, \ldots, x_{n+1}\rangle \in A\})\, K_0^n(x)(d\langle x_0, \ldots x_n\rangle) \\ &= (K_0^n(x) \otimes K_{n+1})(A). \end{aligned}$$

Let K_n probabilistically specify the nth state transition of a system, then the probability that the sequence $\langle x_1, \ldots x_n\rangle$ is an element of A is given by $K_0^n(x)(A)$, provided the system was initially in state x.
It is not difficult to see that the sequence $(K_0^n)_{n\in\mathbb{N}}$ forms a projective system in the following sense: for the measurable subset $A \subseteq X^{n+1}$ and for $x \in X$ the equality

$$K_0^{n+1}(x)(A \times X) = K_0^n(x)(A)$$

holds for each $n \in \mathbb{N}$. This is the exact probabilistic counterpart to the property that a tree is closed with respect to prefixes.
Denote the resp. projections $(x_n)_{n\geq 0} \mapsto \langle x_0, \ldots, x_n\rangle$ by $proj_{n+1}$. Standard arguments [6, V.3] show that there exists a uniquely determined probabilistic relation

$$\mathcal{K}_0^\infty : X \rightsquigarrow X^\infty$$

such that for all $x \in X$ the equality

$$\mathbf{P}(proj_{n+1})(\mathcal{K}_0^\infty(x)) = K_0^n(x)$$

holds. Thus $\mathcal{K}_0^\infty(x)(A)$ is the probability that the infinite sequence σ of states the system is running through is an element of A, provided the system starts in x. Averaging out the starting state x through an initial probability $\mu \in \mathbf{P}(X)$, i.e. forming

$$\mathsf{Tree}(\mathcal{K})_\mu(A) := \int_X \mathcal{K}_0^\infty(x)(A)\; \mu(dx)$$

yields a probability measure on X^∞. This is the probabilistic analogue to the body $[\mathcal{R}]$ of the tree formed from the sequence $\mathcal{R}$ of non-deterministic relations. We have shown how to reverse this construction in the non-deterministic case by showing that each tree T yields a sequence of relations $\mathcal{R}$ with $T = \mathsf{Tree}(\mathcal{R})$. Investigating similarities between non-deterministic relations and their probabilistic counterparts, the question arises whether this kind of reversal is also possible

for the probabilistic case. To be more specific: Under which conditions does there exist for a probability measure $\nu^\infty \in \mathbf{P}(X^\infty)$, a sequence $\mathcal{K} = (K_n)_{n\in\mathbb{N}}$ of probabilistic relations $K_n : X \rightsquigarrow X$ and an initial probability $\mu \in \mathbf{P}(X)$ such that the representation $\nu^\infty = \mathsf{Tree}(\mathcal{K})_\mu$ holds?
Since in this case $\mu = \mathbf{P}(proj_0)(\nu^\infty)$ must hold, the question is reduced to conditions under which we can construct for a probabilistic relation $L : X \rightsquigarrow X^\infty$ a sequence $\mathcal{K}$ of probabilistic relations $X \rightsquigarrow X$ such that $L = \mathcal{K}_0^\infty$.

Definition 2. *A transition probability $L : X \rightsquigarrow X^\infty$ is called* memoryless *iff the projection*

$$\mathbf{P}(proj_{n+1})(L(x))$$

can be written for each $n \in \mathbb{N}, x \in X$ as a disintegration

$$\mathbf{P}(proj_n)(L(x)) \otimes J_n$$

with $J_n : X \rightsquigarrow X$, where J_n is independent of x.

The reader may wish to compare the definition of a memoryless probabilistic relation to that of a memoryless tree in Def. 1. Similarly, a comparison of Prop. 1 for the set valued case with Prop. 2 addressing the probabilistic case may be illuminating. Memoryless transition probabilities characterize those relations that arise through refinements.

Proposition 2. *Let $L : X \rightsquigarrow X^\infty$: be a probabilistic relation. Then the following conditions holds: There exists a sequence $\mathcal{K} = (K_n)_{n\in\mathbb{N}}$ of probabilistic relations $K_n : X \rightsquigarrow X$ such that $L = \mathcal{K}_0^\infty$ iff L is memoryless.*

Thus there are in fact striking similarities between non-deterministic relations and their probabilistic counterparts, when it comes to specify reactive, i.e., long running behavior. Both generate memoryless trees, and from these trees the single step behavior can be recovered. This is always true for the non-deterministic case (since we consider here only possibilities without attaching any constraints), it is possible in the probabilistic case under the condition that probabilities on product spaces can be suitably decomposed.
The next section will deal with a transfer between non-deterministic and probabilistic relations: Given is a tree, can we generate it probabilistically?

3 Representing Measurable Trees

A stochastic representation K of a non-deterministic relation R should have the following properties: we have $K(x)(R(x)) = 1$ for each x, indicating that a state transition in state x is guaranteed to lead to a state in $R(x)$, and we want the latter set to be exactly the set of all target states. This latter condition on exact fitting of $R(x)$ is a bit cumbersome to formulate if the space X is not finite or countable. But the topological structure on X comes in helpful now. We want $R(x)$ to be the smallest set of all states for which each open neighborhood U has positive probability $K(x)(U)$.

This is captured through the support of a probability: Given $\mu \in \mathbf{P}(X)$, define $\mathsf{supp}(\mu)$ as the smallest closed subset $F \subseteq X$ such that $\mu(F) = 1$, thus

$$\mathsf{supp}(\mu) := \bigcap\{F \subseteq X | F \text{ is closed and } \mu(F) = 1\}.$$

It can be shown that $\mu(\mathsf{supp}(\mu)) = 1$, and $x \in \mathsf{supp}(\mu)$ iff $\mu(U) > 0$ for each neighborhood U of x. So this is exactly what we want.

Definition 3. *Let Y be a measurable space, and Z be a Polish space, $R \subseteq Y \times Z$ a non-deterministic relation, and $K : Y \rightsquigarrow Z$ a probabilistic one. Then*

$$R \models K \Leftrightarrow \forall y \in Y : R(y) = \mathsf{supp}(K(y)).$$

We say that K represents *R.*

These are some elementary properties of the representation:

Example 3. Assume $R_i \models K_i$ for $i = 1, 2$, where $R_i \subseteq X \times Y$, then

$$R_1 \cup R_2 \models K_1 \oplus_p K_2.$$

Here the union is taken element wise, $0 \leq p \leq 1$, and

$$(K_1 \oplus_p K_2)(x)(A) := p \cdot k_1(x)(A) + (1 - p) \cdot K_2(x)(A)$$

is the convex combination of K_1 and K_2.

This will be generalized considerably in Proposition 7.
Measurable relations will provide a link between non-deterministic and stochastic systems, as we will see. Let us fix some notations first.
Assume that Y is a measurable, and that Z is a Polish space. A relation $R \subseteq Y \times Z$ induces as above a set-valued map through

$$Y \ni y \mapsto R(y) := \{z \in Z | \langle y, z \rangle \in R\} \in 2^Z.$$

If $R(y)$ always takes closed and non-empty values, and if the (weak) inverse

$$(\exists R)(G) := \{y \in Y | R(y) \cap G \neq \emptyset\}$$

is a measurable set, whenever $G \subseteq Z$ is open, then R is called a *measurable relation on $Y \times Z$.*
It is immediate that the support yields a measurable relation for a probabilistic relation $K : Y \rightsquigarrow Z$: put

$$R_K := \{\langle y, z \rangle \in Y \times Z | z \in \mathsf{supp}(K(y))\},$$

then

$$(\forall R_K)(F) = \{y \in Y | K(y)(F) = 1\}$$

is true for the closed set $F \subseteq Z$, and

$$(\exists R_K)(G) = \{y \in Y | K(y)(G) > 0\}$$

holds for the open set $G \subseteq Z$. This set is measurable. It is also plain that $R \models K$ implies that R has to be a measurable relation. Given a set-valued relation R, a probabilistic relation K that satisfies R can be found. For this, R has to take closed values, and a measurability condition is imposed; from [3] we get:

Proposition 3. *Let $R \subseteq Y \times Z$ be a measurable relation for Z Polish. If Z is σ-compact, or if $R(y)$ assumes compact values for each $y \in Y$, then there exists a probabilistic relation $K : Y \rightsquigarrow Z$ with $R \models K$.*

Compactness plays an important role in the sequel, so we state the representation only for this case, leaving aside a more general formulation. The representation of non-deterministic by stochastic relations presented here is used in [4] to construct the probabilistic refinement of a Kripke model.
We are interested in trees. The notion of a measurable tree is introduced as an analogue to measurable relations.

Definition 4. *The tree $T \subseteq X^*$ is called a* measurable tree *iff the following conditions are satisfied:*

1. *T is memoryless,*
2. *$[T] \neq \emptyset$,*
3. *$T^\bullet := \{\langle v, x\rangle \in X^* \times X | vx \in T\}$ constitutes a measurable relation on $X^* \times X$.*

The last condition implies that $T^\bullet(v)$ is a closed subset of X for all $v \in T$. The condition $[T] \neq \emptyset$ makes sure that $\forall v \in T : T^\bullet(v) \neq \emptyset$, so that the tree continues to grow, hence T has the proper range for a measurable relation. Since $T^\bullet$ constitutes a measurable relation the graph of which is just T, it follows that T is a measurable subset of X^*. The first condition constraints our attention to memoryless trees; this is not too restrictive because a tree that is represented through a stochastic relation is memoryless in view of Prop. 1.
We start with a simple observation: If all relations R_n are measurable relations, then $\mathsf{Tree}(\mathcal{R})$ is a measurable tree:

Lemma 2. *Construct $\mathsf{Tree}(\mathcal{R})$ from the sequence $\mathcal{R} = (R_n)_{n\in\mathbb{N}}$ as above, then $\mathsf{Tree}(\mathcal{R}) \subseteq X^*$ is a Borel set, provided each $R_n \subseteq X \times X$ is. Moreover, $\mathsf{Tree}(\mathcal{R}) = \{\langle v, x\rangle | v \in \mathsf{Tree}(\mathcal{R}), x \in X$ so that $vx \in \mathsf{Tree}(\mathcal{R})\}$ is a measurable relation, provided each R_n is.*

Since a probabilistic relation generates a measurable relation, this has as an easy consequence:

Proposition 4. *Let $(K_n)_{n\in\mathbb{N}}$ be a sequence of probabilistic relations $K_n : X \rightsquigarrow X$. Then $\mathsf{Tree}\left((\mathsf{supp}(K_n))_{n\in\mathbb{N}}\right)$ constitutes a measurable tree.*

We can show now that under a compactness condition a measurable tree may be generated from some probabilistic refinement.

Proposition 5. *Let T be a measurable tree on X, and assume that $T \cap X^k$ is compact for each $k \geq 0$. Then there exists a sequence $(K_n)_{n \geq 0}$ of probabilistic relations $K_n : X \rightsquigarrow X$ such that*

$$T = \mathsf{Tree}\left((\mathsf{supp}(K_n))_{n \in \mathbb{N}}\right)$$

holds.

The proof of this statement makes substantial use of some non trivial properties of Borel sets in Polish spaces.

Proof.

1. Define the sequence $(R_k^T)_{k \geq 0}$ of relations for T as above, then there exists for each $k \geq 1$ a measurable subset $D_k \subseteq X$ such that
$$R_k^T : D_k \to \mathcal{K}(X)$$
is a measurable map. This will be shown now. Fix $k \geq 0$, and let $(x_n)_{n \geq 0} \subseteq R_{k+1}^T(x')$ be a sequence, thus we can find $v_n \in T \cap X^k$ with $v_n x' x_n \in T \cap X^{k+2}$. Since the latter set is compact, we can find a convergent subsequence $(v_{n_\ell} x' x_{n_\ell})_{\ell \geq 0}$ and $vx'x \in T$ with $v_{n_\ell} x' x_{n_\ell} \to vx'x$, as $\ell \to \infty$. Consequently, $R_{k+1}^T(x')$ is closed, and sequentially compact, hence compact, since X is Polish. Thus $R_{k+1}^T(x) \in \mathcal{K}(X)$, provided the former set is not empty. The domain D_k of R_k^T is
$$\begin{aligned} D_k &= \pi_X \left[\{\langle v, x\rangle \in T \times X | T(vx) \neq \emptyset\}\right] \\ &= \pi_X \left[\{\langle v, x\rangle \in T \times X | T(vx) \cap X \neq \emptyset\}\right]. \end{aligned}$$
If we can show that $(\forall R_{k+1}^T)(F)$ is Borel in X whenever $F \subseteq X$ is closed, then measurability of D_k will follow (among others).
2. In fact, if $F \subseteq X$ is closed, then the compactness assumption for T implies that
$$\{\langle v, x\rangle \in T \times X | T(vx) \cap F \neq \emptyset\}$$
is closed, consequently,
$$H^{(F)} := \{\langle v, x\rangle \in T \times X | T(vx) \subseteq F\}$$
is a G_δ set, since F is one. Hence $H^{(F)}$ is Borel. Because the section $H_x^{(F)}$ is compact for each $x \in X$, the Novikov Theorem [7, Th. 5.7.1] implies now that
$$\pi_X \left[H^{(F)}\right] = (\forall R_{k+1}^T)(F)$$
is measurable.
3. The map $R_{k+1}^T : D_k \to \mathcal{K}(X)$ is measurable for each $k \geq 0$. Because R_{k+1}^T takes compact and nonempty values in a Polish space we can find by Prop. 3 a probabilistic relation $K_{k+1} : X \rightsquigarrow X$ such that $R_{k+1}^T \models \mathsf{supp}(K_{k+1})$. Hence
$$T = \mathsf{Tree}\left((\mathsf{supp}(K_n))_{n \in \mathbb{N}}\right)$$
is established.

This result makes the rather strong assumption that each slice $T \cap X^n$ of the tree at height n is compact. A little additional work will show that this may be relaxed to σ-compactness. For this fix a measurable tree T over a σ-compact Polish space X (so X may be represented as

$$X = \bigcup_{n \in \mathbb{N}} X_n,$$

where each X_n is compact) such that $T \subseteq X^*$ is closed. Define for $\alpha = n_0 \dots n_{k-1} \in \mathbb{N}^k$ the compact set

$$X_\alpha := X_{n_0} \times \dots \times X_{n_{k-1}},$$

and put

$$S := \{\alpha \in \mathbb{N}^* | T \cap X_\alpha \neq \emptyset\}.$$

Clearly, S is a tree over $\mathbb{N}$. Now let $\sigma \in [S]$, and set

$$T_\sigma := \{v \in X^* | v \in T \cap X_{\sigma|_{|v|}}\}.$$

From the construction it is clear that

$$T = \overline{\bigcup_{\sigma \in [S]} T_\sigma}$$

holds, the bar denoting topological closure.
Since $T_\sigma \cap X^n$ is compact for each $n \in \mathbb{N}$, $T \subseteq X^*$ is closed, and T is a measurable tree over X, the condition of Prop. 5 is satisfied. Thus there exist probabilistic relations $(K_{n,\sigma})_{n \in \mathbb{N}}$ such that

$$T_\sigma = |(\mathsf{supp}(K_{n,\sigma}))_{n \in \mathbb{N}}|$$

is true.
A representation of T will be obtained by pasting the relations $(K_{n,\sigma})_{n \in \mathbb{N}}$ along their index σ. Since $[S]$ may be uncountable, we have probably more than countably many of these families of probabilistic relations, so gluing cannot be done through simply summing up all members. The observation that the body $[S]$ of tree S is a Polish space will come in helpful now: we construct a probability measures on the set of indices and integrate the $(K_{n,\sigma})$ with this measure.
The following Lemma helps with the construction. Call a probability measure on a Polish space *thick* iff it assigns positive probability to each non-empty open set. Construct for example on the real line the probability measure

$$A \mapsto \int_A f(x)\, dx$$

with a strictly increasing and continuous density $f : \mathbb{R} \to \mathbb{R}_+$, then this constitutes a thick measure. But it can be said more:

Lemma 3. *Let P be a Polish space. There exists a thick measure $\mu_\ell \in \mathbf{P}(P)$. Assume that Q is a Polish space, and that $\phi : P \to \mathbf{P}(Q)$ is continuous, where $\mathbf{P}(Q)$ is endowed with the topology of weak convergence. Define $\mu^\bullet(A) := \int_P \phi(p)(A)\ \mu_\ell(dp)$. Then $\mathsf{supp}(\mu^\bullet) = \overline{\bigcup_{p \in P} \mathsf{supp}(\phi(p))}$ holds.*

Note that the continuity condition imposed above for $\phi : P \to \mathbf{P}(Q)$ is satisfied whenever the set $\{p \in P | \phi(p)(U) > 0\}$ is open for an open $U \subseteq Q$. It turns out that

$$[S] \ni \sigma \mapsto K_{n,\sigma} \in \mathbf{P}(X)$$

has this property for fixed $n \in \mathbb{N}$, since it is a continuous map, when $[S]$ has the topology inherited from the Baire space $\mathcal{N}$, and $\mathbf{P}(X)$ carries the weak topology.
Because the body of tree S is a Polish space by Lemma 1, we obtain the generalization of Prop. 5.

Proposition 6. *Let T be a measurable tree on the σ-compact Polish space X such that $T \subseteq X^*$ is closed. Then there exists a sequence $(K_n)_{n \geq 0}$ of probabilistic relations $K_n : X \rightsquigarrow X$ such that*

$$T = \mathsf{Tree}\left((\mathsf{supp}(K_n))_{n \in \mathbb{N}}\right)$$

holds.

The proof takes a thick probability on $[S]$ and pastes the $(K_{n,\sigma})$ along σ, making heavy use of the construction in Lemma 3, because $\sigma \mapsto K_{n,\sigma}(x)$ is always continuous. This is so since $T_\sigma \cap X^n$ depends only on $\sigma |_n$, hence only on a finite number of components of σ.
In the theory of convex cones, the integral over a probability measure is often interpreted as the generalization of a convex combination. We state as a generalization of Example 3 the following continuous version:

Proposition 7. *Let P be a Polish space, and assume that for a family $(R_p)_{p \in P}$ of relations $R_p \subseteq X \times X$ indexed by P the stochastic representation $R_p \models K_p$ holds. Let $\mu_\ell \in \mathbf{P}(P)$ be a thick probability measure on P. Assume that the family of probabilistic relations $K_p : X \rightsquigarrow X$ has the additional property that $p \mapsto K_p(x)$ is weakly continuous for each $x \in X$. Then*

$$\overline{\bigcup_{p \in P} R_p} \models \lambda x \lambda A. \int_P K_p(x)(A) \; \mu_\ell(dp)$$

holds.

4 Conclusion

We show that there are some interesting similarities between non-deterministic and probabilistic ways of describing a system in terms of computation trees. We first relate that the construction of traces for both kinds of relations exhibit the same principles of construction (which could be described in terms of monads, but this does not happen here). Then we give representations of measurable trees in terms of probabilistic relations under some topological conditions.

References

[1] S. Abramsky, R. Blute, and P. Panangaden. Nuclear and trace ideal in tensored *-categories. *Journal of Pure and Applied Algebra*, 143(1 – 3):3 – 47, 1999. 90

[2] E.-E. Doberkat. The demonic product of probabilistic relations. In Mogens Nielsen and Uffe Engberg, editors, *Proc. Foundations of Software Science and Computation Structures*, volume 2303 of *Lecture Notes in Computer Science*, pages 113 – 127, Berlin, 2002. Springer-Verlag. 87, 90

[3] E.-E. Doberkat. The converse of a probabilistic relation. In A. Gordon, editor, *Proc. Foundations of Software Science and Computation Structures'03*, volume 2620 of *Lecture Notes in Computer Science*, pages 233 – 249, Berlin, 2003. Springer-Verlag. 87, 90, 94

[4] E.-E. Doberkat. Stochastic relations interpreting modal logic. Technical Report 144, Chair for Software-Technology, University of Dortmund, October 2003. 94

[5] P. Panangaden. Probabilistic relations. In C. Baier, M. Huth, M. Kwiatkowska, and M. Ryan, editors, *Proc. PROBMIV*, pages 59 – 74, 1998. Also available from the School of Computer Science, McGill University, Montreal. 90

[6] K. R. Parthasarathy. *Probability Measures on Metric Spaces.* Academic Press, New York, 1967. 90, 91

[7] S. M. Srivastava. *A Course on Borel Sets.* Number 180 in Graduate Texts in Mathematics. Springer-Verlag, Berlin, 1998. 87, 90, 95

Pointer Kleene Algebra

Thorsten Ehm

Institut für Informatik, Universität Augsburg
D-86135 Augsburg, Germany
Ehm@informatik.uni-augsburg.de

Abstract. We present an extension of Kleene algebra (KA) that can be used for modeling a record based view of pointer structures. This is achieved by transferring some concepts of fuzzy relation theory to KAs. The defined framework enables us to maintain within a single extended Kleene algebra several equally shaped KAs modeling distinct record selectors.

Keywords: Kleene algebra, embedding, pointer algebra, pointer structures.

1 Introduction

Pointer algorithms are of great importance in the world of programming. Although very error-prone, they are of special interest because of their performance. In modern programming languages, direct pointer manipulations are forbidden, but this only shifts the problem to a higher level of abstraction. Every object-oriented programming language has to manipulate a large net of connected objects. Even those more abstract pointer structures do not support or simplify the formal reasoning about correctness of the used algorithms. On the system side the problem continues. Modern programming languages support the user with automatic memory management. This requests a garbage collection mechanism that also has to handle potentially extremely interlinked object sets on the heap. To formally solve these problems we introduce a Kleene algebra-based axiomatization of a pointer algebra that enables us to formulate properties of pointer structures and formally derive pointer algorithms.

Kleene algebra [3], the algebra build from the regular operations join, composition and finite iteration, has turned out to be an appropriate tool for the development of graph algorithms and the formalization of graphs. Nevertheless, in all Kleene algebra-based considerations, only unlabeled graphs are treated, which results in a separate treatment of distinctly labeled link structures. This is not an appropriate solution for reachability considerations via more than one or all differently labeled edges. This paper presents a calculus based on Kleene algebra that manages to cope with pointer structures represented by labeled graphs. This is achieved by porting results from abstract fuzzy relation theory to Kleene algebra.

R. Berghammer et al. (Eds.): RelMiCS/Kleene-Algebra Ws 2003, LNCS 3051, pp. 99–111, 2004.

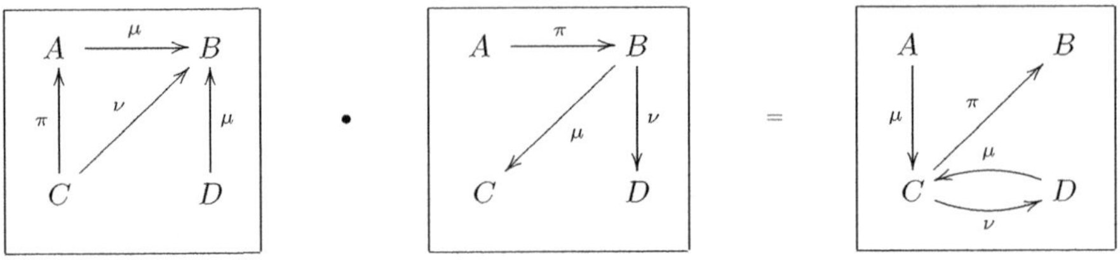

Fig. 1. Example of composition

Fuzzy relation algebra is an abstraction of relation algebra where relations are expressed in a graded way. Comparable to membership grades in fuzzy set theory, object relations are expressed in different gradations. A generalization of these are L-fuzzy relations where the relation grades form a lattice L. If this lattice is atomic we can interpret L-fuzzy relations as consisting of arrows labeled with a set of atoms from L. This again can be seen as an abstract view of labeled graphs which has been shown to be a useful model for a record based view of pointer structures [14].

The idea of the presented formalization is that join represents the union. Composition of two labeled graphs G_1 and G_2 connects nodes S and T with label μ exactly if there is an μ-labeled edge from S to U in G_1 and from U to T in G_2 (Figure 1). So composition should yield all equally labeled paths of length two where the first step is performed in G_1 and the second step in G_2. Algebraically, graphs only consisting of equally labeled self-links are scalars and equally labeled completely connected graphs correspond to ideals (Figure 2). In graph theory, intersection with these ideals is used to have access to μ-labeled subgraphs. As there is no meet in Kleene algebra, we establish a bijective correspondence between scalars and ideals and use composition with scalars to achieve the same result.

There still remains the question why we do not use fuzzy relation algebra straight away. Almost all areas that have to be treated formally demand a powerful but also concise calculus. As these two desires affect each other we are forced to find a compromise between them. Kleene algebras have turned out to be an algebraic system that is simple in its treatment on the one hand and of high expressive power on the other. It has been shown that many algebraic problems previously treated with a relational framework do not really need a converse operation and make do with sequential composition, choice, iteration and some minor extensions [1, 2, 5, 17].

This paper is structured as follows: Section 2 defines Kleene algebra, extensions and operators. A short overview shows in Section 3 which extra axioms are needed to establish a bijection between scalars and ideals. In Section 4 the notion of crispness is introduced and it is explained how L-fuzzy relations can be used to model pointer structures. Section 5 shows how the previously defined framework can be used to algebraically treat pointer structures. Finally, Section 6 summarizes and points out future work.

2 Extensions of Kleene Algebra

This section presents extensions of Kleene algebra and all the operations that are used to achieve an embedding of several Kleene algebras, each modeling a uniquely labeled graph, into one Kleene element. A complete axiomatization with all used extensions is given in Definition 10. As basis we use the axiomatization as given by Kozen [13].

Definition 1 (Kleene Algebra). *A* Kleene algebra $(\mathcal{K}, +, \cdot, 0, 1, ^*)$ *is an idempotent semiring with star, viz.* $(\mathcal{K}, +, 0)$ *and* $(\mathcal{K}, \cdot, 1)$ *are monoids,* $+$ *is commutative and idempotent, composition distributes through joins and* 0 *is an annihilator for composition. For star the following properties are postulated:*

$$1 + a^* \cdot a = a^* \quad \text{(l-star)} \qquad b + a \cdot c \leq c \rightarrow a^* \cdot b \leq c \quad \text{(l-star-ind)}$$
$$1 + a \cdot a^* = a^* \quad \text{(r-star)} \qquad b + c \cdot a \leq c \rightarrow b \cdot a^* \leq c \quad \text{(r-star-ind)}$$

Like in Kleene algebra with tests [13] we also need a Boolean sort which here represents the nodes in a graph. In contrast to Kozen we identify the tests with the set of *predicates* $\mathcal{P} = \{s : s \leq 1\}$ and demand that they form a Boolean algebra $(\mathcal{P}, +, \cdot, \neg, 0, 1)$, with $\neg$ denoting the complement in $\mathcal{P}$. We will denote predicates by symbols s and t. This approach is a special view of dynamic algebra [16], embedding the Boolean sort into the KA such that the module function coincides with the image operation presented in Definition 4. The composition with a predicate can be interpreted as restriction of the graph to a set of nodes.

A helpful tool to gain knowledge about the internal structure of elements but nevertheless staying in an abstract framework is residuation. Residuals characterize largest solutions of certain linear equations and are defined by Galois connections:

Definition 2 (Residuals). $b \leq a \backslash c \overset{\text{def}}{\Leftrightarrow} a \cdot b \leq c \overset{\text{def}}{\Leftrightarrow} a \leq c / b$

We will call a Kleene algebra where all residuals exist a *residuated KA*. In contrast to our earlier approaches, we do not assume anymore the existence of arbitrary residuals. The reasons are, first, that existence of residuals is a rather strong demand and, second, that for our purpose we only need residuals with respect to predicates. These can be defined using a greatest element $\top$. Therefore we assume subsequently that the KAs considered have a top element. Note that in all residuated KAs there is a top element, viz. $0 \backslash 0$.

Lemma 1. *In residuated KAs the residuals of predicate* s *can be expressed as*

$$s \backslash a = a + \neg s \cdot \top \qquad\qquad a / s = a + \top \cdot \neg s \tag{1}$$

Proof. We only show the first equality. Assume a residuated KA, then

"$\geq$": $a + \neg s \cdot \top \leq s \backslash a \Leftrightarrow s \cdot (a + \neg s \cdot \top) \leq a \Leftrightarrow s \cdot a \leq a$
"$\leq$": $s \backslash a = s \cdot (s \backslash a) + \neg s \cdot (s \backslash a) \leq a + \neg s \cdot \top$ □

So, subsequently we only assume (non-residuated) KAs extended with $\top$ and use $\backslash$ as well as $/$ as abbreviation defined by terms (1). In this structure, Lemma (1) implies the Galois connections with respect to predicates, so that subsequently we can use all standard results about residuals restricted to predicates.

Lemma 2. *$s \cdot a \leq b \Leftrightarrow a \leq s\backslash b$ and $a \cdot s \leq b \Leftrightarrow a \leq b/s$*

Proof. Assume $s \cdot a \leq b$. Then $a = s \cdot a + \neg s \cdot a \leq b + \neg s \cdot \top = s\backslash b$. Now assume $a \leq s\backslash b$. Then $s \cdot a \leq s \cdot (s\backslash b) = s \cdot (b + \neg s \cdot \top) = s \cdot b \leq b$ □

For an abstract treatment of domain and codomain we use predicates defined equationally based on an axiomatization given in [4].

Definition 3 (Domain). *The domain $\ulcorner$ of an element is required to satisfy:*

$$a \leq {}^{\ulcorner}a \cdot a \qquad\qquad {}^{\ulcorner}(s \cdot a) \leq s$$

Codomain $\urcorner$ is defined symmetrically. For practical applications these two laws have to be supplemented by a law called left-locality (${}^{\ulcorner}(a \cdot b) = {}^{\ulcorner}(a \cdot {}^{\ulcorner}b)$). This describes the fact that calculation of the domain of a composition only depends on the domain of the right argument. We will show that locality follows from a later added extension and therefore do not demand this law right here.

To work with pointer structures one needs two other essential operations. To follow the links we require a possibility to calculate all the direct successors of a node. This is achieved by the image operator : that calculates the successors of a set of nodes. To be able to modify a pointer structure we also need an update operator | that selectively alters links. Both can be defined using the framework presented so far:

Definition 4 (Image/Update).

$$s : a \stackrel{\text{def}}{=} (s \cdot a)^{\urcorner} \qquad\qquad b \mid a \stackrel{\text{def}}{=} b + \neg{}^{\ulcorner}b \cdot a$$

We assume that $\cdot$ binds stronger than : or | and they again bind stronger than +. Abstract properties of the image operator are investigated in the setting of Kleene modules in [9]. Since the image operator is build from two monotone functions, that both distribute through joins, these properties are directly inherited. The induction principle for the star operator in Kleene algebra (Definitions l-star-ind and r-star-ind), can be lifted to the image under a starred Kleene element.

Lemma 3. *$s + t : a \leq t \Rightarrow s : a^* \leq t$*

The properties of selective updates are shown in detail in [6].

3 Ideals and Scalars

Since we are interested in a record based view of pointer structures, a single sort of links is not sufficient. We need distinct selectors for each record component. A record-based pointer structure then can be represented by a labeled

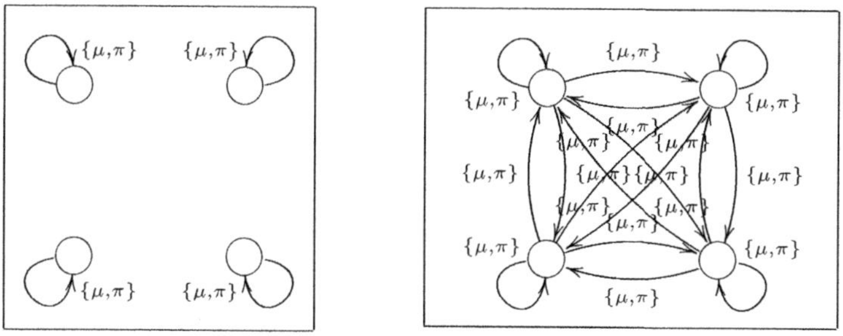

Fig. 2. Representation of scalars and ideals

graph. Each node is identified with the address of a record whereas links marked with selector names represent the record components. The basic trick to model *labeled* graphs is to identify each label with a relation grade of an L-fuzzy relation. These grades form a lattice L which can be interpreted as the lattice of selector handles. Although the reader may think of a matrix algebra with matrix entries from set L as a sort of standard model, our approach is more general. Goguen categories [18], an extension of Dedekind categories, also known as locally complete division allegories in [10], have shown to be a suitable axiomatization for L-fuzzy relations. To represent labeled graphs we port some results from Dedekind categories to KAs.

The access to differently labeled links or subgraphs is based on a bijective connection between ideals and scalars, which does not hold in all Kleene algebras.

Definition 5 (Ideal). *A right ideal is an element $j \in \mathcal{K}$ that satisfies $j = j \cdot \top$. Symmetrically we define the notion of a left ideal. An ideal then is an element that is a left and a right ideal (that fulfills $\top \cdot j \cdot \top = j$).*

An ideal corresponds to a completely connected graph where all arrows are identically labeled with the same selector (see Figure 2). Each of these graphs plays the rôle of a top element in the sublattice for a fixed selector.

Definition 6 (Scalar). *An element $\alpha \in \mathcal{P}$ is called a* scalar *iff $\alpha \cdot \top = \top \cdot \alpha$.*

The set of scalars will be denoted by $\mathcal{S}$. Scalars are similar to ideals except that they are not completely connected. There are only pointers from each node to itself (Figure 2). Scalars not only commute with top but also show some other nice commutativity properties:

Lemma 4. *Let $\alpha \in \mathcal{S}$ and $a \in \mathcal{K}$. Then $\alpha \backslash a = a / \alpha$ and $\alpha \cdot a = a \cdot \alpha$*

In the sequel we denote ideals by j and scalars by α, β. We now show in a short overview how the bijective correspondence between scalars and ideals can be established in KAs. The two mutually inverse functions that relate ideals and scalars bijectively in a Dedekind category with $\sqcap$ denoting the meet are [12]:

$$i_{\mathcal{SJ}}(\alpha) = \alpha \cdot \top \qquad i_{\mathcal{JS}}(j) = j \sqcap 1 \tag{2}$$

The bijection follows from the validity of the modular laws in Dedekind categories. In fact, one only needs the weaker versions that use $\top$ as the conversed element so that the converse is not needed explicitly in the proof (we assume that $\cdot$ binds stronger than $\sqcap$):

$$\top \cdot a \sqcap b \leq \top \cdot (a \sqcap \top \cdot b) \qquad a \cdot \top \sqcap b \leq (a \sqcap b \cdot \top) \cdot \top \tag{3}$$

Although we got rid of conversion, there is no meet operation in KA. But we can find a remedy by replacing the meet with domain. In the case that the restricted modular laws (3) hold, there is a closed formula for domain:

$$\ulcorner a = a \cdot \top \sqcap 1 \tag{4}$$

As a consequence $i_{\mathcal{JS}}$ on ideals simplifies to

$$i_{\mathcal{JS}}(j) = j \sqcap 1 = j \cdot \top \sqcap 1 = \ulcorner j$$

This is an operation also present in Kleene algebra. Now we can give axioms not using meets that are weaker than the modular laws but also imply Equation (4):

Lemma 5. *In Dedekind categories the following conditions are equivalent:*

1. $\ulcorner a \leq a \cdot \top$
2. $\ulcorner a \cdot \top = a \cdot \top$
3. $\ulcorner a = a \cdot \top \sqcap 1$

It is easy to show that 1. and 2. are equivalent even in KAs with $\top$ and domain. Symmetrically, the corresponding formulas for codomain are equivalent. Motivated by Equation 5.1 we call the equations of Lemma 5 and the corresponding codomain versions *subordination* of domain and codomain, respectively. Under the assumption of subordinated domain and codomain we can define the two mappings from (2) in KA with top and domain/codomain by

$$i_{\mathcal{SJ}}(\alpha) = \alpha \cdot \top \qquad\qquad i_{\mathcal{JS}}(j) = \ulcorner j$$

Indeed both functions are injective and the bijection between ideals and scalars can be shown by: $i_{\mathcal{JS}}(i_{\mathcal{SJ}}(\alpha)) = i_{\mathcal{JS}}(\alpha \cdot \top) = \ulcorner(\alpha \cdot \top) = \alpha$

$$i_{\mathcal{SJ}}(i_{\mathcal{JS}}(j)) = i_{\mathcal{SJ}}(\ulcorner j) = \ulcorner j \cdot \top = j \cdot \top = j$$

In the sequel we will assume an extended Kleene algebra which is a KA with top, domain and codomain as well as subordination of domain and codomain.

As we mentioned earlier, locality follows from subordination. So for example left-locality holds in Kleene algebra extended with one of the equations from Lemma 5:

Lemma 6. *Assume subordination of domain. Then* $\ulcorner(a \cdot \ulcorner b) = \ulcorner(a \cdot b)$

Proof. $\ulcorner(a \cdot \ulcorner b) \leq \ulcorner(a \cdot b \cdot \top) = \ulcorner(\ulcorner(a \cdot b) \cdot \top) = \ulcorner(a \cdot b)$, and the opposite direction holds in all Kleene algebras. □

Conversely, right-locality follows from one of the properties that hold symmetrically for codomain. So a KA with subordinated domain and codomain shows locality. Locality of composition is directly inherited by the image operator. The corresponding equality is:

Lemma 7. *Locality of the image operator* $(s : a) : b = s : (a \cdot b)$.

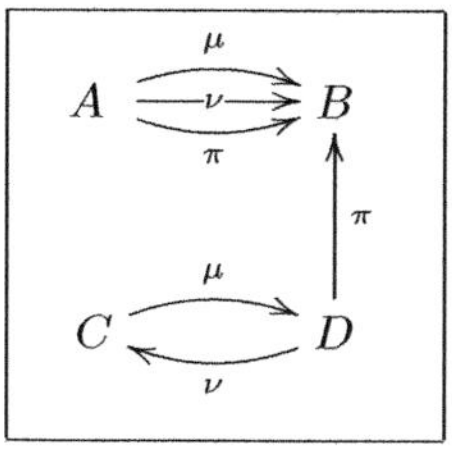

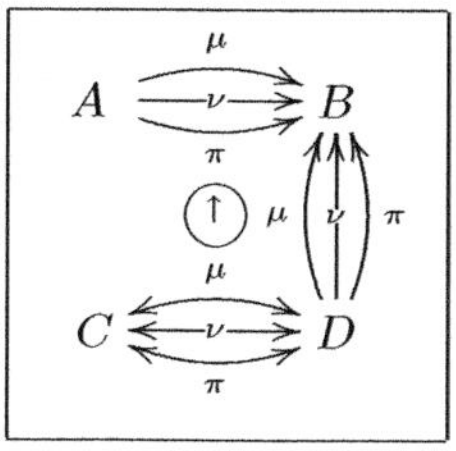

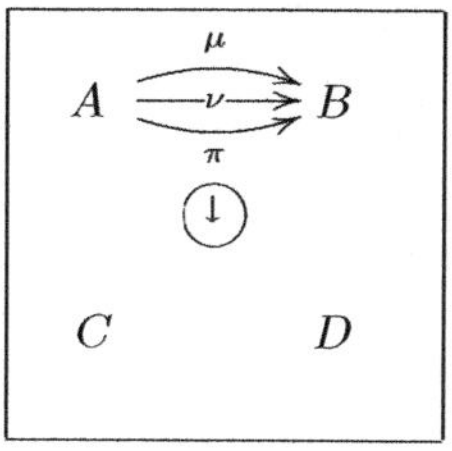

Fig. 3. Example graph and application of $^\uparrow$ and $^\downarrow$

4 Formalizing Crispness

To formalize L-fuzzy relations, a notion of crispness is needed. Crispness describes the total absence of uncertain information. So a crisp relation relates two elements a hundred percent or not at all. To model crispness we define two new operators $^\uparrow$ and $^\downarrow$ that send an element to the least crisp element it is included in and to the greatest crisp element it includes, respectively. The effect of these operations carried over to labeled graphs is depicted in Figure 3.

In the original graph on the left side a crisp connection exists from node A to node B as they are connected by all types of links. Applying $^\uparrow$ results in the graph in the middle, where all previously partially linked nodes are linked totally. Application of $^\downarrow$ yields the graph on the right side in which only the previously crisp parts remain. As $^\uparrow$ and $^\downarrow$ produce related least and greatest elements, we can use a Galois connection to define them. The following equations have to hold:

Definition 7 (Up and Down).

1. *$(^\uparrow, ^\downarrow)$ form a Galois connection, viz:* $a^\uparrow \leq b \Leftrightarrow a \leq b^\downarrow$
2. $(a \cdot b^\downarrow)^\uparrow = a^\uparrow \cdot b^\downarrow$ *and* $(a^\downarrow \cdot b)^\uparrow = a^\downarrow \cdot b^\uparrow$
3. *If α is a scalar and $\alpha \neq 0$, then $\alpha^\uparrow = 1$*

From the Galois connection there follows immediately distributivity of the lower adjoint $^\uparrow$ through joins and the cancellation laws. Further $^\uparrow$ and $^\downarrow$ show the following properties:

Lemma 8.

1. $1^{\uparrow} = 1$
2. $a^{\downarrow\uparrow} = a^{\downarrow}$ *and* $a^{\uparrow\downarrow} = a^{\uparrow}$
3. $a^{\uparrow\uparrow} = a^{\uparrow}$ *and* $a^{\downarrow\downarrow} = a^{\downarrow}$
4. $a \leq a^{\uparrow}$ *and* $a^{\downarrow} \leq a$
5. $a^{\uparrow} = a \Leftrightarrow a^{\downarrow} = a$
6. $0^{\uparrow} = 0$ *and* $\top^{\uparrow} = \top$
7. (a) $a^{\uparrow} = 0 \Leftrightarrow a = 0$
   (b) $a^{\downarrow} = \top \Leftrightarrow a = \top$
   (c) $s^{\downarrow} = 1 \Leftrightarrow s = 1$
8. $(a \cdot b^{\uparrow})^{\uparrow} = a^{\uparrow} \cdot b^{\uparrow} = (a^{\uparrow} \cdot b)^{\uparrow}$

We can now define crisp elements as those that are not changed by $^{\uparrow}$:

Definition 8 (Crisp). *An element $a \in \mathcal{K}$ is called* crisp, *if $a^{\uparrow} = a$.*

Crisp elements are closed under join and composition. All the constants 0,1 and $\top$ are crisp.

To retrieve a desired element from a graph we need a unique identifier for the embedded subgraphs. To achieve this, scalars which represent the labeling of links are used as handles to access the subgraphs labeled with these marks. Then we have to calculate a sort of projection that provides the embedded parts. By validity of $\alpha \backslash a = a + \neg\alpha \cdot \top$ we can see that the residual under a scalar completes the graph with links labeled with marks that are not in α. So two nodes are completely connected in $\alpha \backslash a$ if and only if they were at least linked before via all pointers described by α. Application of $^{\downarrow}$ yields a graph completely connecting all the nodes that were previously connected at least via the α links. By restricting this result to α we get a graph consisting of all the α-links of the original graph. So we define the projection function:

Definition 9 (Projection). $P_{\alpha}(a) = \alpha \cdot (\alpha \backslash a)^{\downarrow}$

Calculation of the image of m under $P_{\alpha}(a)$ gives us the α-successor of m. The sets $\mathcal{K}_{\alpha} = \{\alpha \cdot (\alpha \backslash a)^{\downarrow} \mid a \in \mathcal{K}\}$ of all elements of a scalar α form a Kleene algebra $(\mathcal{K}_{\alpha}, +, \cdot, 0, \alpha, ^{*})$ itself [1]. The ideal $j = \alpha \cdot \top$ corresponding to α is the top element. This shows that the Kleene algebras for each selector are embedded into the KA.

For some proofs we additionally need a property called the *resolution form* of a fuzzy relation.

$$a \leq \sum_{\alpha \in \mathcal{S}} P_{\alpha}(a) \qquad \text{(fuzzy-res)}$$

This demand is quite natural and essentially says that a fuzzy relation is covered by the union of all its projections. To summarize we define:

Definition 10 (Enriched KA). *An enriched Kleene algebra (EKA) is a Kleene algebra $(\mathcal{K}, +, \cdot, 0, 1, ^{*})$ where the set of predicates forms a Boolean lattice, a top element $\top$, subordination of domain/codomain and $^{\uparrow}$, $^{\downarrow}$ satisfying the axioms in Definition 7 and Equation fuzzy-res.*

EKAs are our basis to calculate with properties of pointer structures and labeled graphs.

[1] To be correct, α has to be atomic in the set of scalars [8].

5 Pointer Algebra

This section shortly explains how enriched Kleene algebra elements can be utilized to define operations of a pointer algebra. It is shown that we can express all the operations defined and used in [14]. Some nice and important rules are mentioned. More properties and all the proofs can be found in [8].

As there is no separate sort for addresses, we have to model them as particular elements of the Kleene algebra. In an unlabeled graph they could be represented by predicates which, in turn, are nodes with only a single self-link. For labeled graphs crisp predicates have to be used. These are nodes that have self-pointers with all possible labels.

Definition 11 (Address). *A crisp element* $m \leq 1$ *is called an* address.

In the sequel we will use letters m and n to denote addresses. As addresses are crisp they are closed under join and composition. Additionally they are closed under complement and so form a lattice.

An important requirement for pointer algorithms is the existence of a special address $\diamond$ that models a terminal element for data structures. In contrast to Hoare and Jifeng [11] who propose to model it by an address with all links pointing to itself we choose $\diamond$ to be a special node from which no links start. This is a natural condition for $\diamond$ expressing that $\diamond$ can not be dereferenced. So for all elements a that represent stores it has to hold that there is no image of $\diamond$ under a which is expressed by $\diamond : a = 0$. We will refer to a pair (m, a) consisting of an entry address m and a store a by the notion *pointer structure*. The most important properties in pointer structures are based on reachability observations. We are especially interested in addresses or nodes reachable from a set of starting addresses as well as in the corresponding part of the store. So we define two reachability operators. The first calculates all reachable addresses starting from m in store a. The second operator restricts the store to the part that contains all the links and addresses that are reachable from the entry address. This is a sort of projection of the accessible part of the store.

Definition 12 (Reach/From).

$$\mathit{reach}(m,a) \stackrel{\text{def}}{=} m : (a^{\uparrow})^*$$

$$\mathit{from}(m,a) \stackrel{\text{def}}{=} \mathit{reach}(m,a) \cdot a$$

We immediately notice that *reach* by definition is a closure operator in its first argument. Likewise *from* forms an interior operator in the second argument. Both operators are connected since equivalence of the *from* part implies equivalent sets of reachable nodes. The opposite direction does not have to hold because there can be different stores with the same reachability behaviour.

Lemma 9. $\mathit{from}(m,a) = \mathit{from}(m,b) \Rightarrow \mathit{reach}(m,a) = \mathit{reach}(m,b)$

This shows that the *from* operator really singles out the accessible part of the store. So the reachable addresses of the live part are exactly the ones reachable in the original store:

Corollary 1. $\mathit{reach}(m, \mathit{from}(m,a)) = \mathit{reach}(m,a)$

On the other hand, if we know which addresses are allocated, we are able to define a complementary operator to *reach* that calculates all the used but not reachable records in a pointer structure. Therefore we define *recs* that returns the allocated addresses of a which are all the elements a pointer link starts from.

Definition 13. $\mathit{recs}(a) \stackrel{\text{def}}{=} (\ulcorner a)^{\uparrow}$

As $\ulcorner$ and $^{\uparrow}$ can be commuted, this is equivalent to $\mathit{recs}(a) = \ulcorner(a^{\uparrow})$. It also immediately follows that the allocated records of $\mathit{from}(m,a)$ are all allocated records that are reachable.

Corollary 2. $\mathit{recs}(\mathit{from}(m,a)) = \mathit{reach}(m,a) \cdot \mathit{recs}(a)$

The allocated but non-reachable records are all those in *recs* without the reachable ones. This is an abstract description of garbage nodes (cells that were in use but are no longer reachable from the roots) in the store.

Definition 14. $\mathit{noreach}(m,a) \stackrel{\text{def}}{=} \mathit{recs}(a) \cdot \neg \mathit{reach}(m,a)$

The previously noticed relation between *reach* and *from* (Lemma 9) immediately can be applied to *noreach*. So we get an alternative definition for *noreach*.

Lemma 10. $\mathit{noreach}(m,a) = \mathit{recs}(a) \cdot \neg \mathit{recs}(\mathit{from}(m,a))$

Most simplifications during reasoning about pointer structures is based on localization properties. So for example parts of the memory do not have any effect for the calculation of pointer structure operators if certain addresses are not reachable. Many of the interesting laws need such a non-reachability precondition and serve to simplify *reach* or *from* expressions. Thus, we define the predicate $\not\vdash$ that expresses that none of the nodes in n is reachable from pointer structure (m,a):

Definition 15. $(m,a) \not\vdash n \stackrel{\text{def}}{=} \mathit{reach}(m,a) \cdot n = 0$

If we assume that the allocated records of a distinct part in a composed store are not reachable one can ignore these regions in reachability calculations. Mostly this is applied on joined and updated stores:

Lemma 11. *Assume* $(m,a) \not\vdash \mathit{recs}(b)$, *then*

1. $\mathit{reach}(m, a+b) = \mathit{reach}(m,a)$
2. $\mathit{reach}(m, b \mid a) = \mathit{reach}(m,a)$
3. $\mathit{from}(m, a+b) = \mathit{from}(m,a)$
4. $\mathit{from}(m, b \mid a) = \mathit{from}(m,a)$

The proofs are rather simple using image induction (Lemma 3) whereas in [14] fixed point induction over a suitable predicate is needed.

More interesting predicates of pointer structures are *sharing* and *acyclicity*. Using these preconditions even more sophisticated laws can be shown. We say two pointer structures do not show sharing if there are no common addresses reachable. As exception we allow sharing of $\diamond$, since this address is used as anchor element in all inductively defined data structures.

Definition 16. $\neg sharing(m, n, a) \Leftrightarrow reach(m, a) \cdot reach(n, a) \leq \diamond$

For the definition of acyclicity we have investigated and examined several characterizations in [8]. The most suitable is a predicate called *progressively finite* on graphs. This is equivalent to acyclicity if the graph is finite. As pointer structures in real programs should be finite, we can choose progressive finiteness as a characterization for acyclicity.

Definition 17. *A store a is called acyclic, if* $\forall m \in \mathcal{P}.\ m \leq m : a^{\uparrow^{+}} \rightarrow m \leq 0$

With the assumption of acyclicity we can show stronger properties of pointer algebra operations.

To have the possibility to define single links from one address to another we can also define a ministore $(m \stackrel{\alpha}{\rightarrow} n) = P_\alpha(m \cdot \top \cdot n)$ that models α-linked addresses from m to n. Using such a ministore and the update operator we are able to describe single changes of the pointer structure. In particular we can show some of the most sophisticated rules that are needed to derive algorithms on pointer structures with selective updates. We abbreviate the α-successor of m in store a by $a_\alpha(m) \stackrel{\text{def}}{=} m : P_\alpha(a)$. With this we immediately get:

Corollary 3. *Set* $c = (m \stackrel{\alpha}{\rightarrow} n) \mid b$. *Then*

1. $(n, b) \not\vdash m \Rightarrow from(n, c) = from(n, b)$
2. $\alpha \cdot \beta = 0 \wedge (b_\beta(m), b) \not\vdash m \Rightarrow from(c_\beta(m), c) = from(b_\beta(m), b)$

The proofs are direct consequences of Lemma 11 and properties of ministores. For the second proposition one needs to show that $c_\beta(m) = b_\beta(m)$ which follows from $\alpha \cdot \beta = 0$. In a C-like notation with implicit store the first proposition can be written:

> *from*(*p*->α) = *from*(*q*) holds after assignment *p*->α=*q*, if address p is not reachable from address q.

This says that if the α-successor of a pointer structure p is set to q then, under the given reachability conditions, selection of the α-successor yields a reachable substructure equal to q.

6 Summary and Outlook

We have presented a formalization of pointer structures and operations in an extended Kleene algebra. This algebra models records and links between them. Such a model has proved to be a useful tool for formally deriving pointer algorithms from an abstract specification [7, 14]. There, the framework is used for ensuring correctness and optimizing the resulting algorithms. A big challenge for the calculus is a formal treatment of a garbage collector. Although this has been done several times before, this will be a testbed for analyzing how powerful and concise the algebra presented here is.

Concerning the particular axiomatization, the following can be observed. Because of the Horn formulas for star induction rule and in the Galois connection for the definition of $^\uparrow$ and $^\downarrow$, the defined class of algebras forms a quasi-variety. Nevertheless there are equational axiomatizations for action logic [15], which corresponds to KAs with existing residuals. There, also induction is defined purely equationally. We have here also used an equational characterization to define domain and codomain in contrast to previous axiomatizations that used a Galois connection. Last but not least we also can give an equational characterization of the up and down operators, which means that the whole algebra can be defined as a variety.

Acknowledgement

I would like to thank M. Winter, B. Möller and G. Struth for valuable discussions.

References

[1] K. Clenaghan. Calculational graph algorithmics: reconciling two approaches with dynamic algebra. Technical report CS-R9518, CWI - Centrum voor Wiskunde en Informatica, March 1995. 100

[2] E. Cohen. Separation and reduction. In R. Backhouse and J. N. Oliveira, editors, *Proceedings of Mathematics of Program Construction, 5th International Conference, MPC 2000*, volume 1837 of *Lecture Notes in Computer Science*, pages 45–59. Springer-Verlag, 2000. 100

[3] J. H. Conway. *Regular Algebra and Finite Machines.* Chapman & Hall, London, 1971. 99

[4] J. Desharnais, B. Möller, and G. Struth. Kleene algebra with a domain operator. Technical report 2003-7, Institut für Informatik, Universität Augsburg, 2003. 102

[5] J. Desharnais, B. Möller, and F. Tchier. Kleene under a Demonic Star. In T. Rus, editor, *Algebraic Methodology and Software Technology, 8th International Conference, AMAST 2000*, volume 1816 of *Lecture Notes in Computer Science*, pages 355–370. Springer-Verlag, 2000. 100

[6] T. Ehm. Properties of overwriting for updates in typed Kleene algebras. Technical report 2000-7, Institut für Informatik, Universität Augsburg, 2000. 102

[7] T. Ehm. Transformational Construction of Correct Pointer Algorithms. In D. Bjørner, M. Broy, and A. V. Zamulin, editors, *Perspectives of System Informatics*, volume 2244 of *Lecture Notes in Computer Science*, pages 116–130. Springer-Verlag, July 2001. 109

[8] T. Ehm. *The Kleene Algebra of Nested Pointer Structures: Theory and Applications.* PhD thesis, Universität Augsburg, 2003. 106, 107, 109

[9] T. Ehm, B. Möller, and G. Struth. Kleene modules. This volume, 2003. 102

[10] P. J. Freyd and A. Scedrov. *Categories, Allegories*, volume 39 of *North-Holland Mathematical Library.* North-Holland, Amsterdam, 1990. 103

[11] C. A. R. Hoare and H. Jifeng. A trace model for pointers and objects. In R. Guerraoui, editor, *ECCOP'99 - Object-Oriented Programming, 13th European Conference, Lisbon, Portugal, June 14-18, 1999, Proceedings*, volume 1628 of *Lecture Notes in Computer Science*, pages 1–17. Springer-Verlag, 1999. 107

[12] Y. Kawahara and H. Furusawa. Crispness and representation theorems in Dedekind categories. Technical report DOI-TR 143, Kyushu University, 1997. 103

[13] D. Kozen. Kleene algebra with tests. *ACM Transactions on Programming Languages and Systems*, 19(3):427–443, May 1997. 101

[14] B. Möller. Calculating with pointer structures. In R. Bird and L. Meertens, editors, *Algorithmic Languages and Calculi*, pages 24–48. Proc. IFIP TC2/WG2.1 Working Conference, Le Bischenberg, Feb. 1997, Chapman & Hall, 1997. 100, 107, 108, 109

[15] V. Pratt. Action logic and pure induction. In J. van Benthem and J. Eijck, editors, *Proceedings of JELIA-90, European Workshop on Logics in AI*, Amsterdam, September 1990. 110

[16] V. Pratt. Dynamic Algebras as a well-behaved fragment of Relation Algebras. In C. H. Bergman, R. D. Maddux, and D. L. Pigozzi, editors, *Algebraic Logic and Universal Algebra in Computer Science*, volume 425 of *Lecture Notes in Computer Science*. Springer-Verlag, 1990. 101

[17] J. von Wright. From Kleene algebra to refinement algebra. In B. Möller and E. Boiten, editors, *Mathematics of Program Construction, 6th International Conference, MPC 2002*, volume 2386 of *Lecture Notes in Computer Science*, pages 233–262. Springer-Verlag, 2002. 100

[18] M. Winter. Relational constructions in Goguen categories. In H. de Swart, editor, *6th International Seminar on Relational Methods in Computer Science (RelMiCS)*, pages 222–236, 2001. 103

Kleene Modules

Thorsten Ehm, Bernhard Möller, and Georg Struth

Institut für Informatik, Universität Augsburg
Universitätsstr. 14, D-86135 Augsburg, Germany
{ehm,moeller,struth}@informatik.uni-augsburg.de

Abstract. We propose axioms for Kleene modules (KM). These structures have a Kleene algebra and a Boolean algebra as sorts. The scalar products are mappings from the Kleene algebra and the Boolean algebra into the Boolean algebra that arise as algebraic abstractions of relational image and preimage operations. KM is the basis of algebraic variants of dynamic logics. We develop a calculus for KM and discuss its relation to Kleene algebra with domain and to dynamic and test algebras. As an example, we apply KM to the reachability analysis in digraphs.

1 Introduction

Programs and state transition systems can be described in a bipartite world in which propositions model their static properties and actions or events their dynamics. Propositions live in a Boolean algebra and actions in a Kleene algebra with the regular operations of sequential composition, non-deterministic choice and reflexive transitive closure. Propositions and actions cooperate via modal operators that view actions as mappings on propositions in order to describe state-change and via test operators that embed propositions into actions in order to describe measurements on states and to model the usual program constructs.

Most previous approaches show an asymmetric treatment of propositions and actions. On the one hand, propositional dynamic logic (PDL) [9] and its algebraic relatives dynamic algebras (DA) [12, 17, 19] and test algebras (TA) [17, 19, 22] are proposition-based. DA has only modalities, TA has also tests. Most axiomatizations do not even contain explicit axioms for actions: their algebra is only implicitly imposed via the definition of modalities. On the other hand, Kleene algebra with tests (KAT) [14]—Kleene algebra with an embedded Boolean algebra—is action-based and has only tests, complementarily to DA. Therefore, action-based reasoning in DA and TA and proposition-based reasoning in KAT is indirect and restricted. In order to overcome these rather artificial asymmetries and limitations, KAT has recently been extended to Kleene algebra with domain (KAD) with equational axioms for abstract domain and codomain operations [6]. This alternative to PDL supports both proposition- and action-based reasoning and admits both tests and modalities. The defining axioms of KAD, however, are quite different from those of DA and TA. Therefore, what is the precise relation between KAD and PDL and its algebraic relatives? Moreover, is the asymmetry and the implicitness of the algebra of actions in DA and TA substantial?

R. Berghammer et al. (Eds.): RelMiCS/Kleene-Algebra Ws 2003, LNCS 3051, pp. 112–124, 2004.

We answer these two questions by extending the above picture with a further intermediate structure (c.f. Figure 1). As already observed by Pratt [19], the definition of DA resembles that of a module in algebra, up to implicitness of the algebra of actions, in which the scalar products define the modalities. When DA was presented, this was reasonable, since there was no satisfactory axiomatization of Kleene algebra. So Pratt could only conjecture that a *Kleene module* (KM) with a Kleene algebra as scalar sort and a Boolean algebra as the other would yield a more natural and convenient axiomatization of DA. Depending on more recent developments in Kleene algebra, our axiomatization of KM verifies Pratt's conjecture and shows that the implicitness of Kleene algebra in DA is in fact unnecessary. KM is also used as a key for answering the first question and establishing KAD as a natural extension of previous approaches.

Our Contributions. First, we axiomatize and motivate the class KM as a straightforward adaptation of the usual modules from algebra [11]. We show that the scalar products abstractly characterize relational image and preimage operations. We outline a calculus for KM, including a duality between left and right scalar products in terms of a converse operation and a discussion of separability, that is, when actions are completely determined by their effects on states. We provide several examples of KM. We also relate our approach to a previous one based on a second-order axiomatization of the star [12].

Second, we relate KM and DA. We show that KM subsumes DA and, using a result of [19], that the equational classes of separable KM and separable DA coincide. This answers Pratt's conjecture. Consequently, the axioms of separable KM are complete with respect to the equational theory of finite Kripke structures.

Third, we relate KAD with KM and TA. We identify KAD with a subclass of TA, but obtain a considerably more economic axiomatization of that class. We show that the equational classes of separable KAD and separable TA coincide, improving a previous related result [10]. Consequently, the axioms of separable KAD are complete for the equational theory of finite Kripke (test) structures; the equational theory of separable KAD is EXPTIME-complete.

Fourth, we present extensions of KM that subsume TA, its above-mentioned subclass and KAD. This clarifies a related axiomatization [10].

Fifth, we demonstrate the expressiveness gap between KM and KAD by defining a basic toolkit for dynamic reachability analysis in directed graphs with interesting applications in the development and analysis of (graph) algorithms.

More generally, our technical comparison establishes KAD as a versatile alternative to PDL. Its uniform treatment of modal, scalar product and domain operators supports the interoperability of different traditional approaches to program analysis and development, an integration of action- and proposition-based views and a unification of techniques and results from these approaches.

Related Work. We can only briefly mention some closely related work. Our semiring-based variants of Kleene algebra and KAT are due to Kozen [13, 14].

DA has been proposed by Pratt [19] and Kozen [12] and further investigated, for instance, in [17, 18]. TA has been proposed by Pratt [19] and further investigated in [17, 22]. With the exception of [12], these approaches implicitly axiomatize the algebra of actions, the explicit Kleene algebra axioms for DA in [12] contain a second-order axiom for the star. More recently, Hollenberg [10] has proposed TA with explicit Kleene algebra axioms. This approach is similar, but less economic than ours. The related class of Kleenean semimodules has recently been introduced by Leiß [15] in applications to formal language theory, with our Boolean algebra weakened to a semilattice. Earlier on, Brink [2] has presented Boolean modules, using a relation algebra instead of a Kleene algebra. A particular matrix-model of KM has been implicitly used by Clenaghan [4] for calculating path algorithms. In the context of reachability analysis, concrete models of Kleene algebras or relational approaches have also be used, for instance, by Backhouse, van den Eijnde and van Gasteren [1], by Brunn, Möller and Russling [3], by Ravelo [21] and by Berghammer, von Karger and Wolf [20]. Ehm [7] uses an extension of KM for analyzing pointer structures.

In this extended abstract we can only informally present selected technical results. More details and in particular complete proofs of all statements in this text can be found in [8].

2 Kleene Algebra

A *Kleene algebra* [13] is a structure $(K, +, \cdot, ^*, 0, 1)$ such that $(K, +, \cdot, 0, 1)$ is an (additively) idempotent semiring and *, the *star*, is a unary operation defined by the identities and quasi-identities

$$1 + aa^* \leq a^*, \qquad (*\text{-}1) \qquad\qquad b + ac \leq c \Rightarrow a^*b \leq c, \qquad (*\text{-}3)$$

$$1 + a^*a \leq a^*, \qquad (*\text{-}2) \qquad\qquad b + ca \leq c \Rightarrow ba^* \leq c, \qquad (*\text{-}4)$$

for all $a, b, c \in K$. The natural ordering $\leq$ on K is defined by $a \leq b$ iff $a + b = b$. We call $(*\text{-}1)$, $(*\text{-}2)$ the *star unfold* and $(*\text{-}3)$, $(*\text{-}4)$ the *star induction* laws.

KA denotes the class of Kleene algebras. It includes, for instance, the set-theoretic relations under set union, relational composition and reflexive transitive closure (the *relational Kleene algebra*), and the sets of regular languages (regular events) over some finite alphabet (the *language Kleene algebra*).

The additive submonoid of a Kleene algebra is also an upper semilattice with respect to $\leq$. Moreover, the operations of addition, multiplication and star are monotonic with respect to $\leq$. The equational theory of regular events is the free Kleene algebra generated by the alphabet [13]. We will freely use the well-known theorems of KA (c.f. [8] for a list of theorems needed). In particular, the star unfold laws can be strengthened to equations.

Kleene algebra provides an algebra of actions with operations of non-deterministic choice, sequential composition and iteration. It can be enriched by a Boolean algebra to incorporate also propositions.

A *Boolean algebra* is a complemented distributive lattice. By overloading, we write $+$ and $\cdot$ also for the Boolean join and meet operation and use 0 and 1 for the least and greatest elements of the lattice. $'$ denotes the operation of complementation. BA denotes the class of Boolean algebras. We will consistently use the letters $a, b, c \ldots$ for Kleenean elements and $p, q, r, \ldots$ for Boolean elements. We will freely use the theorems of Boolean algebra in calculations.

A first integration of actions and proposition is Kleene algebra with tests. A *Kleene algebra with tests* [14] is a two-sorted structure (K, B), where $K \in \mathsf{KA}$ and $B \in \mathsf{BA}$ satisfies $B \subseteq K$ and has minimal element 0 and maximal element 1. In general, B is only a subalgebra of the subalgebra of all elements below 1 in K, since elements of the latter need not be multiplicatively idempotent. We call elements of B *tests* and write $\mathsf{test}(K)$ instead of B. For all $p \in \mathsf{test}(K)$ we have that $p^* = 1$. The class of Kleene algebras with tests is denoted by KAT.

3 Definition of Kleene Modules

In this section we define the class of Kleene modules. These are natural variants of the usual modules from algebra [11]. We replace the ring by a Kleene algebra and the Abelian group by a Boolean algebra.

Definition 1. *A* Kleene left-module *is a two-sorted algebra* $(K, B, :)$, *where* $K \in \mathsf{KA}$ *and* $B \in \mathsf{BA}$ *and where the* left scalar product $:$ *is a mapping* $K \times B \to B$ *such that for all* $a, b \in K$ *and* $p, q \in B$,

$$a : (p + q) = a : p + a : q, \qquad \text{(km1)} \qquad\qquad 1 : p = p, \qquad \text{(km4)}$$

$$(a + b) : p = a : p + b : p, \qquad \text{(km2)} \qquad\qquad 0 : p = 0, \qquad \text{(km5)}$$

$$(ab) : p = a : (b : p), \qquad \text{(km3)} \qquad\qquad q + a : p \leq p \Rightarrow a^* : q \leq p. \qquad \text{(km6)}$$

We do not distinguish between the Boolean and Kleenean zeroes and ones. KM_l denotes the class of Kleene left-modules. In accordance with the relation-algebraic tradition, we call scalar products of KM_l also *Peirce products*. We assign priorities $'$ higher than $:$ higher than $+$ and $-$.

Axioms of the form (km1)–(km4) also occur in algebra. For rings, an analog of (km5) is redundant, whereas for semirings—in absence of inverses—it is independent. Axiom (km6) is of course beyond ring theory. It is the star induction rule ($*$-3) with the semiring product replaced by the Peirce product and the sorts of elements adjusted, that is b and c replaced by Boolean elements. We call such a transformation of a KA-expression to a KM_l-expression a *peircing*.

We define *Kleene right-modules* as Kleene left-modules on the opposite semiring in the standard way (c.f [11]) by switching the order of multiplications. We write $p : a$ for right scalar products. A *Kleene bimodule* is a Kleene left-module that is also a Kleene right-module. Left and right scalar products can be uniquely determined by bracketing. We will henceforth consider only Kleene left-modules.

4 Example Structures

We now discuss the two models of Kleene modules that are most important for our purposes, namely relational Kleene modules and Kripke structures. Further example structures can be found in [6].

Example 1. (*Relational Kleene modules*) Consider the relational Kleene algebra $\mathsf{REL}(A) = (2^{A\times A}, \cup, \circ, \emptyset, \Delta, {}^*)$, on a set A with $2^{A\times A}$ denoting the set of binary relations over A and $\cup$, $\circ$, $\emptyset$ and Δ denoting set-union, relational composition, the empty relation and the identity relation, respectively. Finally, for all $R \in \mathsf{REL}(A)$ the expression R^* denotes the reflexive transitive closure of R.

Of course also $\mathsf{REL}(A) \in \mathsf{KAT}$ with $\mathsf{test}(\mathsf{REL}(A))$ being the set of all subrelations of Δ. This holds, since $\mathsf{test}(\mathsf{REL}(A))$ is a field of sets, whence a Boolean algebra, with $P \sqcap Q = P \circ Q$ and $P' = \Delta - P$, the minus denoting set difference. $\mathsf{test}(\mathsf{REL}(A))$ is isomorphic with the field of sets 2^A under the homomorphic extension of the mapping sending B to $\{(b,b) \mid b \in B\}$ for all $B \subseteq A$.

The *preimage* of a set $B \subseteq A$ under a relation $R \subseteq A \times A$ is defined as

$$R : B = \{x \in A \mid \exists y \in B.(x,y) \in R\}, \tag{1}$$

The definition of image is similar. We extend this definition to $\mathsf{REL}(A)$ via the above isomorphism. Then $(\mathsf{REL}(A), \mathsf{test}(\mathsf{REL}(A)), :)$, with $:$ given by (1), is in KM_l. Therefore the KM_l axioms abstractly model binary relations with a preimage operation. □

Example 2. (*Kripke Structure*) By Example 1, there is an isomorphism between the subsets of a set A and the set of subrelations of the identity relation $\Delta \subseteq A \times A$. A *Kripke structure* on a set A is a pair (K, B), where B is a field of sets on A and K is an algebra of binary relations on A under the operations of union, relational composition and reflexive transitive closure. Finally, a preimage operation on (B, K) is defined by (1).

Every Kripke structure contains the identity relation, since it is presumed in the definition of the reflexive transitive closure operation. However, it need not contain the empty relation. Therefore, not every Kripke structure is a Kleene left-module, but every Kripke structure with the empty relation is.

A *Kripke test structure* on A is a Kripke structure with the additional operation

$$?p = \{(x,x) \mid x \in p\},$$

for all $p \in B$. Kri and Krit denote the class of Kripke structures and Kripke test structures, respectively. The Kripke structure $(2^A, 2^{A\times A})$ is called the *full* Kripke structure on A; it is isomorphic with $\mathsf{REL}(A)$ and has all Kripke structures on A as subalgebras. □

The fact that KM_l contains relational structures and Kripke structures yields a natural correspondence with the semantics of modal logics. More example structures can be found in [6]. These examples are based on Kleene algebra with domain. But by the subsumption result in Proposition 3 below, they can easily be transfered to Kleene modules.

5 Calculus of Kleene Modules

In this section, we list some properties of Kleene modules that are helpful in an elementary calculus. These properties are also needed in the syntactic comparison of KM_l with other structures in Section 8.

The first lemma provides some properties that do not mention the star.

Lemma 1. *Let* $(K, B, :) \in \mathsf{KM}_l$. *For all* $a \in K$ *and* $p, q \in B$, *the scalar product has the following properties.*

(i.) *It is right-strict, that is* $a : 0 = 0$.
(ii.) *It is left- and right-monotonic.*
(iii.) $p \leq 0 \Rightarrow a : p \leq 0$.
(iv.) $a : (pq) \leq (a : p)(a : q)$.
(v.) $a : p - a : q \leq a : (p - q)$.

Here, $p - q = pq'$. Remember that Peirce products are left-strict by (km4).

The following statements deal with peircing the star. The first lemma explains why KM_l has no peirced variants of ($*$-1) and ($*$-2) as axioms.

Proposition 1. *Let* $(K, B, :) \in \mathsf{KM}_l$. *Let* $a \in K$ *and* $p \in B$.

(i.) $p + a : (a^* : p) = a^* : p$,
(ii.) $p + a^* : (a : p) = a^* : p$.

The following statement shows that quasi-identity (km6), although quite natural as a peirced variant of ($*$-3), can be replaced as an axiom by an identity.

Proposition 2. *Let* $(K, B, :) \in \mathsf{KM}_l$ *Then the quasi-identity (km6) and the following identity are equivalent.*

$$a^* : p \leq p + a^* : (a : p - p). \tag{2}$$

Proof. The Galois connection $p - q \leq r \Leftrightarrow p \leq q + r$ implies that $p \leq q \Leftrightarrow p - q \leq 0$ and that the cancellation law $p \leq q + (p - q)$ holds.

(km6) implies (2). Let $p = q$. Then $a : p + p = p$ and $a^* : (a : p + p) \leq p$. Adding $p + a^* : (a : p - p)$ to both sides of this last inequality yields

$$\begin{aligned} p + a^* : (a : p - p) &\geq p + a^* : (a : p - p) + a^* : (a : p + p) \\ &= p + a^* : ((a : p - p) + a : p + p) \\ &\geq p + a^* : (a : p) \\ &= (1 + a^* a) : p \\ &= a^* : p. \end{aligned}$$

The second step uses (km1). The third step uses the cancellation law and Kleene algebra. The fourth step uses (km4) and (km2). The fifth step uses again Kleene algebra.

(2) implies (km6). Let $a : p + q \leq p$, whence $a : p \leq p$ and $q \leq p$ and therefore $a : p - p \leq 0$. Using right-monotonicity and (km5), we calculate

$$a^* : q \leq a^* : p \leq p + a^* : (a : p - p) = p + a^* : 0 = p.$$

□

In [8], we present various additional properties. We show, for instance, that (km6) is also equivalent to $a : p \leq p \Rightarrow a^* : p \leq p$, which reflects an unpeirced theorem of KA, and that (2) can be strengthened to an equality. All these properties can easily be translated to theorems of propositional dynamic logic (c.f [9]), using our consideration in Section 8. In particular, (2) translates to an axiom.

6 Extensionality

In Kleene modules, the algebras of actions and propositions are only weakly coupled. The finer the algebra of propositions, the more precisely can we measure properties of actions. In general, actions are *intensional*; their behavior is not completely determined by measurements on states. Set-theoretic relations, however, are *extensional*, since they are sets. Set-theoretic extensionality can be lifted to Kleene modules to enforce relational models. In analogy to dynamic algebra [12, 19], we call $(K, B, :) \in \mathsf{KM}$ *(left)-separable*, if for all $a, b \in K$

$$\forall p \in B.(a : p \leq b : p) \Rightarrow a \leq b. \tag{3}$$

SV denotes the separable subclass of an algebraic class V with appropriate signature.

An adaptation of a three-element Kleene Algebra from [5] shows that separability is independent in KM_l. (3) is equivalent to $\forall p \in B.(a : p = b : p) \Rightarrow a = b$. Moreover, converse implications also hold by monotonicity. The term *separability* can be explained as follows: Assume (3) and let $a \neq b$ for some $a, b \in K$. Then $a : p \neq b : p$ for some $p \in B$; the witness p separates action a from action b.

Besides this relational motivation, separability can also be introduced algebraically. In [8], we show that the relation $\preceq$ on $(K, A, :) \in \mathsf{KM}_l$ defined by

$$a \preceq b \Leftrightarrow \forall p \in B.(a : p \leq b : p),$$

for all $a, b \in K$ is a precongruence on KM_l, that is, the operation of addition, left and right multiplication and star are monotonic with respect to $\preceq$. Moreover, the relation $\sim \, = \, \preceq \cap \succeq$ is a congruence on KM_l. Therefore, a Kleene module is separable, iff $\sim$ is the identity relation.

The relation $\sim$ introduces a natural notion of *observational equivalence*. For a set A, the preimage $R : \{p\}$ of a singleton set $\{p\} \subseteq A$ under a relation $R \subseteq A \times A$ is the set of all $q \in A$ with $(q, p) \in R$. Intuitively, $R : \{p\}$ scans R point-wise for its input-output behavior. Since relations are extensional, they are completely determined by this scanning. In intensional models, one can distinguish between observable and hidden intrinsic behavior. The congruence $\sim$ then identifies two relations up to intrinsic behavior and therefore via observational equivalence. The freedom of choosing the algebra of propositions in KM with arbitrary coarseness fits very well with this idea of measuring and identifying actions in a more or less precise way.

7 Relatives of Kleene Modules

We now situate the class KM_l within the context of Kleene algebra with domain and algebraic variants of propositional dynamic logic. To this end, we define the classes of dynamic algebras, test algebras à la Hollenberg, test algebras à la Pratt and Kleene algebra with domain.

We obtain the class DA [19] of *dynamic algebras* from Definition 1 by requiring an absolutely free algebra of Kleenean signature K (without 0 and 1) instead of a Kleene algebra, by removing (km4) and (km5), by adding right-strictness (Lemma 1 (i)) and the peirced star unfold law of Proposition 1 and by replacing (km6) by (2). Therefore, the algebra of actions is implicitly axiomatized in DA.

A *test algebra* à la Hollenberg [10] is a structure $(K, B, :, ?)$, where $K \in \mathsf{KA}$ and $B \in \mathsf{BA}$ and the operations : of Peirce product type and ? of type $B \to K$ satisfies the axioms (km2), (km3), (km6) and

$$p? : q = pq, \quad (4) \qquad (pq)? = (p?)(q?), \quad (7)$$

$$0? = 0, \quad (5) \qquad (a : 1)?a = a. \quad (8)$$

$$(p + q)? = p? + q?, \quad (6)$$

TA_H denotes the corresponding class. We show in [8] that ? is an embedding from B into K. In analogy to KAT, the symbol ? can therefore be made implicit; the axioms (5)–(7) and the ? symbol in the axioms (4) and (8) can be discarded. TA_H then reduces to KAT with the remaining axioms. Note that the algebra of actions is explicitly axiomatized in TA_H.

We obtain the class TA_P of *test algebras* à la Pratt [19] from DA by extending the signature with ? and by adding the axiom (4). Therefore, the algebra of actions is again implicitly axiomatized in TA_P.

A *Kleene algebra with domain* [6] is a structure (K, δ), where $K \in \mathsf{KAT}$ and the *domain operation* $\delta : K \to \mathsf{test}(K)$ satisfies, for all $a, b \in K$ and $p \in \mathsf{test}(K)$,

$$a \leq \delta(a)a, \quad \text{(d1)} \qquad \delta(pa) \leq p, \quad \text{(d2)} \qquad \delta(a\delta(b)) \leq \delta(ab). \quad \text{(d3)}$$

KAD denotes the class of Kleene algebras with domain. The impact of (d1), (d2) and (d3) can be motivated as follows. (d1) is equivalent to one implication in each of the statements

$$\delta(a) \leq p \Leftrightarrow a \leq pa, \quad \text{(llp)} \qquad \delta(a) \leq p \Leftrightarrow p'a \leq 0. \quad \text{(gla)}$$

which constitute elimination laws for δ. (d2) is equivalent to the other implications. (llp) says that $\delta(a)$ is the least left preserver of a. (gla) says that $\delta(a)'$ is the greatest left annihilator of a. Both properties obviously characterize domain in set-theoretic relations. (d3) states that the domain of ab is not determined by the inner structure of b or its codomain; information about $\delta(b)$ in interaction with a suffices. All three axioms hold in relational Kleene algebra. Note that in contrast to KM_l, there is no particular axiom for the star. As Lemma 2 (vi) below shows, a variant of the star induction law is a theorem of KAD.

As for Kleene modules, a codomain operation can be defined on the opposite Kleene algebra. Moreover, domain has the following properties.

Lemma 2 ([6]). *Let $K \in \mathsf{KAD}$. For all $a \in K$ and $p \in \mathsf{test}(A)$,*

(i.) *Strictness,* $\delta(a) = 0 \Leftrightarrow a = 0$.
(ii.) *Additivity,* $\delta(a + b) = \delta(a) + \delta(b)$.
(iii.) *Monotonicity,* $a \leq b \Rightarrow \delta(a) \leq \delta(b)$.
(iv.) *Locality,* $\delta(ab) = \delta(a\delta(b))$.
(v.) *Stability,* $\delta(p) = p$.
(vi.) *Induction,* $q + \delta(ap) \leq p \Rightarrow \delta(a^*q) \leq p$.

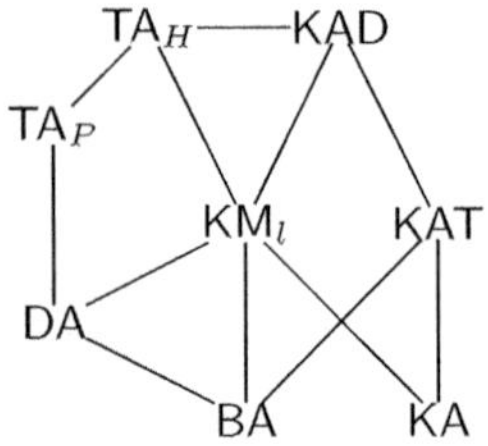

Fig. 1.

Of course, the preimage of a relation R under a set P can also be defined via domain as $\delta(RP)$. We use

$$a : p = \delta(ap), \quad (9) \qquad \delta(a) = a : 1, \quad (10)$$

for defining abstract preimage in KAD and abstract domain in KM_l.

8 Main Results

Our main interest are subsumption relations between the classes introduced in Section 7. Here, we show only the most important ones. A more complete picture can be found in Figure 1. More relations and complete proofs can be found in [8]. We proceed purely calculationally by deriving the axioms in the subsumed class as theorems in the subsuming class. In particular, we use the properties of Peirce products from Section 5, the properties of domain from Section 7 and the translations (10), (9) between Peirce products and domain.

Proposition 3. $\mathsf{TA}_H = \mathsf{KAD} \subseteq \mathsf{KM}_l \subseteq \mathsf{DA}$.

Proof. We first show that $\mathsf{KAD} \subseteq \mathsf{TA}_H$. It follows from the remaining inclusions that all TA_H axioms but (8) are theorems of KAD. Since ? is an embedding, (8) can be written in the form $(a : 1)a = a$. Translating with (9), we see that one inequality is (d1) while the other one holds, since $\delta(a) \leq 1$.

We now show that $\mathsf{TA}_H \subseteq \mathsf{KAD}$, using (10) for translation. By the previous part of the proof it remains to show that axioms (d2) and (d3) are theorems of TA_H. For (d2), we must show that $(pa) : 1 \leq p$ by (10). Using (km3) and (4), which are axioms of TA_H, and Boolean algebra, we calculate

$$(pa) : 1 = p : (a : 1) = p(a : 1) \leq p.$$

For (d3), we must show that $(a(b : 1)) : 1 = (ab) : 1$. We calculate

$$(a(b : 1)) : 1 = a : ((b : 1) : 1) = a : ((b : 1)1) = a : (b : 1) = (ab) : 1.$$

The first step uses (km3). The second step uses (4), the third step uses Boolean algebra, the fourth step uses again (km3).

We now briefly discuss the remaining inclusions. $\mathsf{DA} \subseteq \mathsf{KM}_l$ is immediate from the definition of DA via theorems of KM_l. $\mathsf{KAD} \subseteq \mathsf{KM}_l$ follows from (9) and the results of Lemma 2. □

Proposition 3 shows that the axioms (km2) and (km6) are redundant in TA_H. The axioms (5)–(7) can be made implicit, using KAT for axiomatizing TA_H. Our axiomatization of KAD therefore reduces the number of axioms from eight to three compared to [10]. The reduction to KAT leads to additional economy of expression. Moreover, the axioms of KAD have a natural motivation as abstractions of set-theoretic domain operations, whereas the axiom (8), which does not appear in the traditional axiomatizations of test algebra, is not motivated in [10].

The following consequences of Proposition 3 are not entirely syntactic. They rely on previous semantic considerations [17, 19, 22]. As usual, we write *HSP*(V) for the equational class or variety generated by a class V of algebras. This is the class of homomorphic images of subalgebras of products of algebras in V, according to Birkhoff's theorem. The left equality of the following semantic result is due to Pratt (Theorem 6.4. of [19]); the right equality is an adaptation by Hollenberg of a semantic result by Trnková and Reiterman (Corollary 1 of [22]).

Theorem 1. $HSP(\mathsf{SDA}) = HSP(\mathsf{Kri}) \supseteq HSP(\mathsf{Krit}) = HSP(\mathsf{STA}_H)$.

Based on the left equality of Theorem 1, Pratt conjectures that *HSP*(SDA) *may be defined axiomatically by the dynamic algebra axioms [...] together with an appropriate set of axioms for binary relations.* At the time of writing, Kozen's axiomatization of KA did not yet exist. Hollenberg's axiomatization of TA_H verifies Pratt's conjecture with respect to TA_P. We can show that KM_l verifies it with respect to DA. More interestingly, KAD also verifies it with respect to TA_P. But axiomatically, KAD is a considerable improvement over TA_H.

Corollary 1. $HSP(\mathsf{SKM}_l) = HSP(\mathsf{Kri}) \supseteq HSP(\mathsf{Krit}) = HSP(\mathsf{SKAD})$.

9 Reachability Analysis in Directed Graphs

To demonstrate the applicability of KM and KAD, we present an abstract toolkit based on Kleene algebra for reachability analysis in digraphs. More details and proofs can again be found in [8]. Our toolkit has interesting applications in the development and analysis of graph algorithms, in the analysis of pointer and object structures and in garbage collection algorithms. Here, elements of K denote graphs and elements of $\mathsf{test}(K)$ denote sets of nodes.

The following concepts, for instance, can be defined in KM.

- $\mathsf{reach}(p, a) = p : a^*$ and $\mathsf{nreach}(p, a) = \mathsf{reach}(p, a)'$ denote the set of nodes that is reachable, respectably not reachable, from set p in a.
- $\mathsf{reach\text{-}p}(p, a, q) \Leftrightarrow q \leq \mathsf{reach}(p, a)$ and $\mathsf{nreach\text{-}p}(p, a, q) \Leftrightarrow q \leq \mathsf{nreach}(p, a)$ denote that set q is reachable, respectably non-reachable, from set p in a.
- $\mathsf{final}(p, a) = \mathsf{reach}(p, a)\delta(a)'$ denotes the final nodes with respect to reachability via a from p. When a is a program and p a set of initial states, then $\mathsf{final}(p, a)$ represents the solutions of a.

Note that $\mathsf{nreach\text{-}p}(q, a, p)$ reduces to $paq \leq 0$ in KAD. The following concepts must, however, be defined in KAD.

- $\mathsf{del}(a,b) = \delta(a)'b$ and $\mathsf{ins}(a,b) = a + \mathsf{del}(a,b)$ can be used to model deletions and insertions of edges in a graph.
- $\mathsf{span}(p,a) = \mathsf{reach}(p,a)a$ denotes the subgraph of a that is spanned from p via reachability.

All these definitions can easily be abstracted from the relational model. We can use them for calculating many interesting graph properties in KM or KAD. For instance, we can optimize reachability in KAD.

$$\mathsf{reach}(p,a) = p + \mathsf{reach}(p : a, p'a). \tag{11}$$

Another example are optimization rules for reach and span.

$$\mathsf{nreach\text{-}p}(p,a,\delta(b)) \Rightarrow \mathsf{reach}(p,a+b) = \mathsf{reach}(p,a), \tag{12}$$

$$\mathsf{nreach\text{-}p}(p,a,\delta(b)) \Rightarrow \mathsf{span}(p,a+b) = \mathsf{span}(p,a), \tag{13}$$

$$\mathsf{nreach\text{-}p}(p,a,\delta(b)) \Rightarrow \mathsf{reach}(p,\mathsf{ins}(b,a)) = \mathsf{reach}(p,a), \tag{14}$$

$$\mathsf{nreach\text{-}p}(p,a,\delta(b)) \Rightarrow \mathsf{span}(p,\mathsf{ins}(b,a)) = \mathsf{span}(p,a). \tag{15}$$

These results are applied to pointer analysis in [7]. Also a reconsideration of the previous approaches cited in the introduction seems promising.

10 Conclusion

We have presented an axiomatization of Kleene modules as a complementation to Kleene algebra with domain. This allows a fine-grained comparison with algebras related to propositional dynamic logic. Our results support a transfer between concepts and techniques from set- and relation-based program development methods and those based on modal logics. Although the striking correspondence between scalar products, relational preimage operations and modal operators is not entirely new, we find it still surprising.

In [8], we prove a series of further results. First, we relate KM_l with Kleenean semimodules [15], Boolean modules [2], the dynamic algebras of [12], monotonic predicate transformer algebras and Boolean algebras with operators. In particular, subsumption of the latter shows that Peirce products induce modal (diamond) operators. Second, we establish another duality between left- and right-modules via the operation of converse. Third, we show that separable Kleene bi-modules subsume SKAD. Fourth, our subsumption results allow a translation of previous results for TA to KAD. In particular, the SKAD axioms are complete with respect to the valid equations in Krit. Moreover $HSP(\mathsf{SKAD})$ is EXPTIME-complete. A similar transfer between DA and KM_l is also possible.

At the theoretic side, our results are only first steps of the representation theory for KM_l and KAD. A deeper investigation of these semantic issues is beyond the syntactic analysis of this paper. At the practical side, we have already started considering applications in the development of algorithms (cf. [16]).

References

[1] R.C. Backhouse, J.P.H.W. van den Eijnde, and A.J.M. van Gasteren. Calculating path algorithms. *Science of Computer Programming*, 22(1-2):3–19, 1994. 114

[2] C. Brink. Boolean modules. *Journal of Algebra*, 71:291–313, 1981. 114, 122

[3] T. Brunn, B. Möller, and M. Russling. Layered graph traversals and hamiltonian path problems–an algebraic approach. In J. Jeuring, editor, *Mathematics of Program Construction*, volume 1422 of *LNCS*, pages 96–121. Springer-Verlag, 1998. 114

[4] K. Clenaghan. Calculational graph algorithmics: Reconciling two approaches with dynamic algebra. Technical Report CS-R9518, CWI, Amsterdam, 1994. 114

[5] J.H. Conway. *Regular Algebra and Finite State Machines.* Chapman&Hall, 1971. 118

[6] J. Desharnais, B. Möller, and G. Struth. Kleene algebra with domain. Technical Report 2003-07, Institut für Informatik, Universität Augsburg, 2003. 112, 115, 116, 119

[7] T. Ehm. Pointer Kleene algebra. Technical Report 2003-13, Institut für Informatik, Universität Augsburg, 2003. 114, 122

[8] T. Ehm, B. Möller, and G. Struth. Kleene modules. Technical Report 2003-10, Institut für Informatik, Universität Augsburg, 2003. 114, 117, 118, 119, 120, 121, 122

[9] D. Harel, D. Kozen, and J. Tiuryn. *Dynamic Logic.* MIT Press, 2000. 112, 117

[10] M. Hollenberg. Equational axioms of test algebra. In M. Nielsen and W. Thomas, editors, *Computer Science Logic, 11th International Workshop, CSL '97*, volume 1414 of *LNCS*, pages 295–310. Springer, 1997. 113, 114, 119, 121

[11] N. Jacobson. *Basic Algebra*, volume I,II. Freeman, New York, 1985. 113, 115

[12] D. Kozen. A representation theorem for ∗-free PDL. Technical Report RC7864, IBM, 1979. 112, 113, 114, 118, 122

[13] D. Kozen. A completeness theorem for Kleene algebras and the algebra of regular events. *Information and Computation*, 110(2):366–390, 1994. 113, 114

[14] D. Kozen. Kleene algebra with tests. *Trans. Programming Languages and Systems*, 19(3):427–443, 1997. 112, 113, 115

[15] Hans Leiß. Kleenean semimodules and linear languages. In Zoltán Ésik and Anna Ingólfsdóttir, editors, *FICS'02 Preliminary Proceedings*, number NS-02-2 in BRICS Notes Series, pages 51–53. Univ. of Aarhus, 2002. 114, 122

[16] B. Möller and G. Struth. Greedy-like algorithms in Kleene algebra. This volume. 122

[17] I. Németi. Dynamic algebras of programs. In *Proc. FCT'81 — Fundamentals of Computation Theory*, volume 117 of *LNCS*, pages 281–291. Springer, 1981. 112, 114, 121

[18] V. Pratt. Dynamic logic as a well-behaved fragment of relation algebras. In D. Pigozzi, editor, *Conference on Algebra and Computer Science*, volume 425 of *LNCS*, pages 77–110. Springer, 1988. 114

[19] V. Pratt. Dynamic algebras: Examples, constructions, applications. *Studia Logica*, 50:571–605, 1991. 112, 113, 114, 118, 119, 121

[20] B. von Karger R. Berghammer and A. Wolf. Relation-algebraic derivation of spanning tree algorithms. In J. Jeuring, editor, *Mathematics of Program Construction*, volume 1422 of *LNCS*, pages 23–43. Springer, 1998. 114

[21] J.N. Ravelo. Two graph algorithms derived. *Acta Informatica*, 36:489–510, 1999. 114

[22] V. Trnkova and J. Reiterman. Dynamic algebras with tests. *J. Comput. System Sci.*, 35:229–242, 1987. 112, 114, 121

The Categories of Kleene Algebras, Action Algebras and Action Lattices Are Related by Adjunctions

Hitoshi Furusawa

Laboratory for Verification and Semantics, AIST
3-11-46 Nakoji, Amagasaki, Hyogo, 661-0974, Japan
hitoshi.furusawa@aist.go.jp

Abstract. In this paper we show that the categories of Kleene algebras, action algebras and action lattices are related by adjunctions using the technique of finite limit sketches (FL sketches). This is an answer to one of Kozen's questions which has been open since 1994.

1 Introduction

In [9], Pratt introduced action algebras as an alternative to Kleene algebras [5, 6]. An action algebra is an idempotent semiring equipped with residuals and the iteration operator *. Although, in action algebras, * is characterised purely equationally, they are not closed under the formation of matrices. To make action algebras obtain this good property, which is satisfied by Kleene algebras, Kozen [6] has introduced the largest subvariety of action algebras that are closed under the formation of matrices, called action lattices. They are action algebras forming lattices.

There are several algebraic structures related to Kleene algebras [5, 6]. Kozen has shown relationships between the categories of $*$-continuous Kleene algebras [4, 5], closed semirings and Conway's **S**-algebras in terms of adjunctions. In [6] Kozen has posed the question:

> is there such a relationship between Kleene algebras and action algebras, or between action algebras and action lattices?

This paper gives an answer to this question.

For this purpose, we use knowledge of FL sketches in the sense of Barr and Wells [1, 2]. It is well-known that an FL sketch homomorphism from an FL sketch S to another FL sketch S' induces a functor from the category of models of S' to the category of models of S which has a left adjoint. So we show that Kleene algebras, action algebras and action lattices can be described by FL sketches. Then FL sketch homomorphisms between them will be described. Therefore, the positive answer is derived immediately.

R. Berghammer et al. (Eds.): RelMiCS/Kleene-Algebra Ws 2003, LNCS 3051, pp. 124–136, 2004.

2 FL Sketches

In this section we shall give an overview of FL sketches following Barr and Wells [1, 2].

Definition 1 (Graph, Reflexive Graph). A **graph** G consists of a pair of sets G_0, G_1 together with two functions src, trg: $G_1 \to G_0$. Elements in G_0, G_1 are called **nodes** and **edges**, respectively. Functions src, and trg are called **source** function, **target** function. Graphs are two-sorted algebras with two operations. A **homomorphism of graphs** is defined to be a homomorphism of two-sorted algebras. A **reflexive graph** G is a graph with a function i: $G_0 \to G_1$ satisfying $\mathsf{src} \circ \mathsf{i} = \mathsf{trg} \circ \mathsf{i} = \mathrm{id}_{G_0}$. The function i is called **loop** function. A **homomorphism of reflexive graphs** is a homomorphisms of graphs which preserves operation i.

An edge f whose source and target are a and b is denoted by $f\colon a \to b$.

Definition 2 (Diagram, Commutative Diagram, Cone). Let H and G be a graph and a reflexive graph, respectively. A **diagram** in G of **shape** H is a homomorphism $D\colon H \to G$ of graphs. D is called a **commutative diagram** if H has two distinguished nodes s and d, two paths from s to d, and all edges are part of either paths. If H is equipped with one distinguished node p and a family of edges $P = (\, P_x\colon p \to x \mid x \in H_0 \setminus \{\, p \,\}\,)$, and, for each edge $f \in H_1$ which is not in the set $\{P_x\}_{x \in H_0 \setminus \{p\}}$ determined by P, neither the source nor target is p, the triple $(D\colon H \to G, p, P)$ is called a **cone** in G of **shape** H. The distinguished node p is called the **pivot** and an edge P_x of H is called a **projection**. If each edge of H is a projection, the cone is called **discrete**.

If H is finite, the diagram $D\colon H \to G$ is called a **finite diagram**. Similarly we use terms such as finite commutative diagrams, finite cones and so on. Note that the term "commutative" has no meaning here except as part of our convention since there is no composition of arrows in graphs.

Definition 3 (FL Sketch). An **FL sketch** (finite limit sketch) S is a triple (G, C, Γ) of a reflexive graph G, a set C of commutative diagrams in G, a set Γ of finite cones in G. Edges of G are called **operators** in S.

FL sketches are also known as LE sketches (left exact sketches). An FL sketch (G, C, Γ) whose Γ consists of discrete cones is called an **FP sketch** (finite product sketch). So, an FP sketch is an FL sketch.

Definition 4 (Models in Set, Homomorphism). Let $S = (G, C, \Gamma)$ be an FL sketch. A reflexive graph homomorphism M from G to the underlying reflexive graph of the category **Set** is called a **model** of S if the following conditions hold:

- for each node a of G, M takes the loop $\mathrm{i}(a)$ to the identity map on $M(a)$;
- for each commutative diagram D in C, $M(D(f_n))\circ\cdots\circ M(D(f_1))\circ M(D(f_0)) = M(D(g_m))\circ\cdots\circ M(D(g_1))\circ M(D(g_0))$, where $f_0 f_1 \cdots f_n$ and $g_0 g_1 \cdots g_m$ are the two distinguished paths of D;

– and for each cone $(\gamma\colon H \to G, p, P)$ in Γ, $M \circ \gamma$ is a limit cone in **Set**, for each edge $f\colon x \to y \in H_1$ which is not a projection $M(\gamma(P_y)) = M(\gamma(f)) \circ M(\gamma(P_x))$, and for each set A and a family of maps $F = (F_x\colon A \to M(\gamma(x)) \mid x \in H_0 \setminus \{p\})$, there exists a unique map $k\colon A \to M(\gamma(p))$ such that $F_x = M(\gamma(P_x)) \circ k$.

A **homomorphism** α from M to M' is a G_0-indexed family of maps $(\alpha_x \in \mathbf{Set}(M(x), M(x')) \mid x \in G_0)$ which satisfies $M'(f)\alpha_x = \alpha_y M(f)$ for each edge $f\colon x \to y$ of G. The models of an FL sketch S and homomorphisms between them give rise to a category which we shall denote by $\mathbf{Mod}(S)$.

By replacing **Set** above by any category Z with finite limits, we obtain a more general definition of models, but we shall use only models in **Set**, so the above definition suffices.

Definition 5 (n-ary Operator). Let $f\colon a \to b$ be an operator in an FL sketch $S = (G, C, \Gamma)$. If Γ contains a discrete cone $(\gamma\colon H \to G, p, P)$ for which γ takes p to a and all other nodes to b, f is called an **n-ary operator** on b in S, where n is the number of elements of $H_0 \setminus \{p\}$.

The arity is in general not uniquely determined. We shall use the term "arity" for convenience, although it is not well-established notion.

If M is a model of S, the definition of models forces $M(a)$ to be an n-fold product of $M(b)$ and $M(f)$ to be a function $M(f)\colon M(b)^n \to M(b)$.

Definition 6 (Sketchable). A category C is **FL sketchable** if there exists an FL sketch S, for which the category $\mathbf{Mod}(S)$ of models is equivalent to C.

FL sketches and equational Horn clauses have similar applications. Both are means to express equations which are to hold only under some conditions. In fact, Barr [3] showed the following theorem.

Theorem 1 (Barr [3]). *The category of models of an equational Horn theory is FL sketchable.*

It is known that FP sketches have the same expressive power as purely equational theories. So FL sketches are more expressive than FP sketches. Since we investigate Kleene algebras, we need the expressive power of FL sketches. In the sequel, we do not distinguish FP sketches from FL sketches rigorously.

Semilattices are defined by equations: A **semilattice** is a set equipped with a binary operator $+$ which satisfies the following:

$$(a + b) + c = a + (b + c)$$
$$a + b = b + a$$
$$a + a = a$$

So the category of semilattices is FL sketchable. We develop an FL sketch for semilattices.

Example 1. We define an FL sketch $\mathsf{SL} = (G_{\mathsf{SL}}, C_{\mathsf{SL}}, \Gamma_{\mathsf{SL}})$. The reflexive graph G_{SL} has three nodes we give the names s, s^2 and s^3. The edges in the graph G_{SL} are:

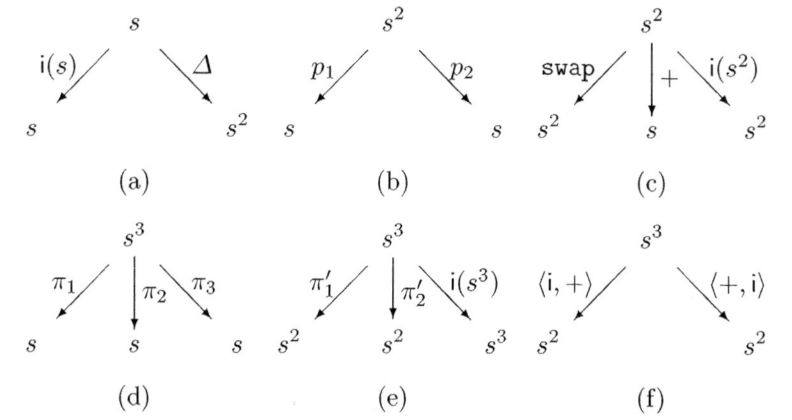

We have given in six pictures, but G_{SL} is just one graph with three nodes and fifteen edges. C_{SL} consists of the following commutative diagrams:

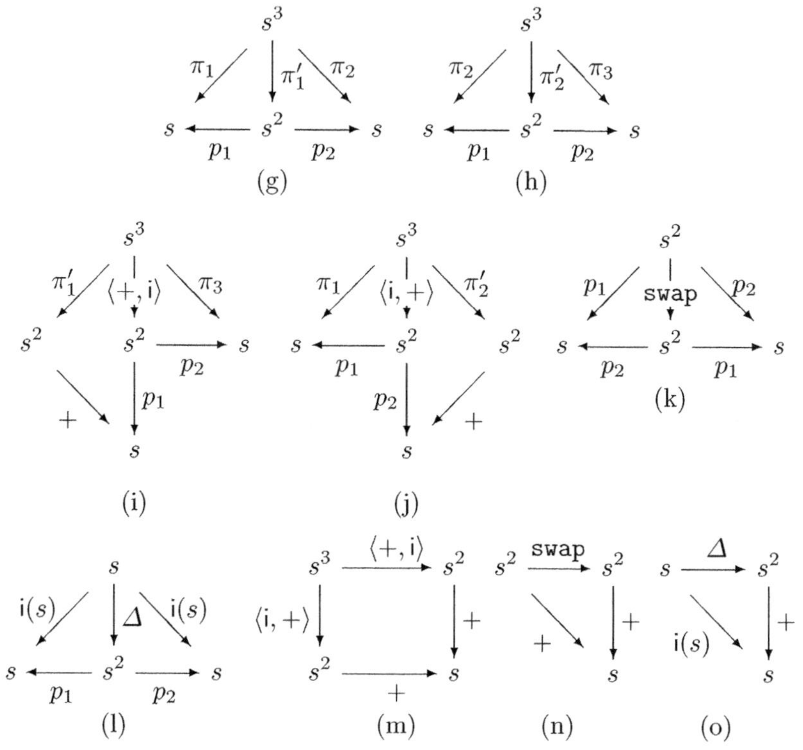

Each (g)-(l) contains two commutative diagrams in one picture. Each drawing should be understood as describing shape graph labelled with the images of its nodes and edges under the commutative diagram. So, if a picture of a diagram contains more than one node with the same label, those nodes are different nodes in its shape graph. Γ_{SL} consists of cones (b) and (d) which appeared in pictures of the edges. As in the case of commutative diagrams, each of (b) and (d) is describing shape graph labelled with the image of its nodes and edges under the diagram of the cone. In (b) we can see two nodes sharing the same labels s. These two nodes are different nodes in the shape graph of the cone. Similarly, the three nodes at the bottom of picture (d) are different nodes in the shape graph of the cone.

Consider a model M of SL which takes s to a set S. Since (b) and (d) are cones, $M(s^2) = S \times S$ and $M(s^3) = S \times S \times S$. In C_{SL} (m), (n) and (o) are main statements, which require associativity, commutativity and idempotency of $M(+)$, respectively. (i) and (j) are needed to define operators which occur in (m), (k) and (l) are needed to define operators which occur in (n) and (o), respectively. (g) and (h) are needed to define operators which occur in (i) and (j).

Definition 7 (Morphism of FL Sketches). Let $S = (G, C, \Gamma)$ and $S' = (G', C', \Gamma')$ be FL sketches. A **morphism** $h\colon S \to S'$ of FL sketches is a homomorphism of reflexive graphs which takes commutative diagrams and cones in G to commutative diagrams and cones in G', respectively, i.e., $h \circ D \in C'$ for each $D \in C$ and $h \circ \gamma \in \Gamma'$ for each $\gamma \in \Gamma$.

The following theorem is central for this paper. Barr and Wells states the result in [1, Chapter 4.4]. See also [8].

Theorem 2. *Let S and T be FL sketches. If $h\colon S \to T$ is a morphism of FL sketches, then the functor $h^*\colon \mathbf{Mod}(T) \to \mathbf{Mod}(S)$ given by composing each model of T with h has a left adjoint $h_\sharp\colon \mathbf{Mod}(S) \to \mathbf{Mod}(T)$.*

3 Action Algebras and Kleene Algebras

In this section we give an overview of action algebras, action lattices, Kleene algebras, and residuated Kleene algebras following Kozen [7]

Definition 8 (Idempotent Semiring). An **idempotent semiring** is a set S equipped with nullary operators 0, 1 and binary operators +, ;, where the triple $(S, 0, +)$ is a semilattice equipped with the unit 0 with respect to +, $(S, 1, ;)$ is a monoid, and these components satisfy the following:

$$x;(y+z) = x;y + x;z$$
$$(x+y);z = x;z + y;z$$
$$x;0 = 0 = 0;x$$

It is well-known that the partial oder $\leq$ defined by

$$x \leq y \iff x + y = y$$

is a semilattice ordering. The partial order $\leq$ defined in this way will be used throughout this paper.

Definition 9 (Residuation Algebra). A **residuation algebra** is an idempotent semiring equipped with binary operators / and \ which satisfies the following:

$$x; a \leq b \iff x \leq b/a$$
$$a; x \leq b \iff x \leq a\backslash b$$

The binary operators / and \ are called left and right residuals, respectively. Although there is a purely equational characterisation of the residual as Pratt and Kozen have shown in [9, 7], we adopt a simpler one because it is not so important in this paper whether purely equational or not.

Definition 10 (Action Algebra, Action Lattice). An **action algebra** is a residuation algebra equipped with unary operator * on A which satisfies the following:

$$1 + a + a^*a^* \leq a^* \qquad (a/a)^* = a/a \qquad (a\backslash a)^* = a\backslash a$$

The category of action algebras will be denoted by **Act**. In particular, an action algebra equipped binary operator $\cdot$ on A and satisfying that the triple $(A, +, \cdot)$ is a lattice is called **action lattice**. The category of action lattices will be denoted by **ActLat**.

Action algebras and action lattices can be axiomatized by equational Horn theory, so the categories **Act** and **ActLat** are FL sketchable by Theorem 1. Even if we adopt purely equational definitions of these two structures, they are FL sketchable.

The FL sketch $\mathsf{ACT} = (G_{\mathsf{ACT}}, C_{\mathsf{ACT}}, \Gamma_{\mathsf{ACT}})$ for action algebras is defined in the following way. G_{ACT} has one distinguished node a. It has two nullary operators 0 and 1, one unary operator * and four binary operators $+$, ;, / and \ (in the sense of Definition 5). The diagrams in C_{ACT} and Γ_{ACT} are determined by these arity conditions and Horn clause axioms, in the way described in [3]. The FL sketch $\mathsf{AL} = (G_{\mathsf{AL}}, C_{\mathsf{AL}}, \Gamma_{\mathsf{AL}})$ for action lattices is quite similar to the FL sketch ACT but it has extra structures. G_{AL} has additional one binary operators $\cdot$ to G_{ACT}. The diagrams in C_{AL} are determined by adding diagrams which corresponds to characterisation of the operators to C_{ACT}. Γ_{AL} is equal to Γ_{ACT}.

Definition 11 (Kleene Algebra). A **Kleene algebra** is a an idempotent semiring equipped with a unary operator * which satisfies the following:

$$1 + (p; p^*) = p^*$$
$$1 + (p^*; p) = p^*$$
$$p; r \leq r \Longrightarrow p^*; r \leq r$$
$$r; p \leq r \Longrightarrow r; p^* \leq r$$

The category of Kleene algebras will be denoted by **Kleene**. In particular, a Kleene algebra equipped with both left and right residuals is called **residuated Kleene algebra**. The category of residuated Kleene algebras will be denoted by **RKleene**.

Kleene algebras and residuated Kleene algebras are axiomatized by equational Horn theory, so, by Theorem 1, the categories **Kleene** and **RKleene** are FL sketchable.

We shall give an outline of the FL sketch KA and RKA for Kleene algebras and Residuated Kleene algebras as we have done for action algebras and action lattices. The FL sketch $\mathsf{KA} = (G_{\mathsf{KA}}, C_{\mathsf{KA}}, \Gamma_{\mathsf{KA}})$ for Kleene algebras is defined in the following way. G_{KA} has one distinguished node a. It has two nullary operators 0 and 1, one unary operator * and two binary operators $+$ and $;$. The diagrams in C_{KA} and Γ_{KA} are determined by these arity conditions and Horn clause axioms. The FL sketch $\mathsf{RKA} = (G_{\mathsf{RKA}}, C_{\mathsf{RKA}}, \Gamma_{\mathsf{RKA}})$ for residuated Kleene algebras is defined by adding extra structures to KA. G_{RKA} has additional two binary operators $/$ and $\backslash$ to G_{KA}. The diagrams in C_{RKA} and Γ_{RKA} are determined by adding diagrams which corresponds to characterisation of the operators to C_{KA} and Γ_{KA}. In Appendix A, we describe an FL sketch for Kleene algebras.

Kozen has shown that residuated Kleene algebras are an alternative characterisation of action algebras in [7].

Lemma 1. *Action algebras are exactly residuated Kleene algebras.*

4 Main Result

This section gives an answer the open question posed by Kozen in [7].

There are trivial inclusions:

$$\mathbf{ActLat} \subseteq \mathbf{Act} \qquad \mathbf{RKleene} \subseteq \mathbf{Kleene} \ .$$

These inclusions determine forgetful functors from **ActLat** to **Act** and from **RKleene** to **Kleene**. They satisfies the following theorem.

Theorem 3.

1. *The forgetful functor from the category* **ActLat** *of action lattices to the category* **Act** *of action algebras has a left adjoint.*
2. *The forgetful functor from the category* **RKleene** *of residuated Kleene algebras to the category* **Kleene** *of Kleene algebras has a left adjoint.*

Proof. Since both of these two are proved in analogous ways, we would like to prove the first. Recall the FL sketches ACT and AL which are described just after Definition 10. By the definitions of the sketches, there is a trivial inclusion i as FL sketch homomorphism from ACT to AL. Since $\mathbf{Mod}(\mathsf{ACT})$ is equivalent to the category **Act** and $\mathbf{Mod}(\mathsf{AL})$ is equivalent to the category **ActLat**, the forgetful functor $i^*\colon \mathbf{ActLat} \to \mathbf{Act}$ induced from i has a left adjoint $i_\#$ by Theorem 2.

In Appendix B, we give an explicit construction of the free residuated Kleene algebra generated by a Kleene algebra.

The following is immediate from the above theorem and Lemma 1.

Corollary 1. *The forgetful functor from the category* **Act** *of action algebras to the category* **Kleene** *of Kleene algebras has a left adjoint.*

As a result, we obtain that the three categories **Kleene**, **Act** and **ActLat** are related by adjunctions as follows:

$$\mathbf{Kleene} \underset{\longleftarrow}{\overset{\longrightarrow}{\bot}} \mathbf{RKleene} \cong \mathbf{Act} \underset{\longleftarrow}{\overset{\longrightarrow}{\bot}} \mathbf{ActLat}$$

By the composition of these two adjunctions, we obtain that **Kleene** and **ActLat** are related by an adjunction again.

Acknowledgement

The author would like to thank Peter Jipsen, Dexter Kozen, Wolfram Kahl, Michael Winter, and the anonymous referees for discussions and helpful comments.

References

[1] M. Barr and C. Wells. *Toposes, Triples and Theories*. Springer, 1984 (updated version available at `http://www.cwru.edu/artsci/math/wells/pub/ttt.html`). 124, 125, 128

[2] M. Barr and C. Wells. *Category theory for computing science, second edition*. Prentice-Hall, 1995. 124, 125

[3] Michael Barr. Models of Horn theory. In J. Gray and A. Scedrov, editors, *Proc. of a Summer Research Conference, Categories in Computer Science and Logic*, number 92 in Contemporary Mathematics, 1–7. AMS, 1980. 126, 129

[4] Dexter Kozen. On induction vs. $*$-continuity. In Kozen, editor, *Proceedings of Workshop in Logic of Programs*, volume 131 of *Springer Lecture Notes in Computer Science*, 167–176, 1981. 124

[5] Dexter Kozen. On Kleene algebras and closed semirings. In Rovan, editor, *Proceedings of Mathematical Foundations of Computer Science*, volume 452 of *Springer Lecture Notes in Computer Science*, 26–47, 1990. 124

[6] Dexter Kozen. A Completeness Theorem for Kleene Algebras and the Algebra of Regular Events. *Information and Computation*, 110:366–390, 1994. 124

[7] Dexter Kozen. On action algebras. In van Eijck and Visser, editors, *Logic and Information Flow*, 78–88. MIT Press, 1994. 128, 129, 130

[8] John W. Gray. Categorical Aspects of Data Type Constructors. *TCS* 50:103–135, 1987. 128

[9] Vaughan Pratt. Action logic and pure induction. In van Eijck, editor, *Proc. Logics in AI: European Workshop JELIA '90*, volume 452 of *Springer Lecture Notes in Computer Science*, 97–120, 1990. 124, 129

Appendix

A Sketch for Kleene Algebra

First we extend SL given in Example 1 to an FL sketch IS for idempotent semirings. KA may be given by extending IS.

We define an FL sketch $\mathsf{IS} = (G_{\mathsf{IS}}, C_{\mathsf{IS}}, \Gamma_{\mathsf{IS}})$. The reflexive graph G_{IS} has two nodes $\mathbf{1}$ and s^4 in addition to the nodes in G_{SL}. The edges in G_{IS} are:

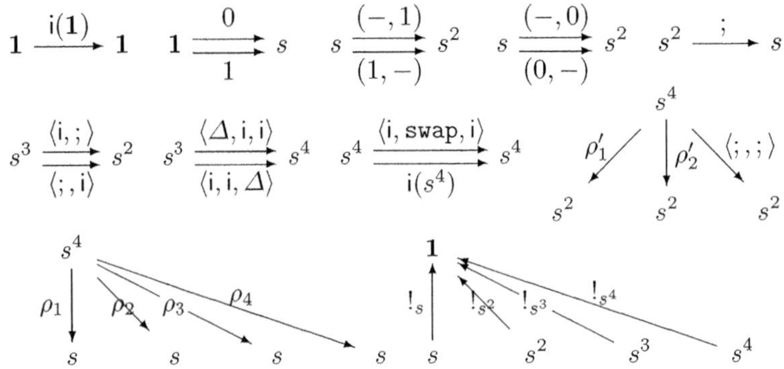

in addition to the edges in G_{SL}. C_{IS} has the following diagrams:

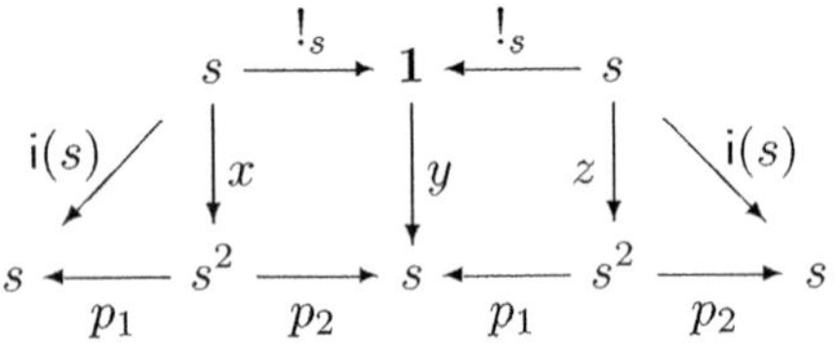

$(x, y, z) = ((-, 0), 0, (0, -)),$
$((-, 1), 1, (1, -))$

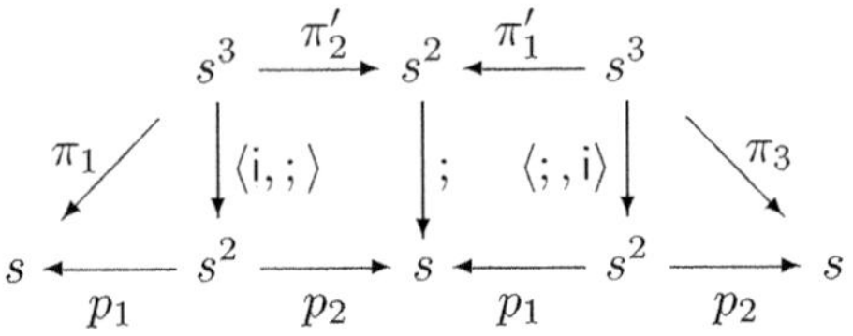

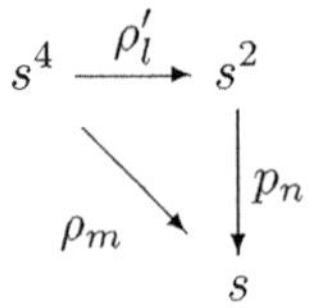

$(l, m, n) = (1, 1, 1), (1, 2, 2),$
$(2, 3, 1), (2, 4, 2)$

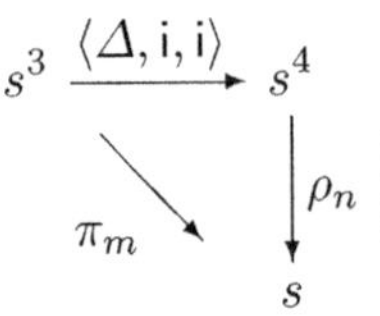

$(m, n) = (1, 1),$
$(1, 2), (2, 3), (3, 4)$

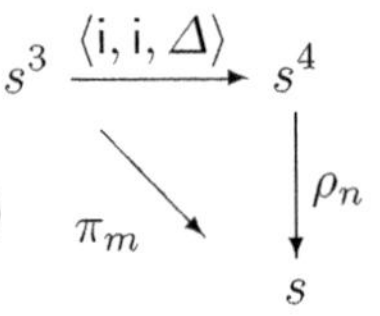

$(m, n) = (1, 1),$
$(2, 2), (3, 3), (3, 4)$

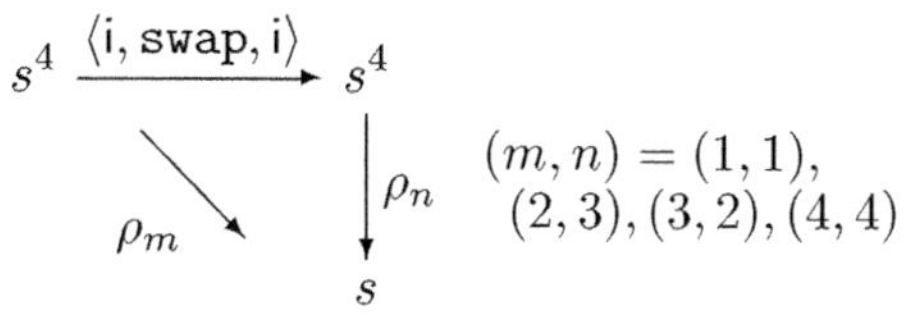

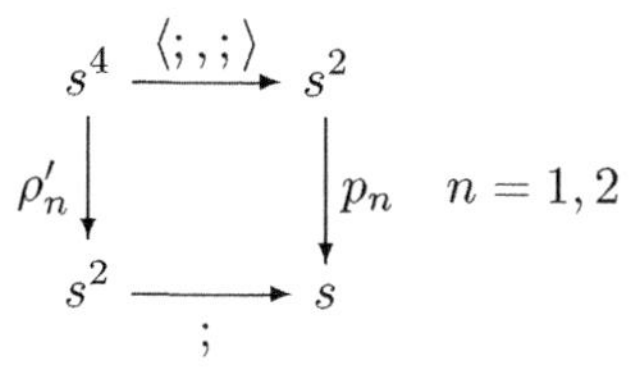

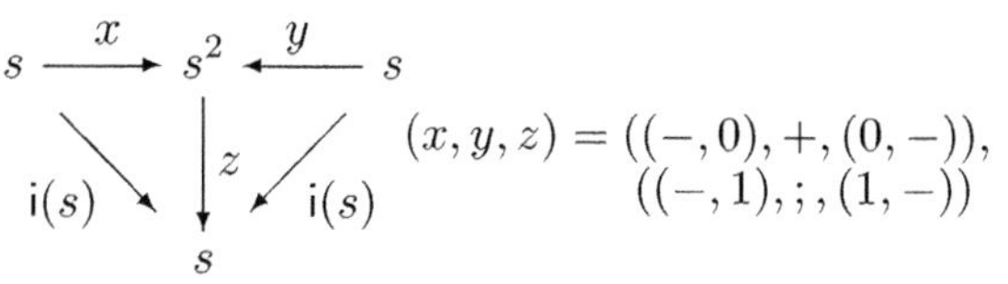

⟨i, swap, i⟩
(m, n) = (1, 1), (2, 3), (3, 2), (4, 4)
⟨; , ;⟩
n = 1, 2

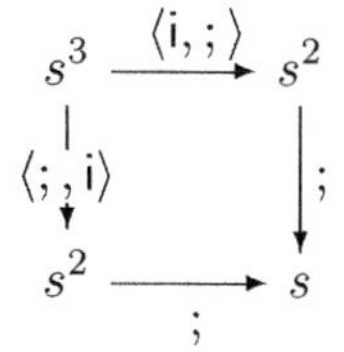

(x, y, z) = ((−, 0), +, (0, −)), ((−, 1), ; , (1, −))
⟨i, ;⟩
⟨; , i⟩

(1) (2)

(x, y) = ((⟨i, +⟩, (Δ, i, i)), ((⟨+, i⟩, (i, i, Δ))
⟨i, swap, i⟩
⟨; , ;⟩
(−, 0)
(0, −)

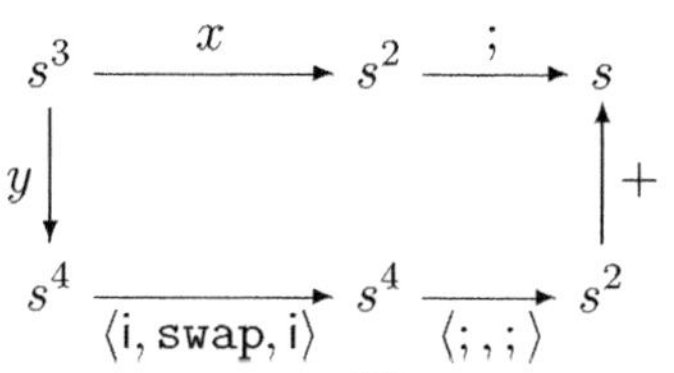

(3) (4)

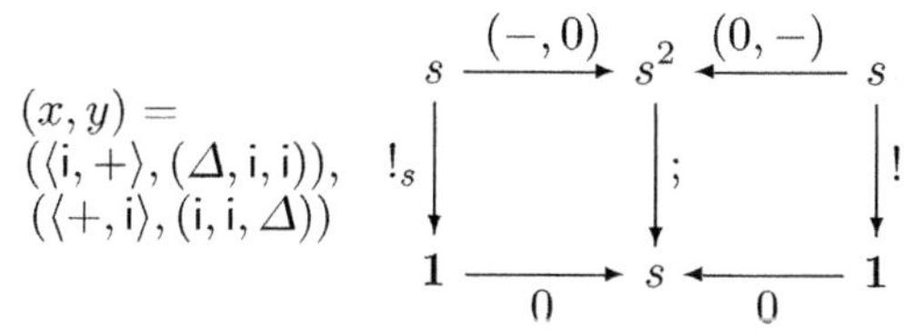

in addition to the diagrams in C_{SL}. Γ_{IS} has (γ) which appeared in pictures in G_{IS}'s edges and $\mathbf{1}$ in addition to the cones in Γ_{SL}.

Consider a model M of IS. Since $\mathbf{1} \in \Gamma_{\mathsf{IS}}$, $M(\mathbf{1})$ is a terminal object, that is, $M(\mathbf{1})$ is a singleton set. In additional pictures of C_{IS} (1)~(4) are main statements. Picture (1) requires that $M(0)$ is the unit with respect to $M(+)$ in the case that $(x, y, z) = ((-, 0), +, (0, -))$, otherwise that so $M(1)$ with respect to $M(;)$. Picture (2) requires associativity of $M(;)$. Picture (3) requires that $M(;)$ distributes over $M(+)$ on both sides. Picture (4) requires that $M(0)$ is the zero with respect to $M(;)$.

We define an FL sketch $\mathsf{KA} = (G_{\mathsf{KA}}, C_{\mathsf{KA}}, \Gamma_{\mathsf{KA}})$. The reflexive graph G_{KA} has two nodes ϵ and ϵ' in addition to the nodes in G_{IS}. The edges in G_{KA} are:

i(ε)
i(ε′)
⟨i, Δ⟩
⟨Δ, i⟩
⟨i, ∗, i⟩
⟨i, i, ∗⟩
⟨i, swap⟩

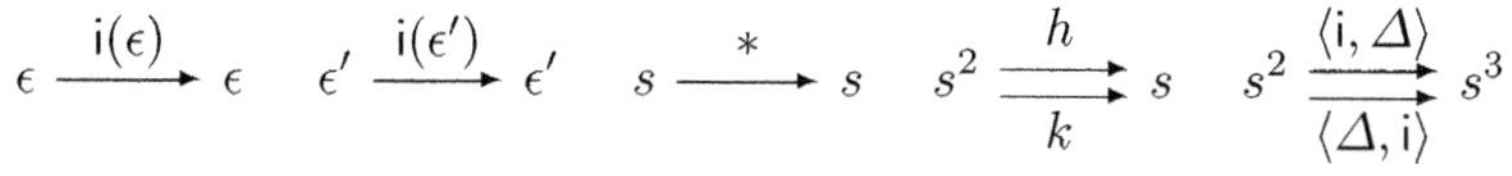

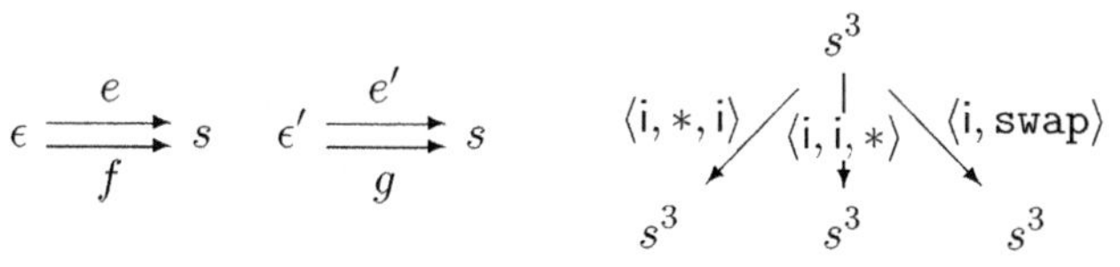

in addition to the edges of G_{IS}. C_{KA} has the following diagrams:

$$
\begin{array}{ccc}
s^2 & \xrightarrow{\langle i, \Delta\rangle} & s^3 \\
& p_m \searrow & \downarrow \pi_n \\
& & s
\end{array}
\quad (m,n)=(1,1),\ (2,2),(2,3)
$$

$$
\begin{array}{ccc}
s^2 & \xrightarrow{\langle \Delta, i\rangle} & s^3 \\
& p_m \searrow & \downarrow \pi_n \\
& & s
\end{array}
\quad (m,n)=(1,1),\ (1,2),(2,3)
$$

$$
\begin{array}{ccc}
s & \xleftarrow{\pi_3} & s^3 \\
* \downarrow & \langle i,i,*\rangle \searrow \pi_n & \downarrow \\
s & \xleftarrow{\pi_3} s^3 \xrightarrow{\pi_n} & s
\end{array}
\quad n = \overline{1,2}
$$

$$
\begin{array}{ccc}
s & \xleftarrow{\pi_2} & s^3 \\
* \downarrow & \langle i,*,i\rangle \searrow \pi_n & \downarrow \\
s & \xleftarrow{\pi_2} s^3 \xrightarrow{\pi_n} & s
\end{array}
\quad n = \overline{1,3}
$$

$$
\begin{array}{ccc}
s^3 & \xrightarrow{\langle i, \mathsf{swap}\rangle} & s^3 \\
& \pi_m \searrow & \downarrow \pi_n \\
& & s
\end{array}
\quad (m,n)=(1,1),\ (2,3),(3,2)
$$

$$
\begin{array}{ccccc}
s^2 & & \xrightarrow{h} & & s \\
\langle \Delta, i\rangle \downarrow & & & & \uparrow + \\
s^3 & \xrightarrow{\langle i, \mathsf{swap}\rangle} & s^3 & \xrightarrow{\langle i, ;\rangle} & s^2
\end{array}
\tag{1}
$$

$$
\begin{array}{ccc}
s^2 & \xrightarrow{k} & s \\
\langle \Delta, i\rangle \downarrow & & \uparrow + \\
s^3 & \xrightarrow{\langle i, ;\rangle} & s^2
\end{array}
\tag{2}
$$

$$
\begin{array}{ccccccc}
s & & & \xrightarrow{*} & & & s \\
(1,-) \downarrow & & & & & & \uparrow + \\
s^2 & \xrightarrow{\langle i, \Delta\rangle} & s^3 & \xrightarrow{x} & s^3 & \xrightarrow{\langle i, ;\rangle} & s^2
\end{array}
\quad x = \begin{array}{c}\langle i,i,*\rangle \\ \langle i,*,i\rangle\end{array}
\tag{3}
$$

$$
\begin{array}{ccccccccc}
\epsilon & & & & \xrightarrow{e} & s^2 & & \xrightarrow{p} & s \\
e \downarrow & & & & & & & & \uparrow + \\
s^2 & \xrightarrow{\langle \Delta, i\rangle} & s^3 & \xrightarrow{\langle i, \mathsf{swap}\rangle} & s^3 & \xrightarrow{\langle i,*,i\rangle} & s^3 & \xrightarrow{\langle i, ;\rangle} & s^2
\end{array}
\tag{4}
$$

$$
\begin{array}{ccccccc}
\epsilon' & & \xrightarrow{e'} & s^2 & & \xrightarrow{p} & s \\
e' \downarrow & & & & & & \uparrow + \\
s^2 & \xrightarrow{\langle \Delta, i\rangle} & s^3 & \xrightarrow{\langle i,i,*\rangle} & s^3 & \xrightarrow{\langle i, ;\rangle} & s^2
\end{array}
\tag{5}
$$

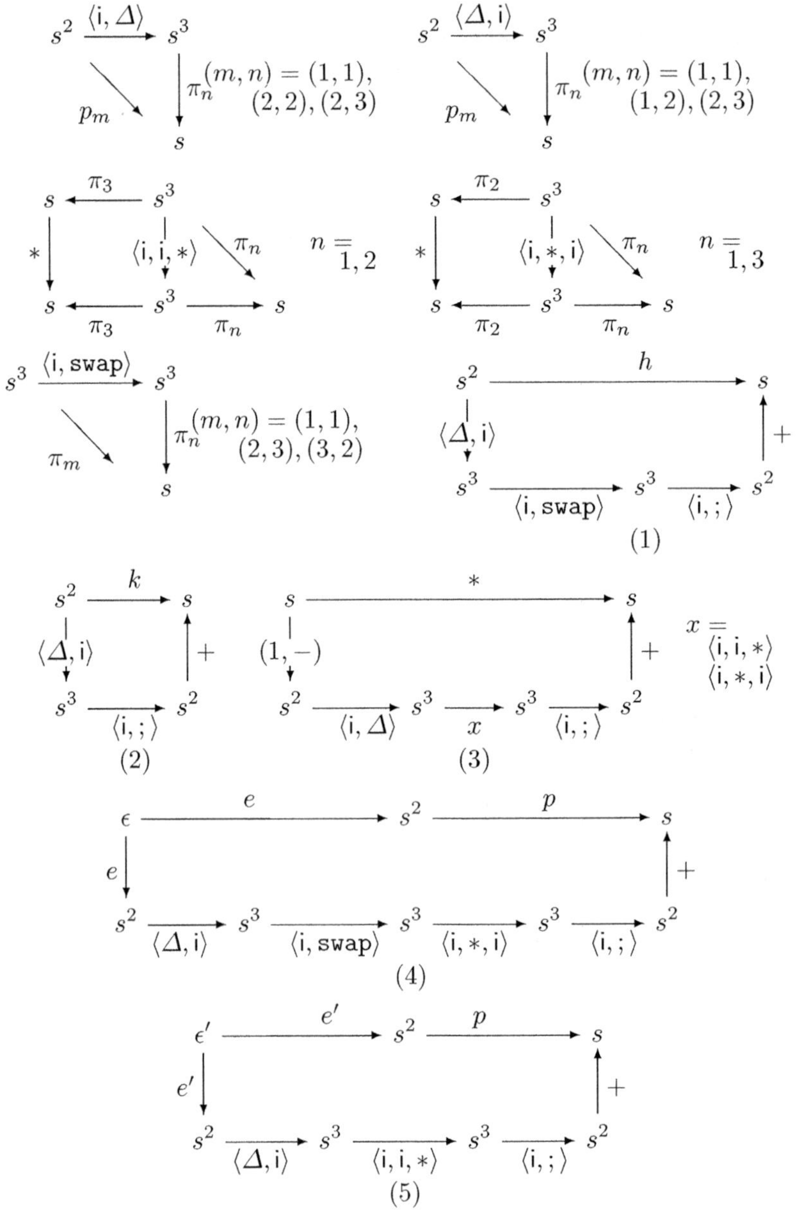

in addition to the diagrams in C_{IS}. Γ_{KA} has the following cones:

$$
\begin{array}{ccc}
& \epsilon & \\
e \swarrow & & \searrow f \\
s^2 \underset{p_1}{\overset{h}{\rightrightarrows}} & & s
\end{array}
\qquad
\begin{array}{ccc}
& \epsilon' & \\
e' \swarrow & & \searrow g \\
s^2 \underset{p_1}{\overset{k}{\rightrightarrows}} & & s
\end{array}
$$

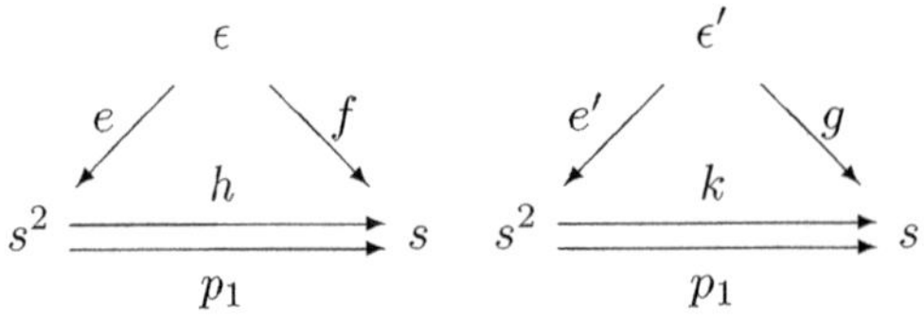

in addition to the cones in Γ_{IS}.

The aim of the additional components in KA is to introduce and to specify only a unary operator $*$ which should be interpreted as a unary operator * of Kleee algebras. Consider a model M. Diagrams (1) and (2) determine h and k that $M(h)(r,p) = M(+)(r, M(;)(p,r))$ and $M(k)(r,p) = M(+)(r, M(;)(r,p))$ for $r, p \in M(s)$, respectively. The additional cones of Γ_{KA} requires that $M(e)$ is an equalizer of $M(h)$ and $M(p_1)$, and so $M(e')$ of $M(k)$ and $M(p_1)$. Therefore, $M(\epsilon)$ and $M(\epsilon')$ are the sets of all pairs (r,p) which satisfy $M(+)(r, M(;)(p,r)) = r$ and $M(+)(r, M(;)(r,p)) = r$, respectively. Indeed these equations are the left-hand sides of two implications in Definition 11. Picture (3) requires two equations, and so (4) and (5) are the right-hand sides of the implications.

B Construction of Free Residuated Kleene Algebra Generated by Kleene Algebra

Given a Kleene algebra $\mathbf{K} = (K, 0, 1, +_K, ;_K, {}^{*_K})$, the set R_0 is defined inductively as follows. C_{KA} has

- If $a \in K$, then $a \in R_0$.
- If $x, y \in R_0$, then so are $x/_0 y$, $x\backslash_0 y$, $x +_0 y$, $x;_0 y$ and x^{*_0}.
- (closing sentence) $x \in R_0$ only if can be obtained by finitely many applications of the above rules.

We define $\equiv$ $(\subseteq R_0 \times R_0)$ to be the least congruence with respect to $+_0$, $;_0$, ^{*_0}, $/_0$ and $\backslash_0$ which satisfies the following.

- If $a = b$, then $a \equiv b$ for $a, b \in K$.
- $a +_K b \equiv a +_0 b$ for $a, b \in K$.
- $a;_K b \equiv a;_0 b$ for $a, b \in K$.
- $a^{*_K} \equiv a^{*_0}$ for $a \in K$.
- $x +_0 (y +_0 z) \equiv (x +_0 y) +_0 z$.
- $x +_0 y \equiv y +_0 x$.
- $x +_0 x \equiv x$.
- $x +_0 0 \equiv 0 +_0 x \equiv x$.
- $x;_0 (y;_0 z) \equiv (x;_0 y);_0 z$.
- $x;_0 1 \equiv 1;_0 x \equiv x$.
- $x;_0 (y +_0 z) \equiv x;_0 y +_0 x;_0 z$.
- $(y +_0 z);_0 x \equiv y;_0 x +_0 z;_0 x$.
- $x;_0 0 \equiv 0;_0 x \equiv 0$.
- $1 +_0 x;_0 x^{*_0} \equiv x^{*_0}$.
- $1 +_0 x^{*_0};_0 x \equiv x^{*_0}$.
- If $x +_0 y;_0 x \equiv x$, then $x +_0 y^{*_0};_0 x \equiv x$.
- If $x +_0 x;_0 y \equiv x$, then $x +_0 x;_0 y^{*_0} \equiv x$.
- $x;_0 y +_0 z \equiv z$ if and only if $x +_0 z/_0 y \equiv z/_0 y$.
- $x;_0 y +_0 z \equiv z$ if and only if $y +_0 x\backslash_0 z \equiv x\backslash_0 z$.

Define $R \stackrel{\text{def}}{=} R_0/\equiv$. The equivalence class of $x \in R_0$ with respect to $\equiv$ is denoted by $[x]$. Defining operators $+$, $;$, *, $/$ and $\backslash$ on R by $[x] + [y] \stackrel{\text{def}}{=} [x +_0 y]$, $[x];[y] \stackrel{\text{def}}{=} [x;_0 y]$, $[x]^* \stackrel{\text{def}}{=} [x^{*_0}]$, $[x]/[y] \stackrel{\text{def}}{=} [x/_0 y]$ and $[x]\backslash[y] \stackrel{\text{def}}{=} [x\backslash_0 y]$, $\mathbf{R} = (R, [0], [1], +, ;, ^*, /, \backslash)$ is the free residuated Kleene algebra generated by a Kleene algebra $\mathbf{K}$, that is, $\mathbf{R}$ is a residuated Kleene algebra which satisfies the following condition together with a Kleene algebra homomorphism $\eta\colon \mathbf{K} \to \mathbf{R}$ taking $a \in K$ to $[a] \in R$.

For each residuated Kleene algebra $\mathbf{M}$ and an Kleene algebra homomorphism $f\colon \mathbf{K} \to \mathbf{M}$, there exists a unique Kleene algebra homomorphism $h\colon \mathbf{R} \to \mathbf{M}$ such that $f = h \circ \eta$.

Towards a Formalisation of Relational Database Theory in Constructive Type Theory

Carlos Gonzalía

Department of Computing Science
Chalmers University of Technology and Gothenburg University
SE-412 96 Gothenburg, Sweden
gonzalia@cs.chalmers.se

Abstract. We offer here an overview of several initial attempts of formalisation of relational database theory in a constructive, type-theoretic, framework. Each successive formalisation is of more generality, and correspondingly more complex, than the previous one. All our work is carried out in the proof editor Alfa for Martin-Löf's monomorphic type theory. Our goal is to obtain a formalisation that provides us with computational content, instead of just being a completely abstract theory.

1 Background

1.1 Type Theory

Per Martin-Löf's *monomorphic type theory* [13] is a formalism based on interpreting the logical constants by means of the Curry-Howard isomorphism between propositions and sets (a proposition being identified with the set of its proofs). Type theory can also serve as a *logical framework*, i.e. we can express different theories inside it by implementing the theory's constants and rules in terms of the features present inside type theory.

Type theory is an interesting foundation for programming, since it allows us to express both specifications and programs within the same formalism. We can both derive a correct program from its specification and also verify that a certain program has some property. The theory provides us with a typed functional programming language with inductive and dependent types all functions of which terminate, along with a specification language encompassing first-order intuitionistic predicate logic.

1.2 Alfa

Alfa [9] is a graphical proof editor based on the proof checker Agda [4], which in turn is based on the manipulation of explicit proof objects in a constructive monomorphic setting. The formal syntax and semantics of the Agda language is what has been called *structured type theory* [5], a version of type theory where we have dependently typed products with labelled fields (this is called a signature

R. Berghammer et al. (Eds.): RelMiCS/Kleene-Algebra Ws 2003, LNCS 3051, pp. 137–148, 2004.

and a value of such a type is called a structure), local definitions and a package mechanism. The concrete syntax of Agda resembles quite closely that of Haskell, and is based on Cayenne (a dependently typed functional programming language) [1].

On top of these features, the Alfa editor allows the user to define the way proof terms are presented on screen with much flexibility. Fundamental to the use of Alfa as an interactive proof assistant is how it allows us to leave some parts of our proofs incomplete, by using a notion of *meta-variable* with a well-defined behaviour. These meta-variables can then be filled in by the user, with the Agda engine checking the correctness of all introduced or refined expressions. We remark that currently there is a new standard library for Alfa, developed by Michael Hedberg [10], of which we make abundant use.

1.3 Relational Database Model

The relational database model was introduced by Codd [3], and has seen a highly successful development and use since then. For our purposes, we deal only with its core basic notions, which we informally describe here.

A *domain* is a (usually finite) set of atomic values. A *relation* in this context is any subset of a Cartesian product of a finite number of domains, this number being called the arity of the relation. A database in this model is a finite set of relations in this sense. One usually thinks of a relation in a database as a table of values, where each row corresponds to a tuple. Quite naturally, names are associated with the columns of these tables. Such names are called *attributes*. A relation then has a *relation scheme*, which is a set of attributes I along with a typing A_i (for $i \in I$) associating a set A_i to each attribute i. Relations in databases are typically finite, and we will assume so for the rest of the present work.

Looking at relations in the preceding way makes use of an ordering of the components of a tuple that is not quite natural, once one considers the presence of attributes. The usual way to refer to one of such components for a tuple t would be by the attribute name of the desired component. That is, a tuple can be thought of as a record, the order of the field labels being irrelevant. We would use for instance the notation $t[A]$ for this.

This becomes more of a real problem when it comes to defining database operations. Many useful operations depend on the attributes present in a scheme, and not on the order those attributes are listed in the scheme. One could then choose to either restrict the operations to those in which the schemes are in some particular order, or take into account all permutations of the scheme. For instance, thinking about the union of two relations (seen as sets of tuples), we could only allow it if the relations are over the same scheme (which would imply the attributes are in the same order); or we could allow it for any relations whose schemes are permutations of each other. The first way is very common in actual database manager systems, but many concepts of relational database theory are actually better expressed in an unordered-scheme way. This kind of concern will actually need to be addressed when carrying out a formalisation in type theory.

Typical database operations include the set-theoretic ones, for instance union and intersection. There are also projection functions, selection operations to get tuples with particular attribute values, and a kind of relation composition called join. A large part of relational database theory deals with *data dependencies*, constraints expressing structural properties of relations and allowing relations to be decomposed in a safe way to improve their storage efficiency and consistency.

2 Previous Work

In our previous work, we addressed the issue of how to define and formalise a constructive, categorical calculus of relations inside Martin-Löf's type theory, and using Agda [7, 8]. We thus set up what we called the LT-allegory of E-relations, a notion based both on the theory of allegories and type theory, and making heavy use of the notion of setoids [2] (a set with an equivalence relation on it) to deal with the equalities provided for each set over which the relations are defined.

In [8] we showed that our formalisation was able to deal with database relations by using filter relations, much like in [11] and [15]. As much as this showed the feasibility of doing so using our existing formalisation, it becomes inconvenient and somewhat contrived. It would be much preferable if we could formalise relational database concepts directly into type theory, skipping the layer provided by the constructive version of allegories. Hence in the current work we have approached the problem by attempting a direct formalisation. We view this work as naturally following on our exploration of type-theoretical relational systems.

3 Type-Theoretic Building Blocks

The type of all sets is denoted **Set**, and the type of all propositions **Prop** (this being the same as **Set** by the Curry-Howard isomorphism, as usual). For a set A, the type of predicates over A is denoted **Pred** A. The usual notations of type theory are used for dependent function types (Π) and function types ($\rightarrow$). We will use $\Rightarrow$ instead of the more common $\supset$ for the implication logical constant, to avoid confusion with the standard notation for set containment, which is used a lot to speak of relational database concepts. **Bool** denotes the set of Boolean values $\{\mathbf{true}, \mathbf{false}\}$.

Definition 1. *A* relation *over a set A is a function of type $A \rightarrow A \rightarrow$* **Prop**.

Definition 2. *A relation over a set A is called* reflexive *if $\Pi x : A \,.\, xRx$ holds.*

Definition 3. *A relation over a set A is called* symmetric *if $\Pi x : A \,.\, \Pi y : A \,.\, xRy \Rightarrow yRx$ holds.*

Definition 4. *A relation over a set A is called* transitive *if $\Pi x : A \,.\, \Pi y : A \,.\, \Pi z : A \,.\, xRy \Rightarrow yRz \Rightarrow xRz$ holds.*

Definition 5. *A relation over a set A is called an* equivalence *if it is reflexive, symmetric, and transitive.*

Definition 6. *A relation R over a set A is called* substitutive *if* $\Pi x : A\ .\ \Pi y : A\ .\ \Pi P : \mathbf{Pred}\ A\ .\ xRy \Rightarrow Px \Rightarrow Py$ *holds.*

Definition 7. *A* Boolean relation *over a set A is a function of type* $A \rightarrow A \rightarrow \mathbf{Bool}$.

To apply the same concepts of reflexivity and such also to Boolean relations, we do it by referring to their 'raised' versions. That is, for a Boolean relation of type $A \rightarrow A \rightarrow \mathbf{Bool}$ we define a relation of type $A \rightarrow A \rightarrow \mathbf{Prop}$ simply by **if** $R\,x\,y$ **then** $\top$ **else** $\bot$ (this kind of definition is called 'by large elimination', since we produce sets as results of the pattern matching). So by abuse of language we will call a Boolean relation reflexive, for instance, when its 'raised' version is reflexive.

Definition 8. *A* setoid *is a set equipped with an equivalence relation over it. For a setoid A, we denote its underlying set (called its* carrier*) by* $|A|$, *and its equivalence relation by* $=_A$.

Definition 9. *A* datoid *is a set equipped with a Boolean relation over it that is reflexive and substitutive. For a datoid A, we denote its underlying set (called its* carrier*) by* $|A|$, *and its relation by* $=_A$.

In the actual implementation of all these notions in Alfa/Agda, all functions and function types are pretty much as in our definitions. Those notions that result from combining several properties (that is, equivalence relations, setoids and datoids) are implemented by means of signatures, dependently typed records that will contain all required proof objects as fields.

For the rest of this work, our notation will be mainly the one used by Alfa, and for obvious reasons of space we will only present essential (or some representative) fragments of our formalisation's proof code.

4 Formalisations of Relational Database Theory

Relational database notions are defined in terms of naive set theory, and when we move to constructive type theory, the notions of set to be used are different. Some hints of the potential usefulness of applying type theory to databases were pointed out by [12]. One purpose of our carrying out these formalisations, quite naturally, is setting up a precise and complete account of relational database theory inside type theory. The other purpose is exploring the issue of proving correctness of particular database implementations, at the level of data structures and algorithms, with respect to relational database theory. Since the actual implementations use very complex such structures and algorithms, we approach the problem through abstract views of implementations. It is our hope that this simpler exploration will make clear how one should attempt, in the close future, a pretty realistic proof of correctness of data structures and algorithms used in actual implementations.

In our setting, members of the universe **Set** are typically inductively defined sets, and by the Curry-Howard isomorphism we can refer to **Set** equivalently as **Prop**, the type of all propositions. To speak of subsets of A we can view them as predicates of type $A \rightarrow$ **Prop**. One standard way of dealing with sets that have an equality is through the already mentioned notion of setoids, denoted by **Setoid**. When we need such equalities to be decidable, we can assert them to be so by requiring to always have a proof of $x = y \vee x \neq y$, for any x, y in the setoid. A different approach is by having the equality produce a Boolean result. In the Alfa library, a notion called **Datoid** is available: a set with such a Boolean-valued equality that is reflexive and substitutive. It is also established in the library that one can always construct a setoid from a datoid, and many standard datoids, setoids, and properties about both notions are provided.

4.1 General Indexed Relations

In this approach a scheme will be some element of **Set**:

$$\mathbf{Scheme} \equiv \mathbf{Set}$$

No assumptions about its structure are needed, in particular nothing is said about a scheme being ordered or not.

The domains associated with attributes will be setoids. This is required, instead of they being simply sets, because the definitions of several fundamental operations (like projections and joins) involve equality comparisons between tuples and subtuples, and these in turn depend on equality comparisons between values of corresponding positions in the tuples. Hence, each and every attribute domain must provide such an equality.

The association of attributes and their domains is not done globally, as one would expect, but at the level of the schemes. Hence, if I is a scheme, for each element of I there will be a setoid A_i. We give a name to the type of all such associations:

$$\mathbf{Typing}\ (I \in \mathbf{Scheme}) \equiv I \rightarrow \mathbf{Setoid}$$

We need to explain a bit our notation for Alfa proof code: the preceding is a definition, as will be denoted from now on by the use of an equivalence sign. The definition has a parameter I which is a **Scheme**, as shown in parenthesis on the left-hand side of the equivalence sign. It is also possible to shift parameters between the left and right-hand sides of the equivalence sign, for instance we could have defined instead

$$\mathbf{Typing} \equiv \lambda I \in \mathbf{Scheme}.\, I \rightarrow \mathbf{Setoid}$$

Finally, we will sometimes just quote the left-hand side of a definition (the header) along with its result type, due to lack of space.

A tuple over a scheme takes as argument not just the scheme, but also the whole association function just described. The tuple itself is a function from elements of the scheme to elements of the associated carrier:

$$\textbf{Tuple}\ (I \in \textbf{Scheme})(T \in \textbf{Typing}\ I) \equiv (i \in I) \to |T\, i|$$

Tuple equality is simply the element-wise equality at the level of the associated setoids.

In this way we avoid most of the issues involved in handling schemes in detail. However, this means for once admitting some relations that are not what one usually thinks of as database relations: I can certainly be infinite. Also, representing schemes in this way will mean their manipulation will be different than is usual in the naive set theory definitions.

Database relations are simply propositional functions over tuples:

$$\textbf{Rel}\ (I \in \textbf{Scheme})(T \in \textbf{Typing}\ I) \equiv \textbf{Tuple}\ I\, T \to \textbf{Prop}$$

This gives us immediately the set operations over relations, in the usual way (union by disjunction, intersection by conjunction, difference by intersection and negation). For instance,

$$\textbf{unionRel}\ (I \in \textbf{Scheme})(A \in \textbf{Typing}\ I)(R, S \in \textbf{Rel}\ I\, A) \equiv \lambda t.R\, t \vee S\, t$$

Selection is simply defined as the conjunction of the predicates for the relation and the selection condition:

$$\begin{aligned}\textbf{selectRel}\ &(I \in \textbf{Scheme})(A \in \textbf{Typing}\ I)\\ &\qquad (P \in \textbf{Pred}\ (\textbf{Tuple}\ I\, A))(R \in \textbf{Rel}\ I\, A) \equiv \lambda x.R\, x \wedge P\, x\end{aligned}$$

Accordingly, properties of set-theoretic relational operations and selections will follow easily from the properties of the logical constants.

For projections, the essential idea is that the scheme gets split into the part we want to project to, and the part that is removed. This can be done by seeing the scheme as a sum type $I+J$, and its typing as having type **Typing** I+**Typing** J. Then two projections, one producing tuples on subscheme I and the other on subscheme J, can be defined by means of existential quantification. For instance,

$$\begin{aligned}\textbf{projRel1}\ &(I, J \in \textbf{Scheme})(A \in \textbf{Typing}\ I)\\ &(B \in \textbf{Typing}\ J)(R \in \textbf{Rel}\ (I+J)(A+B))\\ &\equiv \lambda x.\exists y \in \textbf{Tuple}\ (I+J)(A+B).\, R\, y \wedge (x = y[A])\end{aligned}$$

(Some massaging of the tuples involved is necessary for slicing a tuple, but we omit the details here for brevity.)

The join operation follows a similar pattern of definition, except that we use a type for triple sums, $I + J + K$. The definition is:

$$\begin{aligned}\textbf{joinRel}\ &(I, J, K \in \textbf{Scheme})\\ &(A \in \textbf{Typing}\ I)(B \in \textbf{Typing}\ J)(C \in \textbf{Typing}\ K)\\ &(R \in \textbf{Rel}\ (I+J)(A+B))\ (S \in \textbf{Rel}\ (J+K)(B+C))\\ &\equiv \lambda x.(\exists y \in \textbf{Tuple}\ (I+J)(A+B).R\, y \wedge (y = x[A+B]))\\ &\wedge(\exists z \in \textbf{Tuple}\ (J+K)(B+C).S\, z \wedge (z = x[B+C]))\end{aligned}$$

(Again, we omit all the required auxiliary definitions for massaging the tuples and obtaining the desired slices.) Proofs of properties involving projection and join are not trivial anymore, but quite straightforward and mostly depending on the standard logical constants and quantifiers.

As our representation of schemes doesn't have much structure, dealing with concepts of data dependency theory can become a bit cumbersome. We can define the notion of key if we see the scheme once again as a sum type (the two parts being the key and the non-key attributes), but to deal in general with dependencies we would need the ability to manipulate sets of attributes with more flexibility than this. For instance, to define the notion of candidate key we would need the ability to talk about subschemes in general, so we could then talk about a minimal such subscheme.

4.2 Decidable Relations and Finite Index Sets

Real database relations are defined over finite schemes, so the next formalisation will deal with index sets that are finite. In type theory we can express this by requiring that I is $N_k = \{0, 1, \ldots, k-1\}$. This restricts the attributes to natural numbers only, serving as positional indices. In the Alfa library N_k is called **Fin** k. Additionally, we wish to have operations that are computable, so now relations will be decidable: for this, we use Boolean relations. Since some operations (for instance, selection) need to compare tuples, and hence elements in the domains, this means that the equality over every domain must also be decidable and Boolean-valued.

A scheme will be a particular **Fin** n, where n is the arity of the tuples that form part of relations over this scheme. We only need to know n for knowing which scheme it is:

$$\mathbf{Scheme} \equiv \mathbf{Nat}$$

The domains associated with attributes are now datoids. As in the preceding approach, the association is not done globally but at the level of the schemes. The type of all such associations is thus represented:

$$\mathbf{Typing}\ (n \in \mathbf{Scheme}) \equiv (i \in \mathbf{Fin}\ n) \to \mathbf{Datoid}$$

As in the preceding approach, a tuple over a scheme also takes a typing as an argument, and is a function taking attributes of the scheme to elements of the associated carrier:

$$\mathbf{Tuple}\ (n \in \mathbf{Scheme})(T \in \mathbf{Typing}\ n) \equiv (i \in \mathbf{Fin}\ n) \to |T\,i|$$

Tuple equality is again the element-wise equality at the level of the associated datoids. But its type is now boolean:

$$\mathbf{eqTuple}\ (n \in \mathbf{Scheme})(T \in \mathbf{Typing}\ n)(t_1, t_2 \in \mathbf{Tuple}\ n\,T) \in \mathbf{Bool}$$

Its definition involves the use of a universal quantifier over **Fin** n to express element-wise equality.

Database relations could now be boolean functions over tuples:

$$\mathbf{Rel}\ (n \in \mathbf{Scheme})(T \in \mathbf{Typing}\ n) \equiv \mathbf{Tuple}\ n\, T \to \mathbf{Bool}$$

This gives us immediately the set operations over relations, by using the boolean operators in the usual way (union by or, intersection by and, difference by and and complement). Their type is similar for all of them, for instance intersection is:

$$\mathbf{interRel}\ (n \in \mathbf{Scheme})(T \in \mathbf{Typing}\ n)(R, S \in \mathbf{Rel}\ n\, T) \in \mathbf{Rel}\ n\, T$$

Also as in the preceding approach, selection is quite simple to define, by just taking the and of the (now boolean) selection predicate and the boolean function that represents the relation. Its type is:

$$\begin{gathered}\mathbf{selRel}\ (n \in \mathbf{Scheme})(T \in \mathbf{Typing}\ n)\\ (P \in (\mathbf{Tuple}\ n\, T) \to \mathbf{Bool})(R \in \mathbf{Rel}\ n\, T) \in \mathbf{Rel}\ n\, T\end{gathered}$$

The expected properties of these operations can be established by resorting to properties of the Boolean operators.

However, defining projections and join would require a decidable existential quantification. The library provides us with such, but only for a domain of quantification of the form **Fin n**. As we need to quantify over tuples, some kind of isomorphism would be needed between one such finite set and the type of tuples on a scheme. This would quickly become very cumbersome, and perhaps even not possible in all cases.

We could instead change the approach a bit, and define a relation as a list of tuples over a finite scheme. This now requires us to define database operations in terms of list manipulations (appends, filters, maps and such). Properties of the operations would need to resort to the usual theory of lists. Also, as we deal with schemes in terms of finite sets, a concept analogous to sum types is used for projections: $\mathbf{Fin}\ (m + n)$ types. These types need several properties regarding finite sets (for instance, that an isomorphism allows us to move between $\mathbf{Fin}\ (m + n)$ and $\mathbf{Fin}\ m + \mathbf{Fin}\ n$). Join can be defined by resorting to the corresponding idea for three finite set types, and its type is:

$$\begin{gathered}\mathbf{joinRel}\ (m, n, p \in \mathbf{Scheme})(T_1 \in \mathbf{Typing}\ m)(T_2 \in \mathbf{Typing}\ n)\\ (T_3 \in \mathbf{Typing}\ p)(R \in \mathbf{Rel}\ m + n\ \ T_1 + T_2)(S \in \mathbf{Rel}\ n + p\ \ T_2 + T_3)\\ \in \mathbf{Rel}\ m + n + p\ \ T_1 + T_2 + T_3\end{gathered}$$

(Here, we should remark that the sums of the typings need to be defined as operations that use the isomorphism mentioned above, and not as sum types as in the previous subsection.)

The situation regarding data dependencies would be pretty much as in the previous formalisation: some basic notions can be captured (as keys), but formalising a general theory of functional dependencies will be cumbersome. Also, as the attributes are similar to positional indexes in the local scheme, we lose the possibility of using the same name for an attribute that appears in different schemes of the database. Some kind of lookup tables could be implemented as part of the formalisation to recover such capabilities, but it will clearly make the whole effort quite more complex.

4.3 Relations via Collections

A useful notion that has sometimes been used in databases is that of *collection types*, also called *bulk types*. A collection in this sense can be seen intuitively as a data type corresponding to sets over some base type. Related work has shown it is possible to use the notion of collections as an abstraction layer on which to build constructive formalisations of some relational database theory concepts [14]. However, in that work the authors aim for a generic kind of constructive algebra for databases, while we intend to arrive at a fairly concrete (and much simpler) basis on top of which to build a full formalisation of relational database theory.

We formalise an addition to the Alfa libraries to deal with collections. Our representation is based on a signature, with one field being a list, and the only other field being a proof object asserting that no repeated elements appear in that list. We do this in two steps, since the notion of generic collection may be useful in our future work:

$$GenColl\ (A \in Set, P \in Pred\ (List\ A)) \in Set$$

$$GenColl\ A\ P \equiv \begin{array}{l} \textbf{sig} \\ elems \in List\ A \\ ppty \in P\ elems \end{array}$$

(with different A and P, many useful notions could be captured: ordered lists, association lists, etc.). We remark that **sig** is the Alfa/Agda notation for a dependently typed record, and that we use visual layout (like in Haskell) to establish scopes of definitions.

$$Coll\ (A \in Setoid) \in Set$$

$$Coll\ A \equiv GenColl\ (|\ A\ |)\ (noDupls\ A)$$

Hence collections can be seen as a version of finite sets over a base data type, and having computational content. Heavy use is made of properties of lists to define operations over collections (for instance, intersecting collections requires a proof that the manipulation, doing by a list filtering, actually produces a list without duplicates). The definitions of these operations require a decidable equality over the base type, and we have chosen to use setoids with decidable equality, where the equality is not Boolean valued: instead we always have a proof of $x = y \vee x \neq y$. This makes the proofs simpler to deal with. The set of collections over a base setoid A is a setoid with decidable equality, so we can form collections of collections when needed.

As an example of the operations provided, and the kind of formalisation effort involved, consider the addition of a single element to a collection:

$addElemColl\ (A \in Setoid, dEq \in decEq\,A, a \in |\,A\,|, C \in Coll\,A) \in Coll\,A$

$$addElemColl\,A\,dEq\,a\,C \equiv \begin{array}{l}\textbf{struct}\\ \left[\begin{array}{l} elems \equiv addElem\,A\,dEq\,a\,C.elems \\ ppty \equiv thNoDupAddElem\,A\,dEq\,a \\ \qquad\qquad C.elems\,C.ppty \end{array}\right.\end{array}$$

($thNoDupAddElem$ is the proof that the *addElem* operation, when applied to a collection, gives a collection). We remark that **struct** is the Alfa/Agda notation for defining an instance of a dependently typed record. Here, we make use of one operation for lists to achieve the desired result:

$addElem\ (A \in Setoid, dEq \in decEq\,A, a \in |\,A\,|, L \in List\,(|\,A\,|)) \in List\,(|\,A\,|)$

$$addElem\,A\,dEq\,a\,[\,] \equiv a : [\,]$$

$$addElem\,A\,dEq\,a\,(x : xs) \equiv \begin{array}{l} ifD \\ (dEq\,a\,x) \\ \lambda\,h \to L \\ \lambda\,h \to x : addElem\,A\,dEq\,a\,xs \end{array}$$

The reader will notice two new things that require explanation: *decEq* gives for a setoid a proof that the equality on that setoid is decidable in the sense explained previously in this subsection; *ifD* is a variation of the usual conditional expression, except that here it works on a value of the type $P + \neg P$ by case analysis. Briefly, *ifD cond thenFun elseFun* looks at at a value of type $P + \neg P$ provided by *cond*, then branches through the appropiate function (*thenFun* if P is the case, *elseFun* if $\neg P$ is the case), and passes the corresponding proof object to the branch so selected. While for this example operation the proof object passed to the branch function is not necessary, there are situations in which it is actually used for obtaining correct definitions.

An additional benefit of using collections is that we can now provide enough structure in the schemes to allow complete manipulation of their attributes. We set up a global universe of attribute names *Attrib* (for example, strings) along with a typing $T \equiv Attrib \to Setoid$. Equality over attributes is a decidable equivalence, so we can form collections on the setoid of attributes. Also, we require that the setoids onto which attributes are mapped by T have decidable equivalences as equalities. A scheme is now simply a collection of attributes. A tuple is a collection of pairs (A, v) formed by an attribute A and a value v in the setoid $T A$ (equality on such pairs is defined as equality on their attribute parts, so we have a setoid of such pairs and we can collect them). As this allows any kind of tuple to be formed without connection to any scheme, we also define a notion of proper tuple as one that includes exactly those pairs that have as attribute one belonging to the scheme under consideration (this is done by using equality on collections, and a version of map for collections to extract the attributes). (This way of defining the notion of tuple has been used in [6], too.)

A relation is now nothing more than a collection of proper tuples. The database set-theoretic operations are then defined in terms of the set-theoretic collection operations. Selection is defined in terms of the version of filter for collections. Our collection formalisation provides decidable quantification over collections, so we can express the defining property of a projection or a join. As for the construction of the result tuples in these two operations, we use again the version of map for collections. Hence, all properties of database operations reduce directly to properties of the collection operations they are defined upon. We provide such collection properties in our formalisation too.

As for data dependencies, the use of collections to deal with schemes now allows us to capture all the useful notions in a very straightforward way. Since we can manipulate attributes individually and flexibly using collection operations, we are now able to define all the notions of functional dependency theory. Quite naturally, a functional dependency can be represented as a pair of collections of attributes. And we can reason about dependencies too, using collection properties. A goal to reach for in this context would be to formalise soundness and completeness of the standard set of axioms for functional dependencies. We also remark that defining other kinds of dependencies (for instance multi-valued ones) is possible by means of collections.

5 Conclusions and Further Work

We have discussed several ways to define relational database theory concepts in type theory. Doing so gives not just a formal and precise setup for the theory, but also does it in a way that provides computational content. Both as an application of type theory to a practical matter, and as an exploration of the detailed foundations of relational databases, our effort seems worthwhile.

Some work remains to be done providing enough detail on proofs that our definitions behave as the database operations do, but the way to do it seems clear. A more complex effort would be required in the future to deal with dependency theory, particularly for results involving axioms for that theory as already mentioned. Our efforts suggest that an abstract layer corresponding to the notion of finite subset would capture all the common proof structure.

Databases are not implemented on such levels as our collection abstraction, they are instead defined on actual file structures. It would be a desirable goal, though probably a long-term one, to formalise such file structures as are commonly used, along with a transformation from our abstract layer into these structures, and a proof that such a transformation preserves the expected properties. Needless to say, that would be a major effort, but one probably worthwhile both from the type theory and the database theory points of view.

Acknowledgements

The author is grateful to Gothenburg University and the organizers and sponsors of RelMiCS7 for funding with their respective grants my attendance to this meeting. I also thank Peter Dybjer, Michael Hedberg, and Makoto Takeyama, for helpful discussions about many technical issues of type theory and Alfa/Agda. Finally, thanks are due to the anonymous reviewers for their comments and suggestions on improving this work.

References

[1] Augustsson, L.: Cayenne - a language with dependent types. In *Proceedings of the Third ACM SIGPLAN International Conference on Functional Programming, 1998, Baltimore, Maryland, United States*, ACM Press, 1998, pp. 239–250. 138

[2] Barthe, G., Capretta V., Pons, O.: Setoids in type theory. Journal of Functional Programming, Volume 13, Issue 2, March 2003, pp. 261-293. 139

[3] Codd, E. F.: A Relational Model of Data for Large Shared Data Banks. Communications of the ACM, Volume 13, Number 6, pp. 377–387, June 1970. 138

[4] Coquand, C.: Agda home page. `http://www.cs.chalmers.se/~catarina/agda/` 137

[5] Coquand, C., Coquand, T.: Structured Type Theory. In *Workshop on Logical Frameworks and Meta-languages, Paris, France, September 1999, Proceedings.* Part of PLI'99 (Colloquium on Principles, Logics, and Implementations of High-Level Programming Languages). 137

[6] Darwen, H., Date, C. J.: The Third Manifesto. ACM SIGMOD Record, Volume 24, Issue 1 (March 1995), pp. 39–49. 146

[7] Gonzalía, C.: The Allegory of E-Relations in Constructive Type Theory. In *Relational Methods in Computer Science: The Québec Seminar*, Desharnais, J., Frappier, M., MacCaull, W. (Eds.), Methoδos Verlag, April 2002, pp. 19–38. 139

[8] Gonzalía, C.: *Relation Calculus in Martin-Löf Type Theory.* Ph. Lic. thesis, Technical Report no. 5L, Department of Computing Science, Chalmers University of Technology and Göteborg University, February 2002. 139

[9] Hallgren, T.: Alfa home page. `http://www.cs.chalmers.se/~hallgren/Alfa/` 137

[10] Hedberg, M.: New standard library for Alfa. `http://www.cs.chalmers.se/pub/users/hallgren/Alfa/Alfa/Library/New/` 138

[11] Kawahara, Y., Okhuma, H.: Relational Aspects of Relational Database Dependencies. In *RelMICS 2000: 5th International Seminar on Relational Methods in Computer Science, Valcartier, Québec, Canada, January 2000, Proceedings*; Desharnais, J. (Ed.), pp. 185–194. 139

[12] Mäenpää, P., Tikkanen M.: Dependent Types in Databases, Workshop in Dependent Types in Programming, Göteborg, Sweden, 1999. 140

[13] Nordström, B., Petersson, K., Smith, J.: *Programming in Martin-Löf's Type Theory. An Introduction.* The International Series of Monographs on Computer Science, Oxford University Press, 1990. 137

[14] Rajagopalan, P., Tsang, C. P.: A Generic Algebra for Data Collections Based on Constructive Logic. In *Algebraic Methodology and Software Technology, 4th International Conference, AMAST '95, Montreal, Canada, July 1995, Proceedings*; Alagar, V. S., Nivat, M. (Eds.), Lecture Notes in Computer Science, Volume 936, Springer-Verlag, 1995, pp. 546–560. 145

[15] Schmidt, G., Ströhlein, T.: *Relations and Graphs. Discrete Mathematics for Computer Scientists.* EATCS Monographs on Theoretical Computer Science, Springer-Verlag, 1993. 139

SCAN Is Complete for All Sahlqvist Formulae*

V. Goranko[1], U. Hustadt[2], R. A. Schmidt[3], and D. Vakarelov[4]

[1] Rand Afrikaans University, South Africa
vfg@na.rau.ac.za
[2] University of Liverpool, UK
U.Hustadt@csc.liv.ac.uk
[3] University of Manchester, UK
schmidt@cs.man.ac.uk
[4] Sofia University, Bulgaria
dvak@fmi.uni-sofia.bg

Abstract. SCAN is an algorithm for reducing existential second-order logic formulae to equivalent simpler formulae, often first-order logic formulae. It is provably impossible for such a reduction to first-order logic to be successful for every second-order logic formula which has an equivalent first-order formula. In this paper we show that SCAN successfully computes the first-order equivalents of all Sahlqvist formulae in the classical (multi-)modal language.

1 Introduction

One of the most general results on first-order definability and completeness in modal logic is Sahlqvist's theorem [16] where two notable facts are proved for a large, syntactically defined class of modal formulae, now called Sahlqvist formulae: first, the *correspondence result,* which says that Sahlqvist formulae all define first-order conditions on Kripke frames and these conditions can be effectively "computed" from the modal formulae; and second, the *completeness result*, which says that all those formulae are canonical, i.e. valid in their respective canonical frames, hence axiomatise completely the classes of frames satisfying the corresponding first-order conditions. The class of Sahlqvist formulae has been studied extensively, and several alternative proofs and generalisations have been proposed (see [19, 17, 11, 4, 10, 9]).

Computing the first-order equivalent of a modal formula (if it exists) amounts to the elimination of the universal monadic second-order quantifiers expressing the validity in a frame of that formula or, equivalently, the elimination of the existential monadic second-order quantifiers expressing the satisfiability of the

* This work was supported by EU COST Action 274, and research grants GR/M88761 and GR/R92035 from the UK EPSRC. The first author's work was supported by research grants from Rand Afrikaans University. Part of the work by the third author was done while on sabbatical leave at the Max-Planck-Institut für Informatik, Germany, in 2002. We would also like the thank the referees for their helpful comments.

R. Berghammer et al. (Eds.): RelMiCS/Kleene-Algebra Ws 2003, LNCS 3051, pp. 149–162, 2004.

formula. A well-known algorithm for eliminating existential second-order quantifiers is SCAN [8]. SCAN does not succeed for all first-order definable modal formulae, for example, if we take the K axiom $\Box(\varphi \to \psi) \to (\Box\varphi \to \Box\psi)$ and replace ϕ and ψ by two distinct instances of the McKinsey axiom, e.g. $\Box\Diamond p \to \Diamond\Box p$ and $\Box\Diamond q \to \Diamond\Box q$, respectively, then SCAN will not terminate on the resulting formula, although it is equivalent to the first-order formula $\top$. However, SCAN has been proved correct whenever it gives an answer [8].

An alternative algorithm to SCAN is DLS [5, 13, 18]. SCAN and DLS are incomparable concerning their ability to compute first-order correspondences for modal formulae. For example, while SCAN successfully computes a first-order equivalent formula for $(\Diamond\Box(p\vee q)\wedge\Diamond\Box(p\vee\neg q)\wedge\Diamond\Box(\neg p\vee q)) \to (\Box\Diamond(p\vee q)\vee\Box\Diamond p\vee\Box\Diamond(p\wedge q))$, DLS fails. On the other hand, there are also examples where DLS succeeds and SCAN fails.

In this paper we show that SCAN successfully computes the first-order equivalents of all classical Sahlqvist formulae. To our knowledge, this is the first completeness result for a quantifier elimination algorithm.

We assume basic knowledge of the syntax and semantics of modal logic and first-order resolution. State-of-the-art references on modal logic and on automated deduction including resolution are [2, 3] and [1, 15], respectively.

2 Sahlqvist Formulae

We fix an arbitrary propositional (multi-)modal language. For technical convenience we assume that the primitive connectives in the language are $\neg$, $\wedge$, and the diamonds, while the others are definable as usual, e.g. $\varphi \to \psi$ is defined as $\neg(\varphi \wedge \neg\psi)$. Most of the following definitions are quoted from [2].

An occurrence of a propositional variable in a modal formula φ is *positive (negative)* iff it is in the scope of an even (odd) number of negations. A modal formula φ is *positive (negative) in a variable* q iff all occurrences of q in φ are positive (negative). A modal formula φ is *positive* (*negative*) iff all occurrences of propositional variables in φ are positive (negative). A *boxed atom* is a formula of the form $\Box_{k_1} \ldots \Box_{k_n} q$, where $\Box_{k_1}, \ldots, \Box_{k_n}$ is a (possibly empty) string of (possibly different) boxes and q is a propositional variable.

A *Sahlqvist antecedent* is a modal formula constructed from the propositional constants $\bot$ and $\top$, boxed atoms and negative formulae by applying $\vee$, $\wedge$, and diamonds. A *definite Sahlqvist antecedent* is a Sahlqvist antecedent obtained without applying $\vee$ (i.e. constructed from the propositional constants $\bot$ and $\top$, boxed atoms and negative formulae by applying only $\wedge$ and diamonds). A *(definite) Sahlqvist implication* is a modal formula $\varphi \to \psi$ where φ is a (definite) Sahlqvist antecedent and ψ is a positive formula. A *(definite) Sahlqvist formula* is a modal formula constructed from (definite) Sahlqvist implications by freely applying boxes and conjunctions, and by applying disjunctions to formulae without common propositional variables. A *basic Sahlqvist formula* is a definite Sahlqvist formula obtained from definite Sahlqvist implications without applying conjunctions. It can be shown [9]) that that every Sahlqvist formula is semantically equivalent to a conjunction of basic Sahlqvist formulae.

Example 1.

i. $\Diamond(\neg\Box(p \vee q) \wedge \Diamond\Box\Box q) \rightarrow \Box\Diamond(p \wedge q)$ is a Sahlqvist formula.
ii. $\Box\Diamond p \rightarrow \Box p$ and $\Box(p \vee q) \rightarrow \Box p$ are not Sahlqvist formulae, but can be converted into Sahlqvist formulae defining the same semantic conditions by taking their contrapositions and reversing the signs of p and q.
iii. $\Box\Diamond p \rightarrow \Diamond\Box p$ and $\Box(\Box p \rightarrow p) \rightarrow \Box p$ are not Sahlqvist formulae, and cannot be converted into Sahlqvist formulae defining the same semantic conditions, because both are known not to be first-order definable.

Theorem 1 ([16]). *Every Sahlqvist formula φ is locally first-order definable, i.e. there is a first-order formula $\alpha_\varphi(x)$ of the first-order language with equality and binary relational symbols corresponding to the modal operators in φ, such that for every Kripke frame F with a domain W and $w \in W$, $F, w \models \varphi$ iff $F \Vdash_w \alpha_\varphi(x)$ (where $\models$ denotes modal validity, while $\Vdash$ denotes first-order truth).*

The problem whether a given modal formula is frame-equivalent to some Sahlqvist formula is not known to be decidable, and most probably it is not. However, syntactical transformations like the one used in Example 1.ii can extend the applicability of the result we present in this paper.

3 The SCAN Algorithm

SCAN reduces existentially quantified second-order sentences to equivalent first-order formulations. Given a second-order sentence containing existentially quantified second-order variables, the algorithm generates sufficiently many logical consequences, eventually keeping from the resulting set of formulae only those in which no second-order variables occur. The algorithm involves three stages:

(i) transformation to clausal form and (inner) Skolemization;
(ii) C-resolution;
(iii) reverse Skolemization (unskolemization).

The input of SCAN is a second-order formula of the form $\exists Q_1 \ldots \exists Q_k\ \psi$, where the Q_i are unary predicate variables and ψ is a first-order formula. In the first stage SCAN converts ψ into clausal form, written $\mathrm{Cls}(\psi)$, by transformation into conjunctive normal form, inner Skolemization, and clausifying the Skolemized formula [12, 14]. In the second stage SCAN performs a special kind of constraint resolution, called *C-resolution*. It generates all and only resolvents and factors with the second-order variables that are to be eliminated. When computing frame correspondence properties, all existentially quantified second-order variables $Q_1, \ldots, Q_k$ are eliminated.

To allow for a more concise proof of the main result, our presentation of C-resolution differs slightly from [8] in that we have explicitly included purity deletion, subsumption deletion, and condensation in the calculus.

In the following *clauses* are considered to be multisets of literals. A *literal* is either an *atom* $P(t_1, \ldots, t_n)$ or $t_1 \approx t_2$ (also called *positive literal*) where P is an n-ary predicate symbol and $t_1, \ldots, t_n$ are terms, or a *negative literal* $\neg P(t_1, \ldots, t_n)$, or $t_1 \not\approx t_2$. An atom $t_1 \approx t_2$ is also called an *equation*, while a negative literal $t_1 \not\approx t_2$ is called an *inequation*. We consider clauses to be identical up to variable renaming, that is, if C and D are clauses such that there exists a variable renaming σ with $C\sigma = D$, then we consider C and D to be equal. A *subclause* of a clause C is a submultiset of C. A *variable indecomposable clause* is a clause that cannot be split into non-empty subclauses which do not share variables. The finest partition of a clause into variable indecomposable subclauses is its *variable partition*. If $D = C \vee s_1 \not\approx t_1 \vee \ldots \vee s_n \not\approx t_n$ is a clause such that C does not contain any inequations, and σ is a most general unifier of $s_1 \approx t_1 \wedge \ldots \wedge s_n \approx t_n$ (that is, $s_i\sigma \approx t_i\sigma$ for every i, $1 \leq i \leq n$, and σ is the most general substitution with this property), then we say $C\sigma$ is obtained from D by *constraint elimination*. Note that $C\sigma$ is unique up to variable renaming.

A literal with a unary predicate symbol among $Q_1, \ldots, Q_k$, is a **Q**-literal, a literal with a binary predicate symbol (not including equality) is an **R**-literal.

A subclause D of a clause C is a *split component* of C iff (i) if L' is a literal in C but not in D, then L' and D are variable-disjoint and (ii) there is no proper subclause $D' \subset D$ satisfying property (i). If $C \vee L_1 \vee \ldots \vee L_n$, $n \geq 2$, is a clause and there exists a most general unifier σ of $L_1, \ldots, L_n$, then $(C \vee L_1)\sigma$ is called a *factor* of $C \vee L_1 \vee \ldots \vee L_n$. The *condensation* $\mathrm{cond}(C)$ of a clause C is a minimal subclause D of C such that there exists a substitution σ with $L\sigma \in D$ for every $L \in C$. A clause C is *condensed* iff $\mathrm{cond}(C)$ is identical to C. A *clause set* is a set of clauses. With $\uplus$ we denote the *disjoint union* of two sets.

Derivations in the C-resolution calculus are constructed using *expansion rules* and *deletion rules*. The only expansion rule in the C-resolution calculus is the following:

Deduction: $$\frac{N}{N \cup \{C\}}$$
if C is a C-resolvent or C-factor of premises in N.

C-resolvents and C-factors are computed using the following *inference rules*:

C-Resolution: $$\frac{C \vee Q(s_1, \ldots, s_n) \quad D \vee \neg Q(t_1, \ldots, t_n)}{C \vee D \vee s_1 \not\approx t_1 \vee \ldots \vee s_n \not\approx t_n}$$
provided the two premises have no variables in common and $C \vee Q(s_1, \ldots, s_n)$ and $D \vee \neg Q(t_1, \ldots, t_n)$ are distinct clauses. (As usual we assume the clauses are normalised by variable renaming so that the premises of the C-resolution do not share any variables.) The clause $C \vee Q(s_1, \ldots, s_n)$ is called the *positive premise* and the clause $D \vee \neg Q(t_1, \ldots, t_n)$ the *negative premise* of the inference step. The conclusion is called a *C-resolvent* with respect to Q.

C-Factoring: $$\frac{C \vee Q(s_1, \ldots, s_n) \vee Q(t_1, \ldots, t_n)}{C \vee Q(s_1, \ldots, s_n) \vee s_1 \not\approx t_1 \vee \ldots \vee s_n \not\approx t_n}$$
The conclusion is called a *C-factor* with respect to Q.

In addition, the C-resolution calculus includes four deletion rules:

Purity Deletion: $\dfrac{N \uplus \{C \vee Q(s_1, \ldots, s_n)\}}{N}$

if all inferences with respect to Q with $C \vee Q(s_1, \ldots, s_n)$ as a premise have been performed.

Subsumption Deletion: $\dfrac{N \uplus \{C, D\}}{N \uplus \{C\}}$

if C subsumes D, i.e. there is a substitution σ such that $C\sigma \subseteq D$.

Constraint Elimination: $\dfrac{N \uplus \{C\}}{N \uplus \{D\}}$

if D is obtained from C by constraint elimination.

Condensation: $\dfrac{N \uplus \{C\}}{N \uplus \{\mathrm{cond}(C)\}}$

A *derivation* in the C-resolution calculus is a (possibly infinite) sequence of clause sets $N_0, N_1, \ldots$ such that for every $i \geq 0$, N_{i+1} is obtained from N_i by application of an *expansion rule* or a *deletion rule*. In the following we assume that the condensation rule is applied eagerly, that is, whenever a clause set N_i in a derivation contains a clause C which is not condensed, the condensation rule is applied to N_i to derive N_{i+1} in which C is replaced by $\mathrm{cond}(C)$.

The algorithm generates all possible C-resolvents and C-factors with respect to the predicate variables $Q_1, \ldots, Q_k$. When all C-resolvents and C-factors with respect to a particular Q_i-literal and the rest of the clause set have been generated, purity deletion removes all clauses in which this literal occurs. The subsumption deletion rule is optional for the sake of soundness, but helps simplify clause sets in the derivation.

If the C-resolution stage terminates, it yields a set N of clauses in which the specified second-order variables are eliminated. This set is satisfiability-equivalent to the original second-order formula. If no clauses remain after purity deletion, then the original formula is a tautology; if C-resolution produces the empty clause, then it is unsatisfiable. If N is non-empty, finite and does not contain the empty clause, then in the third stage, SCAN attempts to restore the quantifiers from the Skolem functions by reversing Skolemization. This is not always possible, for instance if the input formula is not first-order definable.

If the input formula is not first-order definable and stage two terminates successfully yielding a non-empty set not containing the empty clause then SCAN produces equivalent second-order formulae in which the specified second-order variables are eliminated but quantifiers involving Skolem functions occur and the reverse Skolemization typically produces Henkin quantifiers. If SCAN terminates and reverse Skolemization is successful, then the result is a first-order formula logically equivalent to the second-order input formula.

The SCAN algorithm as described above differs in two details from the SCAN implementation.[1] First, the implementation does not restrict C-factoring inferences to positive literals, although this is sufficient for the completeness of the

[1] `http://www.mpi-sb.mpg.de/units/ag2/projects/SCAN/index.html`

algorithm. Concerning this aspect the implementation also differs from the SCAN algorithm as described in [8]. Second, the implementation of reverse Skolemization does not take into account that variable-disjoint subclauses can be unskolemized separately, which is crucial for our results. For example, unskolemizing the clause $r(x, f(x)) \vee r(y, g(y))$ should result in the first-order formula $(\forall x \exists u R(x,u)) \vee (\forall y \exists v R(y,v))$. Instead the SCAN implementation will produce a formula involving a Henkin quantifier. Obviously, these two deviations between the SCAN algorithm and its implementation are minor and could be easily incorporated into the implementation.

Since we intend to apply SCAN to Sahlqvist formulae, we now define a translation of modal formulae into second-order logic. Let

$$\Pi(\varphi) = \forall Q_1 \ldots \forall Q_m \forall x \, \mathrm{ST}(\varphi, x),$$

where $\mathrm{ST}(\varphi, x)$ is the (local) standard translation of a modal formula φ with a free variable x, and $Q_1, \ldots, Q_m$ are all the unary predicates occurring in $\mathrm{ST}(\varphi, x)$. The standard translation itself is inductively defined as follows:

$$\begin{aligned} \mathrm{ST}(\bot, x) &= \bot \\ \mathrm{ST}(q_i, x) &= Q_i(x) & \mathrm{ST}(\neg\phi, x) &= \neg\, \mathrm{ST}(\phi, x) \\ \mathrm{ST}(\phi \wedge \psi, x) &= \mathrm{ST}(\phi, x) \wedge \mathrm{ST}(\psi, x) & \mathrm{ST}(\Diamond_{k_i}\phi, x) &= \exists y (x R_{k_i} y \wedge \mathrm{ST}(\phi, y)) \end{aligned}$$

where Q_i is a unary predicate symbol uniquely associated with the propositional variable q_i, R_{k_i} and is a binary predicate symbol representing the accessibility relation associated with $\Diamond_{k_i}$, and x is a first-order variable. The important property of the standard translation is that it preserves the truth of a modal formula at any state w of a Kripke model M, i.e. $M, w \models \phi$ iff $M \models ST(\phi, x)(w/x)$ where M is regarded as a first-order structure for the language of the standard translation. Thus, validity (resp. satisfiability) of a modal formula is expressed by a universal (resp. existential) monadic second-order formula.

Let SCAN* denote the extension of the SCAN algorithm with the preprocessing and postprocessing necessary for computing first-order correspondences for modal logic formulae. The preprocessing involves translating the given modal formula to the second-order formula $\Pi(\varphi)$ and negating the result. The postprocessing involves negating the result output of SCAN, if it terminates.

4 Completeness of SCAN for Sahlqvist Formulae

In the following, we show that SCAN* is complete for Sahlqvist formulae. We need to prove that for any second-order formula ψ obtained by preprocessing from a Sahlqvist formula φ, SCAN can compute a first-order equivalent for ψ. To this end, we have to show two properties:

1. The computation of C-resolvents and C-factors terminates when applied to the set $\mathrm{Cls}(\psi)$ of clauses associated with ψ, i.e. SCAN can generate only finitely many new clauses in the process.

2. The resulting first-order formula, which in general contains Skolem functions, can be successfully unskolemized.

We first consider the case when φ is a definite Sahlqvist implication.

Theorem 2. *Given any definite Sahlqvist implication φ, SCAN* effectively computes a first-order formula α_φ logically equivalent to φ.*

Proof (Sketch). Let $\varphi = A \rightarrow P$. In the preprocessing stage, SCAN* computes $\Pi(\varphi)$ and negates the result. Since $\Pi(\varphi) = \forall Q_1 \ldots \forall Q_m \forall x \, \mathrm{ST}(\varphi, x)$, the negation $\neg\Pi(\varphi)$ is equivalent to $\exists Q_1 \ldots \exists Q_m \exists x \, \mathrm{ST}(\neg\varphi, x) = \exists Q_1 \ldots \exists Q_m \exists x \, \mathrm{ST}(A \wedge \neg P, x)$. So, the initial clause set N we obtain after clausification and Skolemization is given by $\mathrm{Cls}(\mathrm{ST}(A \wedge \neg P, a))$ where a is a Skolem constant. The definite Sahlqvist antecedent A is constructed from propositional constants, boxed atoms and negative formulae by applying only $\wedge$ and diamonds. In addition $\neg P$ is also a negative formula, since P is a positive formula. Thus, $A \wedge \neg P$ is itself a definite Sahlqvist antecedent. Note that

$$\begin{aligned}\mathrm{ST}(\alpha \wedge \beta, a_i) &= \mathrm{ST}(\alpha, a_i) \wedge \mathrm{ST}(\beta, a_i)\\ \mathrm{ST}(\Diamond_{k_i}\alpha, a_i) &= \exists y (a_i R_{k_i} y \wedge \mathrm{ST}(\alpha, y)).\end{aligned}$$

Skolemization will replace the existentially quantified variable y by a new constant a_{i+1} which replaces any occurrence of y in $a_i R_{k_i} y \wedge \mathrm{ST}(\alpha, y)$. Consequently,

$$\begin{aligned}\mathrm{Cls}(\mathrm{ST}(\alpha \wedge \beta, a_i)) &= \mathrm{Cls}(\mathrm{ST}(\alpha, a_i)) \cup \mathrm{Cls}(\mathrm{ST}(\beta, a_i))\\ \mathrm{Cls}(\mathrm{ST}(\Diamond_{k_i}\alpha, a_i)) &= \{a_i R_{k_i} a_{i+1}\} \cup \mathrm{Cls}(\mathrm{ST}(\alpha, a_{i+1}))\end{aligned}$$

It follows by a straightforward inductive argument that we can divide the clause set $N = \mathrm{Cls}(\mathrm{ST}(A \wedge \neg P, a))$ into a set N_n of clauses which stems from negative formulae occurring in $A \wedge \neg P$ and a set N_p of clauses which stems from the translation of the propositional constants $\top$ and $\bot$, and boxed atoms.

The translation of boxed atoms with respect to a constant a_i is given by

$$\begin{aligned}\mathrm{ST}(\Box_{k_1} \ldots \Box_{k_n} q_j, a_i) = \forall x_1 (a_i R_{k_1} x_1 \rightarrow &\\ \forall x_2 (x_1 R_{k_2} x_2 \rightarrow \cdots &\\ \forall x_n (x_{n-1} R_{k_n} x_n \rightarrow Q_j(x_n)) \ldots))&\end{aligned}$$

where $n \geq 0$. Clausification transforms $\mathrm{ST}(\Box_{k_1} \ldots \Box_{k_n} q_j, a_i)$ into a single clause of the form

$$\neg a_i R_{k_1} x_1 \vee \bigvee_{l=1}^{n-1} \neg x_l R_{k_l} x_{l+1} \vee Q_j(x_n),$$

for $n \geq 0$. In the case of $n = 0$, the clause we obtain consists of a single positive ground literal $Q_j(a_i)$. Besides clauses of this form, N_p can only contain the empty clause, which is the result of translation of the propositional constant $\bot$, while the translation of $\top$ will be eliminated during clausification.

Thus, every clause in N_p contains at most one predicate symbol Q_j. Moreover, all clauses in N_p will only contain positive occurrences of unary predicate symbols Q_j. In contrast, by definition, all occurrences of propositional variables q_j in the negative formulae in $A \wedge \neg P$ are negative. So, the corresponding occurrences of unary predicate symbols Q_j in N_n are all negative as well.

We have to establish the following.

The derivation always terminates for the formulae (clauses) under consideration. We define a function μ_1 that assigns to each clause C a triple $\mu_1(C) = \langle n_C^Q, n_C^R, d_C \rangle$ of natural numbers such that n_C^Q is the number of Q_i-literals in C, n_C^R is the number of all remaining literals in C, and d_C is the depth of C. We call $\mu_1(C)$ the *complexity* of clause C. It is straightforward to show that for a given triple $c = \langle n^Q, n^R, d \rangle$ of natural numbers the preimage of c under μ_1 contains only finitely many clauses (up to renaming of variables). We also define an ordering $\succ$ on $\mathbb{N} \times \mathbb{N} \times \mathbb{N}$ by the lexicographic combination of the ordering $>$ on the natural numbers with itself. Obviously, the ordering $\succ$ is well-founded.

We have already established that no clause in N contains a positive Q_i-literal as well as a negative Q_j-literal and the clauses in N_p have the property that each clause which contains a positive Q_i-literal contains exactly one such literal. It follows that in any C-resolution derivation from N no inference steps by C-factoring are possible. Furthermore, any C-resolvent D obtained by C-resolution with positive premise C_p and negative premise C_n will not contain a positive Q_i-literal, and D contains less Q_i-literals than C_n. Thus, $\mu_1(C_n) \succ \mu_1(D)$. Since no other inference steps are allowed in the C-resolution calculus, we have established that the conclusion of any inference step in a derivation from N is of strictly smaller complexity than one of its premises. The application of a deletion rule will only replace a clause C of complexity $\mu_1(C)$ be a clause D with smaller complexity $\mu_1(D)$. It follows that any derivation from N terminates.

The restoration of the quantifiers does not fail. To ensure that the restoration of quantifiers does not fail once the derivation from N has terminated, we can show by induction that for any clause C in a derivation from N, (i) C contains only inequations of the form

$$b \not\approx z, \quad b \not\approx c, \quad b \not\approx f(z), \quad y \not\approx z, \quad y \not\approx c, \quad \text{or} \quad y \not\approx f(z),$$

(ii) there are no two inequations in C of the form $y \not\approx f(z)$ and $y \not\approx g(z)$ with $f \neq g$, and (iii) if C contains negative Q_i-literals then these are of the form $\neg Q_i(z)$, $\neg Q_i(c)$ or $\neg Q_i(f(z))$, where c is a Skolem constant and f a unary Skolem function. An alternative formulation of property (ii) is that for any two inequations $x_1 \not\approx f(y)$ and $x_2 \not\approx g(z)$ in C with $f \neq g$ we have $x_1 \neq x_2$.

Inspection of the unskolemization procedure defined in [6] shows that properties (i) and (ii) are sufficient to ensure that unskolemization is successful, property (iii) enables us to show that the other two properties are preserved in inference steps. □

The theorem can be extended to the case of basic Sahlqvist formulae, obtained from definite Sahlqvist implications by applying boxes and disjunctions to formulae not sharing predicate variables.

In contrast to definite Sahlqvist antecedents, Sahlqvist antecedents can include disjunction as a connective. This makes the proof of completeness of SCAN with respect to Sahlqvist implications much more involved. The cornerstone of our proof is the notion of a *chain*.

Let $(t_1, \ldots, t_n)$ be an ordered sequence of pairwise distinct terms. A *chain* C over $(t_1, \ldots, t_n)$ is a clause containing only literals of the form $(\neg)sR_{k_i}t$ and $(\neg)Q_j(u)$ such that the following three conditions are satisfied:

(1) for every i, $1 \leq i \leq n-1$, either $\neg t_i R_{k_i} t_{i+1}$ or $t_i R_{k_i} t_{i+1}$ is in C;
(2) for every $(\neg)uR_{k_i}v \in C$, $u = t_j$ and $v = t_{j+1}$ for some j, $1 \leq j \leq n-1$;
(3) for every $(\neg)Q_j(u) \in C$, $u = t_j$ for some j, $1 \leq j \leq n$.

Lemma 1. *Let C be a chain over $(t_1, \ldots, t_n)$. Then there does not exist an ordered sequence $(s_1, \ldots, s_m)$ of pairwise distinct terms which is distinct from $(t_1, \ldots, t_n)$ such that C is also a chain over $(s_1, \ldots, s_m)$.*

The *length* of a chain C over $(t_1, \ldots, t_n)$ is n. Note that by Lemma 1 the chain C uniquely determines $(t_1, \ldots, t_n)$. So, the length of a chain is a well-defined notion.

The link between the clauses we obtain from translating Sahlqvist formulae or modal formulae, in general, and chains is not as straightforward as one may hope. For example, consider the Sahlqvist antecedent $\neg q_1 \vee \Box q_2 \vee \Box q_3$. The clausal form of its translation consists of the single clause

$$\neg Q_1(a) \vee \neg a\ Rx \vee Q_2(x) \vee \neg a\ Ry \vee Q_3(y).$$

It is straightforward to check that we cannot arrange the terms a, x, and y in an ordered sequence S such that the whole clause would be a chain over S. Instead the clause consists of at least two chains: $\neg Q_1(a) \vee \neg a\ Rx \vee Q_2(x)$ over (a, x) and $\neg a\ Ry \vee \neg Q_3(y)$ over (a, y), or alternatively, $\neg a\ Rx \vee Q_2(x)$ over (a, x) and $\neg Q_1(a) \vee \neg a\ Ry \vee \neg Q_3(y)$ over (a, y). However, we could also divide the clause into three or more chains, for example, $\neg Q_1(a)$ over (a), $Q_2(x)$ over (x), $Q_3(y)$ over (y), $\neg a\ Rx$ over (a, x) and $\neg a\ Ry$ over (a, y).

In the following we will only consider *maximal chains*. A chain C over $(t_1, \ldots, t_n)$ is *maximal* with respect to a clause D iff C is a variable indecomposable subclause of D and there is no chain C' over $(s_1, \ldots, s_m)$, $m > n$, such that C' is a subclause of D and for every i, $1 \leq i \leq n$, $t_i \in \{s_1, \ldots, s_m\}$.

Under this definition our example clause can only be partitioned into three maximal chains, namely, $\neg Q_1(a)$ over (a), $\neg a\ Rx \vee Q_2(x)$ over (a, x), and $\neg a\ Ry \vee Q_3(y)$ over (a, y). We can see the obvious link between the modal subformula of $\neg q_1 \vee \Box q_2 \vee \Box q_3$ and these three maximals. So, it makes sense to say that the first chain *is associated with* the negative formula $\neg q_1$, the second *is associated with* from the boxed atom $\Box q_2$, and the third from the boxed atom $\Box q_3$. In general, more than one maximal chain can be associated with a single negative formula, while exactly one maximal chain is associated with a boxed atom. We call a maximal chain which is associated with a boxed atom a *positive chain*, while all the chains associated with a negative formula are called *negative chains*.

It turns out that in the case of boxed atoms, the clauses we obtain have another important property: The clauses consist of a single maximal chain which is *rooted*. A chain C over $(t_1, \ldots, t_n)$ is *rooted* iff t_1 is a ground term.

Lemma 2. *Let φ be a Sahlqvist implication. Then any clause C in $\mathrm{Cls}(\neg\Pi(\varphi))$ can be partitioned into a collection $\mathcal{D}$ of maximal chains. For any two maximal chains D and D' in $\mathcal{D}$, either D and D' are identical or they share at most one variable. In addition, if a maximal chain D in $\mathcal{D}$ is associated with a boxed atom in φ, then D is rooted and shares no variables with the other maximal chains in $\mathcal{D}$.*

Theorem 3. *Given any Sahlqvist implication φ,* SCAN* *effectively computes a first-order formula α_φ logically equivalent to φ.*

Proof (Sketch). Let $\varphi = A \to P$. We know that $\mathrm{Cls}(\neg\Pi(\varphi)) = \mathrm{Cls}(\mathrm{ST}(A, a)) \cup \mathrm{Cls}(\mathrm{ST}(\neg P, a))$, where a is a Skolem constant. We also know that all clauses in $N_0 = \mathrm{Cls}(\neg\Pi(\varphi))$ satisfy the conditions stated in Lemma 2, in particular, any clause C in $\mathrm{Cls}(\neg\Pi(\varphi))$ can be partitioned into a collection $\mathcal{D}$ of maximal chains.

We introduce some additional notation for chains. We know that a chain associated with a boxed atom contains exactly one positive **Q**-literal $Q_i(t_i)$, where t_i is either a Skolem constant or a variable, and in the following we denote such a chain by $C^+[Q_i(t_i)]$ or $C_j^+[Q_i(t_i)]$. Chains associated with a negative formula may contain one or more negative **Q**-literals $\neg Q_1(t_1), \ldots, \neg Q_n(t_n)$, where each t_i is either a Skolem constant, a variable, or a Skolem term of the form $f(x)$. We denote these chains by $C^-[\neg Q_1(t_1), \ \ldots, \neg Q_n(t_n)]$ or $C_j^-[\neg Q_1(t_1), \ \ldots, \neg Q_n(t_n)]$. By $C^+[\top]$ we denote the clause we obtain by removing the **Q**-literal $Q_i(t_i)$ from the chain $C^+[Q_i(t_i)]$. Analogously, $C^-[\top, \ \ldots, \neg Q_n(t_n)]$ denotes the clause obtained by removing $\neg Q_1(t_i)$ from the chain $C^-[\neg Q_1(t_1), \ \ldots, \neg Q_n(t_n)]$.

Unlike in the case of definite Sahlqvist implications, since a clause in N_0 can contain more than one positive **Q**-literal, inference steps by C-factoring are possible. Such an inference step would derive a clause $D_1 \vee C_1^+[Q(t_1)] \vee C_2^+[\top] \vee t_1 \not\approx t_2$ from a clause $D_1 \vee C_1^+[Q(t_1)] \vee C_2^+[Q(t_2)]$. Since t_1 and t_2 are either variables or constants, the constraint $t_1 \not\approx t_2$ can take the forms

$$b \not\approx z, \quad b \not\approx c, \quad y \not\approx z, \quad \text{or} \quad y \not\approx c.$$

In all cases, except where b and c are distinct constants, a most general unifier σ of t_1 and t_2 exists, and constraint elimination replaces $D_1 \vee C_1^+[Q(t_1)] \vee C_2^+[\top] \vee t_1 \not\approx t_2$ by $(D_1 \vee C_1^+[Q(t_1)] \vee C_2^+[\top])\sigma$. Note that this clause is identical to $D_1 \vee (C_1^+[Q(t_1)] \vee C_2^+[\top])\sigma$, that is, the subclause D_1 is not affected by the inference step nor does it influence the result of the inference step. A problem occurs in the following situation: The clause $Q(a) \vee \neg R(a, x) \vee Q(x)$ is a chain over (a, x) and a C-factoring step is possible which derives $Q(a) \vee \neg R(a, x) \vee a \not\approx x$. This C-factor is simplified by constraint elimination to $Q(a) \vee \neg R(a, a)$. However, an **R**-literal like $\neg R(a, a)$ with two identical arguments is not allowed in a chain. We could modify the definition of a chain to allow for these literals, but it is simpler to consider a clause like $Q(a) \vee \neg R(a, a)$ as shorthand for $Q(a) \vee \neg R(a, x) \vee a \not\approx x$.

It is important to note that the condition that a maximal chain associated with a boxed atom does not share any variables with other chains may no longer be true for C-factors. For example, consider the clause $\neg R(a, x) \vee$

$Q(x) \vee \neg R(a,u) \vee \neg R(u,v) \vee Q(v)$, obtained from $\neg\Pi(\Box q \vee \Box\Box q \to \top)$, which can be partitioned into two maximal chains, $\neg R(a,x) \vee Q(x)$ over (a,x) and $\neg R(a,u) \vee \neg R(u,v) \vee Q(v)$ over (a,u,v). This clause has the C-factor $\neg R(a,x) \vee Q(x) \vee \neg R(a,u) \vee \neg R(u,v) \vee v \not\approx x$. Constraint elimination replaces this C-factor by $\neg R(a,x) \vee Q(x) \vee \neg R(a,u) \vee \neg R(u,x)$ which can be partitioned into two maximal chains $\neg R(a,x) \vee Q(x)$ over (a,x) and $\neg R(a,u) \vee \neg R(u,x)$ over (a,u,x) that share the variable x. Let us call such clauses *factored positive chains.*

Note that the length of chains in a C-factor is the length of chains in the premise clause. Also, the depth of terms in a C-factor is the same as in the premise clause. Let there be c_p positive chains in the clauses of N_0. Then we can potentially derive $2^{c_p} - 1$ factored positive chains.

A C-resolvent can only be derived from a clause $D_1 \vee C^+[Q_i(t_i)]$ and a clause $D_2 \vee C^-[\neg Q'_1(t'_1), \ \ldots, \neg Q'_n(t'_n)]$ where one of the Q'_j, $1 \leq j \leq n$, is identical to Q_i. Without loss of generality, we assume that $Q_i = Q'_1$. Then the resolvent is $D_1 \vee C^+[\top] \vee D_2 \vee C^-[\top, \neg Q'_2(t'_2), \ \ldots, \neg Q'_n(t'_n)] \vee t_i \not\approx t'_1$. The term t_i will either be a Skolem constant b or a variable y, while the term t'_1 can either be a Skolem constant c, a variable z or a Skolem term $f(z)$. Thus, $t_i \not\approx t'_1$ has one of the following forms:

$$b \not\approx z, \quad b \not\approx c, \quad b \not\approx f(z), \quad y \not\approx z, \quad y \not\approx c, \quad \text{or} \quad y \not\approx f(z),$$

If t_i and t'_1 are unifiable by a most general unifier σ, then constraint elimination replaces the C-resolvent by $(D_1 \vee C^+[\top] \vee D_2 \vee C^-[\top, \neg Q'_2(t'_2), \ \ldots, \neg Q'_n(t'_n)])\sigma$, which is identical to $D_1 \vee D_2 \vee C^+[\top] \vee (C^-[\top, \neg Q'_2(t'_2), \ \ldots, \neg Q'_n(t'_n)])\sigma$. If t_i is a variable, then the t_i and t'_1 must be unifiable, since t_i cannot occur in t'_1. Furthermore, if t_i and t'_1 are unifiable, then the most general unifier is either the identity substitution or a substitution replacing t_i by t'_1. However, if t_i and t'_1 are not unifiable, then the constraint $t_i \not\approx t'_1$ cannot be eliminated. In this case t_i must be a Skolem constant b and t'_1 is either a Skolem constant c distinct from b or a Skolem term $f(z)$. Again, no terms deeper than terms in N_0 occur.

We will focus on the union of negative chains that occur within a single clause and share variables. We call these *joined negative chains.* Joined negative chains are variable indecomposable subclauses of the clauses in which they occur and they are variable-disjoint from the rest of the clause. A C-resolution inference step involves one such joined negative chain and one factored positive chain, and the result is again a joined negative chain. Let c_n be the number of joined negative chains in N_0 and let there be at most n_Q occurrences of $\mathbf{Q}$-literals in any joined negative chain. Then at most $c_n \times n_Q \times 2^{c_p}$ joined negative chains can be derived by one or more C-resolution inference steps.

Each clause in a derivation $N_0, N_1, \ldots$ is a collection of factored positive chains and joined negative chains without duplicates modulo variable renaming. Any clause containing duplicates modulo variable renaming would immediately be replaced by its condensation. Given there are $2^{c_p} - 1$ factored positive chains and $c_n \times n_Q \times 2^{c_p}$ joined negative chains. We can derive at most $2^{c_n \times n_Q \times 2^{c_p} + 2^{c_p} - 1}$ distinct clauses. This ensures the termination of the derivation.

As to reverse Skolemization, again the Q_i-literals in N have one of the forms

$$Q_i(b), \quad Q_i(y), \quad \neg Q_i(z), \quad \neg Q_i(c) \quad \text{or} \quad \neg Q_i(f(z)),$$

where b denotes any Skolem constant, f denotes any Skolem function, and y and z arbitrary variables. The possible forms of inequality literals in all C-resolvents and C-factors are then

$$b \not\approx z, \quad b \not\approx c, \quad b \not\approx f(z), \quad y \not\approx z, \quad y \not\approx c, \quad \text{or} \quad y \not\approx f(z).$$

I.e. the form of inequality literals is $s \not\approx t$, where s and t are variable-disjoint, s is either a variable or a constant, and t is either a variable, a constant or a Skolem term of the form $f(x)$. What is again crucial is that no derived clause contains two inequations $y \not\approx s$ and $y \not\approx t$, where s and t are compound terms. This cannot happen. Consequently, restoration of the quantifiers can always be successfully accomplished during reverse Skolemization [6].

Finally, the general case of a Sahlqvist formula φ obtained from Sahlqvist implications by freely applying boxes and conjunctions, as well as disjunctions to formulae sharing no common propositional variables can be dealt with by induction of the number of applications of such disjunctions. The basis of the induction (with no such applications) has been established by Theorem 3.

We can now state the main theorem.

Theorem 4 (Completeness with Respect to Sahlqvist Formulae). *Given any Sahlqvist formula φ,* SCAN* *effectively computes a first-order formula α_φ logically equivalent to φ.*

5 Conclusion

Sahlqvist's theorem [16] can be seen as a milestone for a particular kind of classification problem in modal logics, namely, for which modal formulae can we establish correspondence and completeness results.

At first sight the work on SCAN seems to be unrelated to this classification problem, since SCAN tries to provide a general automated approach to establishing correspondence results while, as already noted, it is not complete for all modal formulae which have a first-order correspondent.

It is interesting to compare this situation to that in first-order logic. The classification problem in first-order logic is to identify fragments of first-order logic which are decidable/undecidable. It has been shown in recent years that general automated approaches to the satisfiability problem in first-order logic, in particular approaches based on the resolution principle, also provide decision procedures for many decidable fragments [7]. They can also provide a good starting point for establishing new decidable fragments of first-order logic.

In this paper we have applied similar approach to SCAN, showing that this general automated approach computes first-order correspondents for the class of Sahlqvist formulae. In fact, the class of modal formulae for which SCAN is successful is strictly larger than the class of Sahlqvist formulae (a witness being the formula given in the introduction). An important open problem is to find a purely logical characterization of that class.

Interestingly, since SCAN is based on a variation of the resolution principle, we were able to employ techniques used for establishing decidability results based on the resolution principle in first-order logic, to obtain our result.

It can be expected that corresponding results can also be obtained for other classes of modal formulae, and even classes for which until now no correspondence and completeness results have been established.

References

[1] W. Bibel and P. H. Schmitt, editors. *Automated Deduction – A Basis for Applications, Vol. I–III.* Kluwer, 1998. 150

[2] P. Blackburn, M. de Rijke, and V. Venema. *Modal Logic.* Cambridge Univ. Press, 2001. 150

[3] A. Chagrov and M. Zakharyaschev. *Modal Logic*, volume 35 of *Oxford Logic Guides.* Clarendon Press, Oxford, 1997. 150

[4] M. de Rijke, and Y. Venema Sahlqvist's Theorem For Boolean Algebras with Operators with an Application to Cylindric Algebras, *Studia Logica*, 54:61–78, 1995. 149

[5] P. Doherty, W. Lukaszewics, and A. Szalas. Computing circumscription revisited: A reduction algorithm. *Journal of Automated Reasoning*, 18(3):297–336, 1997. 150

[6] T. Engel. Quantifier Elimination in Second-Order Predicate Logic. MSc thesis, Saarland University, Saarbrücken, Germany, 1996. 156, 160

[7] C. G. Fermüller, A. Leitsch, U. Hustadt, and T. Tammet. Resolution decision procedures. In *Handbook of Automated Reasoning*, pp. 1791–1849. Elsevier, 2001. 160

[8] D. M. Gabbay and H. J. Ohlbach. Quantifier elimination in second-order predicate logic. *South African Computer Journal*, 7:35–43, 1992. 150, 151, 154

[9] V. Goranko and D. Vakarelov. Sahlqvist formulas unleashed in polyadic modal languages. In *Advances in Modal Logic*, vol. 3, pp. 221-240. World Scientific, 2002. 149, 151

[10] B. Jónsson. On the canonicity of Sahlqvist identities. *Studia Logica*, 53:473–491, 1994. 149

[11] M. Kracht. How completeness and correspondence theory got married. In *Diamonds and Defaults*, pp. 175–214. Kluwer, 1993. 149

[12] A. Nonnengart. Strong skolemization. Research Report MPI-I-96-2-010, Max-Planck-Institut für Informatik, Saarbrücken, 1996. 151

[13] A. Nonnengart, H. J. Ohlbach, and A. Szalas. Quantifier elimination for second-order predicate logic. In *Logic, Language and Reasoning: Essays in honour of Dov Gabbay.* Kluwer, 1999. 150

[14] A. Nonnengart and C. Weidenbach. Computing small clause normal forms. In *Handbook of Automated Reasoning*, pp. 335–367. Elsevier, 2001. 151

[15] A. Robinson and A. Voronkov, editors. *Handbook of Automated Reasoning.* Elsevier Science, 2001. 150

[16] H. Sahlqvist. Completeness and correspondence in the first and second order semantics for modal logics. In *Proc. of the 3rd Scandinavian Logic Symposium, 1973*, pp. 110–143. North-Holland, 1975. 149, 151, 160

[17] G. Sambin and V. Vaccaro. A new proof of Sahlqvist's theorem on modal definability and completeness. *Journal of Symbolic Logic*, 54(3):992–999, 1989. 149

[18] A. Szałas. On the correspondence between modal and classical logic: An automated approach. *Journal of Logic and Computation*, 3(6):605–620, 1993. 150

[19] J. van Benthem. *Modal Logic and Classical Logic*. Bibliopolis, 1983. 149

Relations and GUHA-Style Data Mining II

Petr Hájek*

Institute of Computer Science, Academy of Sciences of the Czech Republic
Pod Vodárenskou věží 2, 182 07 Prague 8, Czech Republic

Abstract. The problem of representability of a (finite) Boolean algebra with an additional binary relation by a data matrix (information structure) and a binary generalized quantifier is studied for various classes of (associational) quantifiers. The computational complexity of the problem for the class of all associational quantifiers and for the class of all implicational quantifiers is determined and the problem is related to (generalized) threshold functions and (positive) assumability.

1 Introduction

Orlowska and Düntsch [1] study information relations given by data (e.g. objects x, y are similar, etc.), Boolean algebras with operators, and their representability by data ("information structures" in their terminology).

GUHA (General Unary Hypotheses Automaton) is a method (existing and developed since the 1960's) of what is presently called data mining (see e.g. [2, 4, 13, 14], for theoretical foundations see particularly [3] and for recent contributions see e.g. [6, 7, 10, 11]). Here we shall be only interested in some aspects of theoretical foundations of GUHA (which may be useful as foundations of other data mining methods as well); namely, in generalized predicate calculi for expressing dependencies hidden in data, notably a sort of finite model theory with generalized quantifiers (mostly two-dimensional), and Boolean algebras with additional relations given by a data set and a generalized quantifier. Our main problem is to find conditions under which a (finite) Boolean algebra with an additional binary relation is (isomorphic to) an algebra given in this way (by some data and quantifier). This is hoped to be a possible way of mutual influence of the GUHA approach and the approach of Orlowska et al.

2 Main Notions

For the reader's convenience we survey here the main notions of the GUHA-style theory of observational logical calculi (calculi for speaking about observed data). The reader familiar with some GUHA theory (in particular with my Relmics 6 paper [5]) may skip this section.

Fix a finite set of unary predicates $P_1, \ldots, P_n$ – names of attributes of objects (for simplicity we shall deal only with two-valued, i.e. Boolean attributes). Also

* Partial support by the COST Action 274 – TARSKI is acknowledged.

R. Berghammer et al. (Eds.): RelMiCS/Kleene-Algebra Ws 2003, LNCS 3051, pp. 163–170, 2004.

fix an object variable x and consider only formulas with no variable except x. Open formulas are Boolean combinations of atomic formulas. Thus for example, $(P_1(x)\&\neg P_3(x))\vee P_7(x)$ is an open formula. A *data matrix* is a rectangular matrix $\mathbf{M} = (u_{ij})_{i=1,\dots m}^{j=1,\dots n}$ of zeros and ones with n columns; its i-th row is understood as information on i-th object; $u_{ij} = 1$ says "the i-th object has the j-th (atomic) property", in other words, "the i-th object satisfies the atomic formula $P_j(x)$". Using the obvious semantics of connectives we define *satisfaction* of an open formula $\varphi(x)$ by the i-th object (thus open formulas define atomic and composed properties of objects).

For each open formula φ and each data matrix $\mathbf{M}$, let X_φ be the set of all objects satisfying φ in $\mathbf{M}$. Let $Fr_{\mathbf{M}}(\varphi)$ be the cardinality of X_φ. A subset of $\mathbf{M}$ is *definable* if it is X_φ for some open formula φ. Definable subsets of $\mathbf{M}$ form a Boolean formula (with respect to union, intersection and complement); call it $Def(\mathbf{M})$. Each pair of open formulas φ, ψ defines its four-fold table
$a = Fr(\varphi\&\psi), b = Fr(\varphi\&\neg\psi)$
$c = Fr(\neg\varphi\&\psi), d = Fr(\neg\varphi\&\neg\psi)$
($\mathbf{M}$ omitted); put also $r = a + b, s = c + d, k = a + c, l = b + d$ (marginal sums), $m = a + b + c + d$ (cardinality of $\mathbf{M}$).

The semantics of a binary quantifier $\sim$ (applied to a pair of formulas) is given by a four-argument function $Tr_\sim$ (or just Tr) assigning to each quadruple (a, b, c, d) of natural numbers with $a + b + c + d > 0$ the truth value 1 or 0. The quantifier $\sim$ describes a relation between two (definable) properties; the corresponding formula may be written $(\sim x)(\varphi(x), \psi(x))$ (in the analogy to classical unary quantifier $(\forall x)\varphi(x)$) or simply $\sim(\varphi, \psi)$ or (most often) $\varphi \sim \psi$.

Caution: Don't think of $\sim$ as of a connective; $\varphi \sim \psi$ does not express any property of individual objects but a relation between φ, ψ defined as follows: let (a, b, c, d) be the four-fold table of φ in $\mathbf{M}$. Then $\varphi \sim \psi$ is true in $\mathbf{M}$ iff $Tr_\sim(a, b, c, d) = 1$; $\varphi \sim \psi$ is false in $\mathbf{M}$ iff $Tr_\sim(a, b, c, d) = 0$. We further define: $\sim$ is *monotone* if Tr is monotone in each variable. Furthermore, the quantifier $\sim$ is *associational* if Tr is non-decreasing in a, d and non-increasing in b, c.
In particular: the quantifier is *implicational* if it is associational and its truth function Tr is independent of c, d. An associational quantifier expresses positive association; if $\sim$ is associational and $\varphi \sim \psi$ is true in your data then it means that for (sufficiently) many objects the truth values of φ, ψ coincide and for (sufficiently) few objects they differ. If $\sim$ is implicational then we only pay attention to objects satisfying φ; from those objects (sufficiently) many objects satisfy ψ and (sufficiently) few objects satisfy $\neg\psi$.

We recall some "classical" examples:

(1) Implicational quantifiers.
$FIMPL_{p,s}$: $Tr(a, b, c, d) = 1$ iff $a/(a + b) \geq p$ and $a \geq s$.
$LIMPL_{p,\alpha}$: $Tr(a, b, c, d) = 1$ iff $\sum_{i=a}^{r} \binom{r}{i} p^i (1-p)^{r-i} \leq \alpha$
If $\sim$ is $FIMPL_{p,s}$ then $\varphi \sim \psi$ just says that at least s objects satisfy $\varphi\&\psi$ and the relative frequency of objects satisfying ψ among those satisfying φ is at least p. ($FIMPL$ is called the founded implicational quantifier, see

already [2], and was much later rediscovered by Agrawal in what he calls "association rules".)

$LIMPL_{p,x}$ (lower critical implicational quantifier) is statistically motivated: if $\sim$ is $LIMPL_{p,x}$ then $\varphi \sim \psi$ says that the hypothesis that for objects satisfying φ the probability $P(\psi)$ is a least p can be accepted with the significance (risk) α (α small).

(2) Symmetric associational quantifiers:

$SIMPLE_h$: $a \cdot d \geq h \cdot b \cdot c \quad (h \geq 1)$

$FISHER_\alpha : \sum_{i=0}^{\min(b,c)} \frac{(a+b)!(c+d)!(a+c)!(b+d)!}{m!(a+i)!(b-i)!(c-i)!(d+i)!} \leq \alpha$ and $ad > bc$

For $\sim$ being $SIMPLE_1$, $\varphi \sim \psi$ says that $ad > bc$, or equivalently, $\frac{a}{a+b} > \frac{c}{c+d}$, thus the relative frequency of ψ among objects satisfying φ is bigger than the relative frequency of ψ among objects satisfying $\neg\varphi$. $SIMPLE_h$ for $h > 1$ is still stronger. The truth function of $FISHER_\alpha$ is based on the statistical test of positive dependence: $\varphi \sim \psi$ says that the statistical hypothesis saying $P(\psi|\varphi) > P(\psi)$ can be accepted on the significance level α. See [3] for the corresponding theory, in particular for "frame assumptions" of statistical hypothesis testing involved.

Let us mention also unary quantifiers expressing in some way "many", i.e. Quantifiers μ such that $(\mu x)\varphi$ (or just $\mu\varphi$) says "many objects have φ". In fact, each implicational quantifier (denote it $\Rightarrow^*$) defines a quantifier "many" if $(\mu x)\varphi$ is defined as $\top \Rightarrow^* \varphi$, where $\top$ stands for a formula "true" satisfied by each object of each data set. The truth function of this μ is defined from the truth functions of $\Rightarrow^*$ as $Tr_\mu(a,b) = Tr_{\Rightarrow^*}(a,b,0,0)$. Call each such quantifier μ a (unary) *multitudinal* quantifier.

3 Problem(s) of Representability

A given associational quantifier $\sim$ defines a binary relation of association $R_\sim$ on $Def(\mathbf{M})$ as follows: for any open formulas φ, ψ, $R_\sim(X_\varphi, X_\psi)$ holds iff $\varphi \sim \psi$ is true in $\mathbf{M}$. For any Boolean algebra $\mathbf{B}$ define the following partial order $\prec$ between pairs of elements of $\mathbf{B}$: $(u_1, v_1) \prec (u_2, v_2)$ iff

$$u_2 \wedge v_2 \geq u_1 \wedge v_1, u_2 \wedge -v_2 \leq u_1 \wedge -v_1,$$

$$-u_2 \wedge v_2 \leq -u_1 \wedge v_1, -u_2 \wedge -v_2 \geq -u_1 \wedge -v_1$$

Now a binary relation R on $\mathbf{B}$ is called *weakly associational* if, for each $u_1, v_1, u_2, v_2 \in \mathbf{B}$, $R(u_1, v_1)$ and $(u_1, v_1) \prec (u_2, v_2)$ implies $R(u_2, v_2)$. Evidently, if $\sim$ is associational and $\mathbf{B} = Def(\mathbf{M})$ then $R_\sim$ defined above is weakly associational.

The problem reads as follows: Given a (finite) Boolean algebra $\mathbf{B}$ and a binary weakly associational relation R on B, is it isomorphic to $(Def(\mathbf{M}), R_\sim)$ for some data $\mathbf{M}$ and some associational quantifier $\sim$?

It has several variants:

- $\sim$ uniquely given in advance
- $\sim$ may be from a parametric class, e.g. $FIMPLE_{p,s}$
- $\sim$ may be from a broad class, e.g. any implicational
- R may be only partially defined: one gives two disjoint sets of pairs – R^+ is the set of positive pairs, R^- set of negative pairs.

  R may be any weakly associational relation containing R^+, disjoint from R^-.

Remark. Recall that if you have predicates $P_1, \ldots, P_k$ then an *elementary conjunction* is a formula $\lambda_1 \& \ldots \& \lambda_k$ where each λ_i is either $P_i(x)$ or $\neg P_i(x)$. Let $\mathbf{B}$ be a finite Boolean algebra with k atoms (and 2^k elements). Take $\ulcorner \log_2(k) \urcorner$ predicates and identify atoms of $\mathbf{B}$ with k elementary conjunctions of your predicates. Call them *selected.*

Note that if $\mathbf{B}$ is isomorphic to some $Def(\mathbf{M})$ then it is isomorphic to a $Def(\mathbf{M})$ with $Fr(\kappa) > 0$ for all selected EC's and $Fr(\kappa) = 0$ for all other EC's. $\mathbf{M}$ is then fully determined by weights $w_1, \ldots, w_k$ of selected atoms, w_i positive natural numbers.

Observation. Let $u = (u_1, \ldots, u_k) \in \{0,1\}^k$ be the characteristic vector of atoms whose disjunction is φ; then in $\mathbf{M}$, $Fr(\varphi) = w \circ u$ (scalar product).

The representation problem was first formulated in my paper [5] for Relmics 6 meeting in Tilburg.

4 Complexity of Representability

Let $\mathbf{B}$ be a Boolean algebra with k atoms presented by k selected elementary conjunctions of literals built from $\ulcorner \log(k) \urcorner$ predicates as above. Each element of $\mathbf{B}$ can be expressed as a disjunction of $\leq k$ atoms. (Its length is polynomial in k). Assume we have two disjoint sets R^+, R^- of positive and negative pairs of elements of $\mathbf{B}$ as above. In fact: these pairs are given by pairs of open formulas, each of length polynomial in k.

The representability of $(\mathbf{B}, R^+, R^-)$ by a $(Def(\mathbf{M}), R_\sim)$ becomes equivalent to the satisfiability of

$$\bigwedge\{\varphi \sim \psi / (\varphi, \psi) \in R^+\} \& \bigwedge\{\neg(\varphi \sim \psi) | (\varphi, \psi) \in R^-\}.$$

Call the last formula $\Phi(R^+, R^-)$.

Theorem 1. *The present representability problem, for $\sim$ varying over all implicational quantifiers, is NP-complete. Similarly for $\sim$ varying over all associational quantifiers.*

Proof. By [5], the satisfiability problem of formulas of the form $\Phi(R^+, R^-)$ with $\sim$ ranging on all implicational quantifiers is NP-complete.

By [8], the same holds for associational quantifiers.

5 A Counterexample

Recall the ordering $(u_1, v_1) \prec (u_2, v_2)$ defined above. A necessary condition for the representability by an associational quantifier reads that there are no u_1, u_2, v_1, v_2 satisfying $(u_1, v_1) \prec (u_2, v_2)$ such that $R^+(u_1, v_1)$ and $R^-(u_2, v_2)$ (It is impossible that (u_1, v_1) is a positive pair and the (u_2, v_2) is a negative pair.) For representability by an implicational quantifier the necessary condition results by deleting the second half (beginning by $-u_2 \wedge v_2$ from the definition of $\prec$).

For one-dimensional multitudinal quantifier $Many^*$ the condition further simplifies: all u's are 1_B (true) and everything reduces to: if $v_2 \geq v_1$ then $R^+(v_1)$ implies not $R^-(v_2)$.

It is easy to show that this condition is *not* sufficient for representability: assume 5 atoms, let $u = (1,1,1,0,0)$ represent the join of the atoms 0, 1, 2 and let $v = (10011)$ represent the join of atoms 0, 3, 4. Let $R^+ = \{u, v\}$, let R^- consist of all other three-atom elements containing the atom 0. Clearly, our condition is satisfied. Let $w = (w_0, \ldots, w_4)$ be given and let k be such that w, k realize $(\mathbf{B}, R^+, R^-)$, i.e.

$$w_0 + w_1 + w_2 \geq k, w_0 + w_3 + w_4 \geq k,$$

but $w_0 + w_i + w_j < k$ for $(i,j) = (1,3), (1,4), (2,3), (2,4)$. From the first two inequalities we get

$$2w_0 + w_1 + w_2 + w_3 + w_4 \geq 2k,$$

from the last four we get

$$2w_0 + w_1 + w_2 + w_3 + w_4 < 2k,$$

a contradiction.

Thus we get a new algebraic condition: let $x, y_1, \ldots, y_4$ be pairwise disjoint elements of B and let R be the relation given by a unary multitudinal quantifier $(R = R^+; \mathbf{B} - R = R^-)$.

If $[R(x \vee y_1 \vee y_2)$ and $R(x \vee y_3 \vee y_4)]$ where x, y_1, y_2, y_3, y_4 are pairwise disjont, then

$$R(x \vee y_1 \vee y_3) \text{ or } R(x \vee y_1 \vee y_4) \text{ or } R(x \vee y_2 \vee y_3) \text{ or } R(x \vee y_2 \vee y_4).$$

6 Threshold Functions

Next we relate our problem to the problem of threshold boolean functions. Recall that elements of B are represented by vector $u \in \{0,1\}^k$ and that the cardinality of the corresponding definable set in the concerned data M is $w \circ u$ when w is a vector of frequencies of atoms ($\circ$ denotes the scalar product).

We recall some definitions from [9]. A function $f : \{0,1\}^n \to \{0,1\}$ is a *threshold function* if there exists a $t \in Re$ and some $w \in Re^n$ such that for all $u \in \{0,1\}^n$,

$$f(u) = 1 \text{ iff } w \circ u \geq t.$$

f is *summable* if there are finitely many elements ($m > 1$) $a_1, \ldots, a_m$ with $f(a_i) = 1$ and $b_1, \ldots, b_m$ with $f(b_i) = 0$ such that $\sum a_i = \sum b_i$

Theorem 2. (Asumability theorem, Elgot and Chow, see [9] Theorem 7.2.1). f is a threshold function iff f is not summable.

This is almost what we want for *Many*; but we need non-negative (positive) and natural (rational) weights w_i. The existence of rational weights follows from the existence of real weights by continuity.

Let us define: f is a *positive* threshold function if

$$f(u) = 1 \text{ iff } w \circ u \geq t$$

for some *non-negative* t, w (cf. [12]). Analyzing the proof of the asummability theorem in [9] we are lead to the following definition:

A function $f : \{0,1\}^n \to \{0,1\}$ is *positively summable* if $\exists a_1, \ldots, a_m$ with $f(a_i) = 1$, $\exists b_1, \ldots, b_m$ with $f(b_i) = 0$ such that $\sum a_i \leq \sum b_i$. (Note that a_i, b_i are vectors of zeros and ones; the sum is understood coordinatewise.)

Theorem 3. (Positive asummability theorem.) f is a positive threshold function iff it is not positively summable (i.e. it is positively asummable).

Proof. Let f be a positive threshold function w.r.t w, t and let a_i be big, b_i small w.r.t. f; i.e. $f(a_i) = 1, f(b_i) = 0$ thus $w \circ a_i > w \circ b_i, \sum w \circ a_i > \sum w \circ b_i$, $w \circ \sum a_i > w \circ \sum b_i, w > 0$ (recall that all sums are sums of vectors!). Thus $\sum a_i \leq \sum b_i$ is impossible and f is positively asummable.

Conversely, assume that f is not a positive threshold function w.r.t. any w, t, hence the system

$$w \circ (a_p - b_q) > 0, w \circ e_i \geq 0$$

(where a_p enumerates all big, b_q all small vectors w.r.t.f and e_i is the i-th unit vector $(0, \ldots, 0, 1, 0, \ldots, 0)$) has no solution. By [9] Lemma 7.2.1, there are $g_{pq}, g_i \geq 0$, *integers* such that $\sum_{p,q} g_{pq} w \circ (a_p - b_q) + \sum g_i w \circ e_i = 0$ for all w. The coefficient of w_i is $\sum_{p,q} g_{p,q}(a_{p_i} + b_{q_i}) + g_i = 0$. Thus for all i, $\sum_p (\sum_q g_{pq}) a_{p_i} \leq \sum_q (\sum_p g_{pq}) b_{q_i}$, briefly, for some non-negative *integers* λ_p, λ'_q we get $\sum_p \lambda_p a_p \leq \sum_q \lambda'_q b_q$ (coordinatewise), which gives positive summability.

Example 1. $g(x_1, \ldots, x_n) = x_1 x_2 \vee x_3 x_4$.
$a_1 = (1100), a_2 = (0011)$ – big
$b_1 = (1010), b_2 = (0101)$ – small
$\sum a_i = \sum b_i = (2222)$

Let $w_1 + w_2 > t, w_3 + w_4 > t$, then $(w_1 + w_3) + (w_2 + w_4) > 2t$, thus (1010) big or (0101) big; also (1001) or (0110) is big, contradiction. Consequently, f is not a threshold function.

Corollary 1. *Let* $\mathbf{B} = 2^n$ *(algebra with* n *atoms) and* $R \subseteq B$ *be an upper segment (*$u \in R$ *and* $u \leq v$ *imply* $v \in R$*).* $(\mathbf{B}, R)$ *is representable by a data matrix and a unary multitudinal quantifier iff the characteristic function* χ_R *of* R *is a positive threshold function, iff* χ_R *is positively asummable.*

We generalize for implicational quantifier: Observe that if $(\mathbf{B}, R)$ is given by an implicational quantifier then

$$R(u,v) \text{ iff } R(u, u \wedge v)$$

for all u, v.

Thus assume $R^+, R^- \subseteq B \times B$, disjoint, $(u,v) \in R^+ \cup R^-$ implies $v \leq u$. Let $p = h_1/h_2$ be a rational number in $\in (0,1)$. Call $(\mathbf{B}, R^+, R^-)$ strongly representable by $FIMPL_p$ if for some weight vector w, of positive numbers

$$\frac{w \circ v}{w \circ u} > p \text{ for } (u,v) \in R^+ \text{and} \frac{w \circ v}{w \circ u} < p \text{ for } (u,v) \in R^-.$$

Clearly, this is equivalent to the following:
$w \circ (h_2 v - h_1 u) > 0$ for $(u,v) \in R^+$,
$w \circ (h_1 u - h_2 v) > 0$ for $(u,v) \in R^-$,
$w \circ e_i > 0$ where $e_i = (0 \ldots 0, 1, 0 \ldots 0)$ (i-th unit vector).

By Theorem 7.2.1 in [9], we get the following

Theorem 4. Let $\mathbf{B}, R^+, R^-, p = h_1/h_2$ be as above. $(\mathbf{B}, R^+, R^-)$ is *not* representable by $FIMPL_p$ iff there is a finite sequence (u_i, v_i) of elements of R^+ (repetition possible) and a finite sequence of $(u'_j, v'_j) \in R^-$ of the same length such that $\sum_i (h_2 v_i - h_1 u_i) \leq \sum_j (h_2 v'_j - h_2 u'_j)$.

Summarizing, we have contributed to the theory of representability of a finite Boolean algebra with a binary relation by some data and an associational quantifier. We determined the computational complexity of the problem and related it to the theory of assumability of Boolean functions.

This contributes to the foundations of GUHA-style data mining based on information structures understood as finite models (in the sense of finite model theory) and analyzed by logical languages with generalized quantifiers. In particular, our results may give some new deduction rules. Research is to be continued.

References

[1] Düntsch I., Orłowska E.: Beyond modalities: Sufficiency and mixed algebras. In: Orlowska et al. ed., Relational methods for computer science applications, Physica Verlag 2001, p.263-286. 163

[2] Hájek P., Havel I., Chytil M.: The GUHA method of automatic hypotheses determination, Computing 1(1966) 293-308. 163, 165

[3] Hájek P., Havránek T.: Mechanizing Hypothesis Formation (Mathematical Foundations for a General Theory), Springer-Verlag 1978, 396 pp. *Free internet version:* www.cs.cas.cz/ hajek/guhabook. 163, 165

[4] Hájek P., Sochorová A., Zvárová J.: GUHA for personal computers, Comp. Stat., Data Arch. 19, pp. 149-153. 163

[5] Hájek P.: Relations in GUHA style data mining. Proc. Relmics 6, Tilburg (The Netherlands) 91-96. 163, 166

[6] Hájek P.: The GUHA method and mining association rules. Proc. CIMA'2001 (Bangor, Wales) 533-539. 163
[7] Hájek P., Holeňa M.: Formal logics of discovery and hypothesis formation by machine. Theoretical Computer 292 (2003) 345-357. 163
[8] Hájek P.: On generalized quantifiers, finite sets and data mining. In (Klopotek et al., ed.) Intelligent Information Processing and Data Mining, Physica Verlag 2003, 489-496. 166
[9] Muroga S.: Threshold logic and its applications, Wiley 1971. 167, 168, 169
[10] Rauch J., Šimůnek M.: Mining for 4ft association rules. Proc. Discovery Science 2000 Kyoto, Springer Verlag 2000, 268-272. 163
[11] Rauch J.: Interesting Association Rules and Multi-relational Association Rules. Communications of Institute of Information and Computing Machinery, Taiwan. Vol. 5, No. 2, May 2002, pp. 77-82. 163
[12] Servedio R. A.: Probabilistic construction of monotone formulae for positive linear threshold functions, unpublished manuscript from 1999, see `http://citeseer.nj.nec.com/354927.html`. 168
[13] GUHA+– project web site `http://www.cs.cas.cz/ics/software.html`. 163
[14] http://lispminer.vse.cz. 163

A Note on Complex Algebras of Semigroups

Peter Jipsen

Chapman University, Orange CA 92866, USA
jipsen@chapman.edu
http://www.chapman.edu/~jipsen/

Abstract. The main result is that the variety generated by complex algebras of (commutative) semigroups is not finitely based. It is shown that this variety coincides with the variety generated by complex algebras of partial (commutative) semigroups. An example is given of an 8-element commutative Boolean semigroup that is not in this variety, and an analysis of all smaller Boolean semigroups shows that there is no smaller example. However, without associativity the situation is quite different: the variety generated by complex algebras of (commutative) binars is finitely based and is equal to the variety of all Boolean algebras with a (commutative) binary operator.

A *binar* is a set A with a (total) binary operation $\cdot$, and in a *partial binar* this operation is allowed to be partial. We write $x \cdot y \in A$ to indicate that the product of x and y exists. A *partial semigroup* is an associative partial binar, i.e. for all $x, y, z \in A$, if $(x \cdot y) \cdot z \in A$ or $x \cdot (y \cdot z) \in A$, then both terms exist and evaluate to the same element of A. Similarly, a *commutative partial binar* is a binar such that if $x \cdot y \in A$ then $x \cdot y = y \cdot x \in A$.

Let (P)(C)Bn and (P)(C)Sg denote the class of all (partial) (commutative) groupoids and all (partial) (commutative) semigroups respectively. For $A \in \mathsf{PBn}$ the *complex algebra of* A is defined as $\mathrm{Cm}(A) = \langle P(A), \cup, \emptyset, \cap, A, \setminus, \cdot \rangle$, where

$$X \cdot Y = \{x \cdot y \mid x \in X,\ y \in Y \text{ and } x \cdot y \text{ exists}\}$$

is the complex product of $X, Y \in \mathrm{Cm}(A)$. Algebras of the form $\mathrm{Cm}(A)$ are examples of *Boolean algebras with a binary operator*, i.e., algebras $\langle B, \vee, 0, \wedge, 1, \neg, \cdot \rangle$ such that $\langle B, \vee, 0, \wedge, 1, \neg \rangle$ is a Boolean algebra and $\cdot$ is a binary operation that distributes over finite (including empty) joins in each argument. A *Boolean semigroup* is a Boolean algebra with an associative binary operator.

For a class $\mathcal{K}$ of algebras, $\mathrm{Cm}(\mathcal{K})$ denotes the class of all complex algebras of $\mathcal{K}$, $\mathbf{H}(\mathcal{K})$ is the class of all homomorphic images of $\mathcal{K}$, and $\mathbf{V}(\mathcal{K})$ is the variety generated by $\mathcal{K}$, i.e., the smallest equationally defined class that contains $\mathcal{K}$.

The aim of this note is to contrast the equational theory of Cm((C)Bn) with that of Cm((C)Sg). It turns out that the former is finitely based while the latter is not.

Lemma 1. *$\mathbf{V}(\mathrm{Cm}(\mathsf{Sg})) = \mathbf{V}(\mathrm{Cm}(\mathsf{PSg})$, $\mathbf{V}(\mathrm{Cm}(\mathsf{CSg})) = \mathbf{V}(\mathrm{Cm}(\mathsf{PCSg}))$, $\mathbf{V}(\mathrm{Cm}(\mathsf{Bn})) = \mathbf{V}(\mathrm{Cm}(\mathsf{PBn})$ and $\mathbf{V}(\mathrm{Cm}(\mathsf{CBn})) = \mathbf{V}(\mathrm{Cm}(\mathsf{PCBn}))$.*

R. Berghammer et al. (Eds.): RelMiCS/Kleene-Algebra Ws 2003, LNCS 3051, pp. 171–177, 2004.

Proof. We prove the first result and note that the argument for the other results is identical. Since $\mathsf{Sg} \subseteq \mathsf{PSg}$, the forward inclusion is obvious.

Let A be a partial semigroup, and define $A_\infty = \langle A \cup \{\infty\}, \cdot \rangle$ where

$$x \cdot y = \begin{cases} xy \text{ if } xy \in A \\ \infty \text{ otherwise.} \end{cases}$$

It is easy to check that $A_\infty \in \mathsf{Sg}$. Define $h : \mathrm{Cm}(A_\infty) \to \mathrm{Cm}(A)$ by $h(X) = X \setminus \{\infty\}$. Then $(X \cdot Y) \setminus \{\infty\} = (X \setminus \{\infty\})(Y \setminus \{\infty\})$ since

$$z = x \cdot y \text{ and } z \neq \infty \quad \text{iff} \quad z = xy \text{ and } x \neq \infty \text{ and } y \neq \infty.$$

Hence h is a homomorphism, and it follows that $\mathrm{Cm}(\mathsf{PSg}) \subseteq \mathbf{H}\mathrm{Cm}(\mathsf{Sg})$. Therefore $\mathbf{V}(\mathrm{Cm}(\mathsf{PSg})) \subseteq \mathbf{V}(\mathrm{Cm}(\mathsf{Sg}))$.

Let Rel be the class of algebras that are isomorphic to algebras of binary relations closed under Boolean operations $(\cup, \cap, \setminus, \emptyset, T)$ and relation composition. The subclass of algebras that are commutative under composition is denoted by CRel.

Note that the top relation T is always transitive. The proof below shows that the equational theory of Rel does not change even if we assume T is also irreflexive (and hence a strict partial order).

Theorem 1. $\mathsf{(C)Rel}$ *is a variety, and* $\mathbf{V}(\mathrm{Cm}(\mathsf{(C)Sg})) = \mathsf{(C)Rel}$.

Proof. The class $\mathsf{(C)Rel}$ is easily seen to be closed under subalgebras and products. The proof that $\mathsf{(C)Rel}$ is closed under homomorphic images is similar to a proof in [2] Theorem 5.5.10 that shows cylindric-relativized set algebras are a variety (see also [6] Theorem 1.5).

Moreover, it is easy to see that $\mathsf{(C)Rel} \subseteq \mathbf{V}(\mathrm{Cm}(\mathsf{P(C)Sg}))$ since the algebra of all subsets of a transitive relation is the complex algebra of a partial semigroup, with ordered pairs as elements, and $(w, x) \cdot (y, z) = (w, z)$ if $x = y$ (undefined otherwise).

To prove the opposite inclusion, we show that any complex algebra of a semigroup can be embedded in a member of Rel. The commutative case follows since if the semigroup is commutative then the image under this embedding will be a member of CRel.

Let S be a semigroup. We would like to find a set U and a collection $\{R_a \subseteq U^2 \mid a \in S\}$ of **disjoint nonempty** binary relations on U such that $R_a \circ R_b = R_{ab}$. If S is a left-cancellative semigroup, we can simply take the Cayley embedding $R_a = \{(x, xa) \mid x \in S\}$. However, if S is not cancellative then this approach does not give disjoint relations, so we take a step-by-step approach and use transfinite induction to build the R_a. A detailed discussion of this method for representing relation algebras can be found in [3] or [4]. Since our setting is somewhat different, and to avoid lengthy definitions, we take a rather informal approach here. To simplify the argument, we will arrange that all the relations are irreflexive and antisymmetric.

Suppose we have an "approximate embedding", by which we mean a collection of disjoint irreflexive antisymmetric relations $R_{a,\kappa}$ on a set U_κ such that $R_{a,\kappa} \circ R_{b,\kappa} \subseteq R_{ab,\kappa}$.

Using the well-ordering principle, we list all the pairs in $R_{ab,\kappa} \setminus (R_{a,\kappa} \circ R_{b,\kappa})$ for all $a, b \in S$, and proceed to extend U_κ and the $R_{a,\kappa}$ so as to eventually obtain $R_a \circ R_b = R_{ab}$, where R_a is the union of all the $R_{a,\kappa}$ constructed along the way. For each $u \neq v$ with $\langle u, v\rangle \in R_{ab,\kappa} \setminus (R_{a,\kappa} \circ R_{b,\kappa})$, choose $w \notin U_\kappa$ and let

$U_{\kappa+1} = U_\kappa \cup \{w\}$

$R'_z = \bigcup\{R_{x,\kappa} \circ \{\langle u, w\rangle\} : xa = z\} \cup \bigcup\{\{\langle w, v\rangle\} \circ R_{y,\kappa} : by = z\}$

$R_{a,\kappa+1} = R_{a,\kappa} \cup \{\langle u, w\rangle\} \cup R'_a$

$R_{b,\kappa+1} = R_{b,\kappa} \cup \{\langle w, v\rangle\} \cup R'_b$ and

$R_{z,\kappa+1} = R_{z,\kappa} \cup R'_z$ if $z \neq a, b$.

For limit ordinals λ, we let $U_\lambda = \bigcup_{\kappa<\lambda} U_\kappa$ and $R_{x,\lambda} = \bigcup_{\kappa<\lambda} R_{x,\kappa}$.

It remains to check that the new relations are still an approximate embedding. By construction, they are disjoint, irreflexive and antisymmetric since $w \notin U_\kappa$. Checking the inclusion $R_{c,\kappa+1} \circ R_{d,\kappa+1} \subseteq R_{cd,\kappa+1}$ involves several cases, depending on whether $c, d \in \{a, b\}$. Since they are similar, we consider only the case $c, d \notin \{a, b\}$. Let $\langle p, q\rangle \in R_{c,\kappa+1} \circ R_{d,\kappa+1}$. Then there exists $r \in U_{\kappa+1}$ such that $\langle p, r\rangle \in R_{c,\kappa+1}$ and $\langle r, q\rangle \in R_{d,\kappa+1}$. If $r \in U_\kappa$ then the conclusion follows from the assumption that $R_{z,\kappa}$ is an approximate embedding. So we may assume $r = w$ (the unique element in $U_{\kappa+1} \setminus U_\kappa$). By construction $\langle p, u\rangle \in R_{xa,\kappa}$ for some x such that $xa = c$ and $\langle v, q\rangle \in R_{by,\kappa}$ for some y such that $by = d$. Since $\langle u, v\rangle \in R_{ab,\kappa}$ it follows that $\langle p, q\rangle \in R_{x(ab)y,\kappa}$. By associativity we have $R_{x(ab)y,\kappa} \subseteq R_{(xa)(by),\kappa+1} = R_{cd,\kappa+1}$, as required.

Finally, to start the construction take U_0 to be a disjoint union of S and $S' = S \times \{0\}$, and for each $a \in S$ define $R_{a,0} = \{\langle a, a'\rangle\}$, where $a' = \langle a, 0\rangle$.

Now the main result follows easily from the "representation theorem" that we have just established. Previously it was known from [8] that the variety generated by complex algebras of groups (i.e., the variety of *group relation algebras*) is not finitely based. In this case the analogous representation theorem states that every group relation algebra is representable, a result that follows directly from Cayley's theorem for groups.

Corollary 1. $\mathbf{V}(\mathrm{Cm}(\mathsf{Sg}))$ *and* $\mathbf{V}(\mathrm{Cm}(\mathsf{CSg}))$ *are not finitely based.*

Proof. In [1] (Theorem 4) Andreka shows that the class Rel (called $\mathbf{R}(\cup, \cap, |, -)$ in [1]) is not finitely axiomatizable, and by the preceding result $\mathsf{Rel} = \mathbf{V}(\mathrm{Cm}(\mathsf{Sg}))$. Andreka's result is proved using a sequence of finite commutative relation algebras (from [7]) such that the Boolean semigroup reducts of these algebras are not in Rel, but the ultraproduct is in CRel. It follows that $\mathsf{CRel} = \mathbf{V}(\mathrm{Cm}(\mathsf{CSg}))$ is also not finitely axiomatizable.

In fact one can find an 8-element commutative Boolean semigroup that is not in $\mathbf{V}(\mathrm{Cm}(\mathsf{Sg}))$: Let A be the finite Boolean algebra with atoms $\{a, b, c\}$, and define

$\cdot$	a	b	c
a	$b \vee c$	$a \vee b$	$a \vee c$
b	$a \vee b$	$a \vee c$	$b \vee c$
c	$a \vee c$	$b \vee c$	$a \vee b$

It is straight forward to check that this operation is associative. The following identity fails in this algebra but holds in $\mathbf{V}(\mathrm{Cm}(\mathrm{Sg}))$: $s \leq t_1 \vee t_2 \vee t_3 \vee t_4 \vee t_5$ where

$$\begin{aligned}
s &= x_{04} \wedge [(x_{02} \wedge (x_{01} \cdot x_{12})) \cdot (x_{24} \wedge (x_{23} \cdot x_{34}))] \\
t_1 &= x_{04} \wedge (x_{01} \cdot \neg x_{14}) \\
t_2 &= x_{01} \cdot [x_{14} \wedge (x_{12} \cdot x_{24}) \wedge (\neg y_{13} \cdot x_{34})] \\
t_3 &= x_{04} \wedge (\neg x_{03} \cdot x_{34}) \\
t_4 &= [x_{03} \wedge (x_{02} \cdot x_{23}) \wedge (x_{01} \cdot \neg z_{13})] \cdot x_{34} \\
t_5 &= x_{01} \cdot [y_{13} \wedge (x_{12} \cdot x_{23}) \wedge z_{13}] \cdot x_{34}
\end{aligned}$$

This identity was derived from an identity of R. Maddux for a closely related relation algebra (see e.g. [6]). To see that it holds in Cm(Sg), let S be a semigroup, and assume $a_{04} \in s \subseteq S$. Then $a_{04} \in x_{04}$ and there exist $a_{02}, a_{01}, a_{12}, a_{24}, a_{23}, a_{34}$ in S such that $a_{ij} \in x_{ij}$ for the given subscripts, $a_{02} \cdot a_{24} = a_{04}$, $a_{01} \cdot a_{12} = a_{02}$ and $a_{23} \cdot a_{34} = a_{24}$. Suppose $a_{04} \notin t_i$ for $i = 1, 2, 3, 4$. It remains to show that $a_{04} \in t_5$. Let $a_{14} = a_{12} \cdot a_{24}$, $a_{13} = a_{12} \cdot a_{23}$ and $a_{03} = a_{02} \cdot a_{23}$. Then $a_{04} = a_{01} \cdot a_{12} \cdot a_{24} = a_{01} \cdot a_{14} \notin x_{01} \cdot \neg x_{14}$, so $a_{14} \in x_{14}$. Moreover, since $a_{02} \notin t_2$, we have $a_{14} \notin \neg y_{13} \cdot x_{34}$, hence $a_{13} \in y_{13}$. Similarly $a_{03} \in x_{03}$ and $a_{13} \in z_{13}$. But now $a_{04} = a_{01} \cdot a_{13} \cdot a_{34} \in t_5$.

The identity fails in the algebra A if one assigns $x_{02} = x_{04} = a$, $x_{23} = x_{24} = b$, $x_{01} = x_{12} = x_{34} = c$, $x_{03} = x_{14} = z_{13} = a \vee c$, and $y_{13} = a \vee b$ since in this case $s = a \wedge [a \cdot b] = a$, while $t_1 = t_2 = t_3 = t_4 = t_5 = 0$.

The following result shows that there is no smaller example.

Theorem 2. *All four-element Boolean semigroups are in* $\mathbf{V}(\mathrm{Cm}(\mathrm{Sg}))$.

Proof. P. Reich enumerated all four-element Boolean semigroups in [9]. There are a total of 50 (including isomorphic copies), which reduces to 28 if isomorphic copies are excluded. Of these, 6 are non-commutative with a corresponding "opposite" algebra, so only 22 need to be represented. Ten of the 22 algebras are complex algebras of partial semigroups, so by Lemma 1, they are in $\mathbf{V}(\mathrm{Cm}(\mathrm{Sg}))$.

This leaves 12 representation problems. The operation tables for the semigroup operation of these algebras $A_i = \langle \{0 < a, b < 1\}, \vee, 0, \wedge, 1, \neg, \circ_i \rangle$ ($i = 1, \ldots, 12$) are listed below. Reich [9] gives finite representations for 5 of them (A_1– A_5 below), leaving the remaining 7 open.

$\circ_1$	a	b
a	0	0
b	0	1

$\circ_2$	a	b
a	0	0
b	a	1

$\circ_3$	a	b
a	0	a
b	a	1

$\circ_4$	a	b
a	a	a
b	a	1

$\circ_5$	a	b
a	a	b
b	b	1

$\circ_6$	a	b
a	a	b
b	1	b

$\circ_7$	a	b
a	a	b
b	1	1

$\circ_8$	a	b
a	b	1
b	1	1

$\circ_9$	a	b
a	a	a
b	1	1

$\circ_{10}$	a	b
a	a	1
b	1	b

$\circ_{11}$	a	b
a	a	1
b	1	1

$\circ_{12}$	a	b
a	1	1
b	1	1

We now indicate how to construct partial semigroups $S_6, \ldots, S_{12}$ and embeddings $f_i : A_i \to \mathrm{Cm}(S_i)$ for $i = 6, \ldots, 12$. In each case it suffices to specify $f_i(a)$, since $f_i(0) = \varnothing$, $f_i(1) = S_i$, and $f_i(b) = S_i \backslash f_i(a)$.

The algebras A_{10}, A_{11}, and A_{12} are in fact subalgebras of complex algebras of semilattices.

For A_{10} take the chain $\langle \mathbb{N}, \wedge \rangle$ and define $f_{10}(a)$ to be the even numbers.

For A_{12} take a countable binary tree $\langle B_\infty, \wedge \rangle$ (with root as the bottom element) and define $f_{12}(a)$ to be the elements of even height.

For A_{11} we construct a combination of these two semilattices. Let $C_\infty = B_\infty \cup B'_\infty$ where $B'_\infty = B_\infty \times \{0\}$. Each element of height n in B'_∞ is inserted into the order of B_∞ directly below the corresponding element of height n in B_∞ (so the root of B_∞ becomes the root of C_∞), and $f_{11}(a) = B_\infty$.

We note that A_{12} also has a finite representation in the rectangular band $B = \langle \{0,1\}^2, * \rangle$, where $\langle i, j \rangle * \langle k, l \rangle = \langle i, l \rangle$ and $f_{12}(a) = \{\langle 0, 0 \rangle, \langle 1, 1 \rangle\}$.

For the remaining four algebras we are only able to give step-by-step constructions of embeddings into the complex algebra of a partial semigroup defined by a strict dense partial order. The details are similar to the proof of Theorem 1, except that since the atoms of these algebras do not form a semigroup under $\cdot$ ($= \circ_i$), the relation $R_{xy,\kappa}$ is the union of relations $R_{z,\kappa}$ where z ranges over all atoms below $x \cdot y$. Hence the set of pairs $R'_z = R_{z,\kappa+1} \setminus R_{z,\kappa}$ is in general not determined by a definition similar to the one given in Theorem 1. Instead it is convenient to describe the approximate embedding relations $R_{z,\kappa}$ by a partial map $m_\kappa : U_\kappa \times U_\kappa \to \{a, b\}$, where $m_\kappa(p,q) = z$ iff $\langle p, q \rangle \in R_{z,\kappa}$. The definition of $R_{z,\kappa+1}$ is then given by a partial map $m_{\kappa+1}$. This map extends m_κ and on the new pairs $\langle p, w \rangle, \langle w, q \rangle \in U_{\kappa+1} \times U_{\kappa+1}$ it is defined by the following table for the algebra A_6:

1-step completion for A_6	$m_\kappa(p,u)$ $m_\kappa(p,v)$	a a	a b	b a	b b	$m_\kappa(u,q)$ $m_\kappa(v,q)$	a a	a b	b a	b b
$m_\kappa(u,v) = a \leq a \circ_6 a$	$m_{\kappa+1}(p,w)$	a	–	a	**b**	$m_{\kappa+1}(w,q)$	a	b	–	–
$m_\kappa(u,v) = a \leq b \circ_6 a$	$m_{\kappa+1}(p,w)$	b	–	b	b	$m_{\kappa+1}(w,q)$	a	b	–	–
$m_\kappa(u,v) = b \leq a \circ_6 b$	$m_{\kappa+1}(p,w)$	–	a	–	b	$m_{\kappa+1}(w,q)$	a	–	b	b
$m_\kappa(u,v) = b \leq b \circ_6 a$	$m_{\kappa+1}(p,w)$	–	b	–	b	$m_{\kappa+1}(w,q)$	a	–	a	b
$m_\kappa(u,v) = b \leq b \circ_6 b$	$m_{\kappa+1}(p,w)$	–	b	–	b	$m_{\kappa+1}(w,q)$	**a**	–	b	b

This table is to be interpreted as follows. Each row (after the first two) represents a choice of $x, y, z \in \{a, b\}$ and $u, v \in U_\kappa$ such that $m_\kappa(u,v) = z \leq x \circ_6 y$ (hence $\langle u, v \rangle \in R_{z,\kappa}$), but $\langle u, v \rangle \notin R_{x,\kappa} \circ R_{y,\kappa}$. So one chooses $w \notin U_\kappa$ and defines $m_{\kappa+1}(u,w) = x$ and $m_{\kappa+1}(w,v) = y$. To complete the definition of $m_{\kappa+1}$, for each $p \in U_\kappa \setminus \{u\}$ the value of $m_{\kappa+1}(p,w)$ is given by the first half of the row, and depends on the values of $m_\kappa(p,u)$ and $m_\kappa(p,v)$ (listed in the first two rows). The table has a dash (–) as entry if $m_\kappa(p,u) \circ_6 m_\kappa(u,v) \not\geq m_\kappa(p,v)$.

The definition of $m_{\kappa+1}(w,q)$ is similar and uses the second half of the row. The entries in these rows are largely determined by the operation table for $\circ_6$, but in those places where a choice needed to be made, the chosen atom is listed in boldface. The appropriate choices were found by a backtrack search algorithm. It remains to check that the given definition produces relations $R_{z,\kappa+1}$ that are again an approximate embedding. This involves a tedious but straight forward case analysis. The process of refining approximate embeddings in this step-by-step way is iterated in a suitable countable sequence to ensure that $R_z = \bigcup_{\kappa\leq\omega} R_{z,\kappa}$ is indeed an embedding of A_6 into the complex algebra of a partial semigroup. Readers familiar with representing relation algebras by games as in [3] or [4], may note that the table above specifies a winning strategy for the existential player in such a game.

For the algebra A_7, the procedure is identical, except that the definition of $m_{\kappa+1}$ is determined by the following table:

1-step completion for A_7	$m_\kappa(p,u)$ / $m_\kappa(p,v)$	a / a	a / b	b / a	b / b	$m_\kappa(u,q)$ / $m_\kappa(v,q)$	a / a	a / b	b / a	b / b
$m_\kappa(u,v) = a \leq a \circ_7 a$	$m_{\kappa+1}(p,w)$	a	–	a	**b**	$m_{\kappa+1}(w,q)$	a	b	–	–
$m_\kappa(u,v) = a \leq b \circ_7 a$	$m_{\kappa+1}(p,w)$	b	–	a	b	$m_{\kappa+1}(w,q)$	a	b	–	–
$m_\kappa(u,v) = a \leq b \circ_7 b$	$m_{\kappa+1}(p,w)$	b	–	b	b	$m_{\kappa+1}(w,q)$	a	a	–	–
$m_\kappa(u,v) = b \leq a \circ_7 b$	$m_{\kappa+1}(p,w)$	–	a	b	**b**	$m_{\kappa+1}(w,q)$	a	a	b	b
$m_\kappa(u,v) = b \leq b \circ_7 a$	$m_{\kappa+1}(p,w)$	–	b	a	**b**	$m_{\kappa+1}(w,q)$	a	b	a	b
$m_\kappa(u,v) = b \leq b \circ_7 b$	$m_{\kappa+1}(p,w)$	–	b	b	b	$m_{\kappa+1}(w,q)$	**a**	**a**	**b**	**b**

For A_8 one can give a similar 1-step completion table, but in this case the information in the table can be summarized by $m_{\kappa+1}(p,w) = b = m_{\kappa+1}(w,q)$ for all $p,q \in U_\kappa \setminus \{u,v\}$. This definition produces an approximate embedding at each step since $b \leq x \circ_8 y$, $x \leq b \circ_8 y$ and $x \leq y \circ_8 b$ for all $x,y \in \{a,b\}$.

Finally, for A_9 the 1-step completion table is almost as easy to describe as for A_8. Here we set $m_{\kappa+1}(p,w) = b$ if $m_\kappa(p,u) = b = m_\kappa(p,v)$ and otherwise let $m_{\kappa+1}(p,w) = a = m_{\kappa+1}(w,q)$ for all $p,q \in U_\kappa \setminus \{u,v\}$. As before it is tedious, but not difficult to check that this definition of $m_{\kappa+1}$ again produces an approximate embedding.

It is not known whether the algebras $A_6, \ldots, A_{11}$ can be embedded in complex algebras of *finite* semigroups.

Finally we contrast the equational theory of complex algebras of semigroups with the following result adapted from [5] (Theorem 3.20).

Theorem 3. *Every Boolean algebra with a binary operator can be embedded in a member of* Cm(PBn). *If the operator is commutative, then the algebra can be embedded in a member of* Cm(PCBn).

Corollary 2. **V**((C)Bn)) *is the variety of Boolean algebras with a (commutative) binary operator, and hence is finitely based.*

References

[1] H. Andreka, *Representations of distributive lattice ordered semigroups with binary relations*, Algebra Universalis, 28 (1991), 12–25. 173

[2] L. Henkin, J. D. Monk and A. Tarski, *Cylindric algebras, Part II*, North-Holland, Amsterdam, 1985. 172

[3] R. Hirsch and I. Hodkinson, *Step by step — building representations in algebraic logic*, J. Symbolic Logic 62 (1997), 816–847. 172, 176

[4] R. Hirsch and I. Hodkinson, *Relation algebras by games*, Studies in Logic and the Foundations of Mathematics, Vol 147, Elsevier Science, North-Holland, 2002. 172, 176

[5] P. Jipsen, *Computer aided investigations of relation algebras*, dissertation, Vanderbilt University, 1992, `www.chapman.edu/~jipsen/dissertation/`. 176

[6] P. Jipsen and R. D. Maddux, *Nonrepresentable sequential algebras*, Log. J. IGPL 5 (1997), no. 4, 565–574. 172, 174

[7] R. D. Maddux, *Nonfinite axiomatizability results for cylindric and relation algebras*, J. Symbolic Logic 54(3), (1989), 951–974. 173

[8] D. Monk, *On representable relation algebras*, Michigan Math. J. 11 (1964), 207–210. 173

[9] P. Reich, *Complex algebras of semigroups*, dissertation, Iowa State University, 1996. 174

[10] A. Tarski, *On the calculus of relations*, J. Symbolic Logic 6, (1941), 73–89.

Calculational Relation-Algebraic Proofs in Isabelle/Isar

Wolfram Kahl

Department of Computing and Software
McMaster University

Abstract. We propose a collection of theories in the proof assistant Isabelle/Isar that support calculational reasoning in and about heterogeneous relational algebras and Kleene algebras.

1 Introduction and Related Work

Abstract relational algebra is a useful tool for high-level reasoning that, through appropriate models, provides theorems in fields such as data mining, fuzzy databases, graph transformation, and game theory. Frequently, once an application structure is identified as a model of a particular relation-algebraic theory, that theory becomes the preferred reasoning environment in this application area. Since relation-algebraic reasoning typically follows a very calculational style, and, due to the expressive power of its constructs and rules, also proceeds in relatively formal steps, one would expect that computer support for this kind of reasoning should be relatively easy to implement. Since the number of rules that can be applied in any given situation tends to be quite large, and expressions can become quite complex, computer support also appears to be very desirable.

Some applications, such as fuzzy relations [Fur98], or graph transformation [Kah01, Kah02], involve structures where complements in particular may not be available. These structures therefore require weaker formalisations, such as Dedekind categories, or other kinds of allegories [FS90]. Besides the category structure encompassing composition and identities, allegories are equipped with meet and converse, and are closely related with data-flow graphs. The dual view of control-flow graphs corresponds to Kleene algebras which besides composition and identities feature join and iteration (via the Kleene star). Recent years have seen a rapid growth of interest in computer science applications of Kleene algebras and related structures, studied from a relational perspective, see e.g. [DM01]. Since all of these structures still share a considerable body of common theory, it appears desirable to structure the theory support for relation-algebraic reasoning in such a way that the organisation of results reflects the necessary premises in an intuitive way. On the "data-flow side", the different kinds of allegories proposed in [FS90] offer themselves naturally for this structuring; on the "control-flow side", we will use Kleene algebras and several extensions like Kleene algebras with tests [Koz97] and Kleene algebras with domain [DMS03].

R. Berghammer et al. (Eds.): RelMiCS/Kleene-Algebra Ws 2003, LNCS 3051, pp. 178–190, 2004.

While allegories, being extensions of categories, correspond to the *heterogeneous* approach to relation algebras of [SS93, SHW97], Kleene algebras are mostly studied in a *homogenous* setting. Since we believe that the heterogeneous approach with its strongly-typed flavour has significant advantages in particular when it comes to computer science applications, we will adopt it throughout. Therefore, in the remainder of this paper, "Kleene algebra" always means "typed Kleene algebra" in the sense of [Koz98], and then, both allegories and typed Kleene algebras are considered as kinds of categories.

Furthermore, since complex applications will require the possibility to reason about relational algebras built from other relational algebras via certain construction principles — for example product algebras or matrix algebras — it becomes a necessity to allow reasoning not only *within* a single relational algebra, but also *about* several structures and the connections between them, using operations from both *in a single formula.* Therefore, relational algebras must become addressable as objects, as in the relational programming framework RATH [KS00].

Quite a few projects in the recent past have strived to provide computer-aided proof assistance for abstract relation-algebraic reasoning, each with its particular motivation and priorities, all quite different from the approach we are proposing in this paper.

The relation-algebraic formula manipulation system and proof checker RALF [BH94, HBS94, KH98] was designed as a special-purpose proof assistant for abstract (heterogeneous) relation algebras with the goal of supporting proofs in the calculational style typical for relation-algebraic reasoning. RALF has a graphical user interface presenting goal formulae in their tree structure, a feature that allows easy interactive selection of the subexpressions to be transformed by proof steps. During interaction, only the current sub-goal is visible; document output is generated in the calculational style of proof presentation. RALF is based on a fixed axiomatisation, and only supports reasoning *within* a single relation algebra.

Math∫pad is a flexible quasi-WYSIWYG syntax-directed editing environment for mathematical documents that has been designed to support calculational proof presentation. It has been connected with the theorem prover PVS to enable checking of relation-algebraic proofs contained in Math∫pad documents [VB99]. Although the infrastructure is general, the system has been used only *within* concrete relation algebra; there appears to be no provision so far for working in weaker theories, nor for reasoning *about* multiple relation-algebraic structures in the same context.

An interesting related experiment is "PCP: Point and Click Proofs", a proof assistant based on a small JavaScript rewriting engine that allows users to interactively construct proofs of properties in a wide range of mathematical structures, characterised by equational and quasi-equational theories [Jip03, Jip01]. Currently, (homogeneous) relation algebra and Kleene algebra are already supported by the system, which is extensible and still under development. It appears not to be geared to the addition of a type system as required for heterogeneous relation algebras, and is also limited to reasoning *within* structures.

A previous formalisation of heterogeneous relation algebras in Isabelle, RALL [vOG97], uses the Isabelle/HOL type system to support reasoning in abstract *heterogeneous* relation algebras with minimal effort, but at the cost of limiting itself to reasoning *within* a *single* relation algebra, as well. Besides interactive tactical proving, with special tactics allowing to approach the calculational style, RALL also explores automatic proving using the isomorphism of *atomisation* from relation-algebraic formulae into predicate logic, which allows Isabelle's classical reasoner to tackle the atomised versions of relation-algebraic formulae. This approach is not available for weaker structures like allegories or Dedekind categories, which were also not considered in RALL.

Struth realised a formalisation in Isabelle-1999 of untyped Kleene algebras via a hierarchy of axiomatic type classes, and used this to fully formalise Church-Rosser proofs in Kleene algebras [Str02]. Although Struth did all his reasoning within a single structure, axiomatic type classes [Wen97] do support reasoning "between" several structures; but they impose severe limitations to reasoning *about* structures.

It turns out that our objectives, namely *calculational reasoning* both *within* and *about* models of *different theories* with a relation-algebraic flavour are quite nicely supported by recent developments in the theorem prover Isabelle [NPW02]:

For reasoning *within* abstract algebraic structures in essentially the same way as it is done in pencil-and-paper mathematics, the concept of *locales* has been introduced into Isabelle [KWP99]. If such locales are based on records, this also allows reasoning *about* algebraic structures in the sense that several instances of the same structure can be used simultaneously in a single context.

While internally Isabelle still is a tactical theorem prover, the addition of the interpreted "Isar" language for "Intelligible semi-automated reasoning" allows proofs to be structured in the same way as in traditional mathematical proof presentation [Nip03, Wen02].

One additional aspect of Isar is its support for calculational reasoning [BW01]; this has been designed to also support user-defined transitive relations, such as the inclusion ordering of relations.

These features together allow us to provide a collection of theories extending from categories via different kinds of allegories and Kleene algebras up to heterogeneous relation algebras, geared towards calculational reasoning *in and about* relational algebras in a way that allows easy connection with concrete application theories.

Filling in these theories to a degree that they will become a useful starting point for applications is still ongoing work. From the examples we show in the next section, the reader may obtain a first flavour of working with the current state of our theory collection, which is not yet optimally tuned towards the automatic proving support provided by Isabelle. Once this is achieved, we will turn to extending the capabilities of Isabelle/Isar for the purpose of further streamlining of calculational proof support for relational reasoning, for example to eliminate certain repetitive patterns, but also to include special-purpose decision procedures. The main purpose of this paper is to make a case that user-friendly mechanised support for fully formal calculational reasoning in relational algebras appears to become realistically achievable in the near future.

After the examples in the next section, we proceed to give an overview over the organisation of our theories in Sect. 3, and in Sect. 4 we discuss some technical details underlying the possibility of reasoning about and between structures.

2 Example Proofs

Essentially following [FS90], we define allegories as a special kind of categories. However, while Freyd and Scedrov treat categories and allegories as one-sorted algebraic structures with morphisms only, we follow the more usual approach of providing a separate sort for objects which serve as *source* and *target* of morphisms.

The inclusion ordering $\sqsubseteq$ in allegories is defined from the meet (or intersection) operation $\sqcap$:

$$\textit{incl-def}[\textit{iff}]\text{: } [\![\ R : a \leftrightarrow b;\ S : a \leftrightarrow b\ ]\!] \Longrightarrow (R \sqsubseteq S) = (S \sqcap R = R)$$

In the higher-order logic HOL, terms of type *bool* are used *en lieu* of propositions, so equality also takes on the role of propositional equivalence. The presence of the assumptions $[\![\ R : a \leftrightarrow b;\ S : a \leftrightarrow b\ ]\!]$ implies that the equivalence $(R \sqsubseteq S) = (R \sqcap S = R)$ only needs to hold if R and S belong to the same *homset* $a \leftrightarrow b$, i.e., if R and S are well-defined morphisms between two well-defined objects a and b.

As a first example proof, we show $R \sqcap S \sqsubseteq R$ via the above definition of inclusion from the algebraic properties of $\sqcap$:

$$\begin{aligned} R \sqcap (R \sqcap S) &= (R \sqcap R) \sqcap S && \text{associativity} \\ &= R \sqcap S && \text{idempotence} \end{aligned}$$

The Isar proof consists of the same steps. Since in our design, the well-typedness for relation-algebraic expressions is not dealt with by HOL's type system, but rather by homset membership, we also need to aid the system to explicitly discharge all well-typedness conditions — the automatic reasoning tactic *auto* is sufficient in most cases, aided by introducing the typings as introduction rules with [*intro*].

```
lemma (in Allegory) meet-decr1:
  assumes R-t[intro]: R : a ↔ b
  assumes S-t[intro]: S : a ↔ b
  shows R ⊓ S ⊑ R
proof –
 have R ⊓ (R ⊓ S) = (R  ⊓ R) ⊓ S        by (rule meet-assoc [THEN sym], auto)
 also have ... = R ⊓ S                   by (subst meet-idem, auto)
 finally show ?thesis                    by (rule-tac incl-contract, auto)
qed
```

Obviously, we were able to transfer our proof into Isar without significant loss of readability or conciseness.

As a slightly more complicated example, we show that for univalent relations F, the equality $R \sqcap S{:}F = (R{:}F^{\smile} \sqcap S){:}F$ holds — this can be shown by a cyclic inclusion chain:

$$\begin{aligned} R \sqcap S{:}F &\sqsubseteq (R{:}F^{\smile} \sqcap S){:}F && \text{modal rule} \\ &\sqsubseteq R{:}F^{\smile}{:}F \sqcap S{:}F && \text{meet-subdistributivity} \\ &\sqsubseteq R \sqcap S{:}F && F \text{ univalent} \end{aligned}$$

Here, quite a few implicit steps were hidden, and some of these technical details need to be made explicit in the Isar proof. Some remain hidden; for example, modal rules come in four shapes (equivalent by conversion and commutativity of meet), so the Isabelle theory for allegories binds the theorem reference *modal* to the set containing all four shapes of modal rules. Also, since longer inclusion chains do not uniquely determine to which inclusions the antisymmetry rule should be applied, we need to split the calculation into two separate inclusion chains:

```
lemma (in Allegory) unival-meet-escape-1:
  assumes F-u: univalent F
  assumes F-t[intro]: F : b ↔ c
  assumes R-t[intro]: R : a ↔ c
  assumes S-t[intro]: S : a ↔ b
  shows     R ⊓ S ; F = (R ; F˘ ⊓ S) ; F
proof -
 have R ⊓ S ; F ⊑ (R ; F˘ ⊓ S) ; F                  by (rule modal, auto)
 moreover have (R ; F˘ ⊓ S) ; F ⊑ R ⊓ S ; F
  proof -
  have (R ; F˘ ⊓ S) ; F ⊑ (R ; F˘) ; F ⊓ S ; F      by (rule meet-cmp, auto)
  also have (R ; F˘) ; F = R ; (F˘ ; F)            by (rule cmp-assoc, auto)
  also from F-u have F˘ ; F ⊑ Id c                 by (rule univalent, auto)
  also (incl-mon-trans) have R ; Id c = R          by (rule right-id, auto)
  also show ?thesis                               by (rule calculation, best+)
  qed
 ultimately show ?thesis                          by (rule incl-antisym, best+)
qed
```

For this proof, a few additional points need explanation. While in most cases, *auto* can discharge all well-typedness conditions, here there are two cases where *auto* does not succeed, and we used the best-fit-first reasoning tactic *best* instead. Since several conditions had to be discharged, application of *best* had to be iterated — *auto* attempts to discharge all open subgoals, so never needs iteration. One of the "**also**" occurrences has an explicitly specified transitivity rule *incl-mon-trans* — this is necessary for cases where the transitivity rule cannot be uniquely determined by Isabelle.

Furthermore, the conclusion of the second inclusion chain looks strange: instead of "**finally show** *?thesis*", which is an abbreviation for "**also from** *calculation* **show** *?thesis*" (where *calculation* is the name of the local register that

Isar uses to accumulate the result of calculational proofs), we now have "**also show** *?thesis* **by** ... *calculation*" — the reason for this is that *calculation* as produced by the previous chain is not a simple fact, but a meta-logic quantification carrying many (well-typedness) assumptions, so that it is not easily accessible to the automatic proof tools and instead needs to be explicitly applied to *?thesis*, the statement to be shown, as a rule.

The typing assumptions have, as always, again been flagged with [*intro*] as introduction rules, so *auto* will use them to discharge the type correctness conditions that all the other rule applications introduce. The assumption about univalence has not been flagged in this way — if it had been, we would not have needed to refer to it via "from *F-u*" for making use of univalence of *F*. The decision not to make this assumption available to Isabelle's automated proof tools therefore contributes to traceability.

Finally, experienced Isar users will note that instead of showing the theorem in the last line of the proof by "*rule incl-antisym*", we might have declared this as the rule organising the whole proof by inserting it after the opening "proof" in the place where "–" explicitly excludes any proof structuring rule. However, *incl-antisym* has not only the two inclusions as premises, but also two well-typedness conditions, which then would have to be dealt with explicitly, while with our approach they are implicitly discharged by the last "*best+*".

Although this second proof employed a larger number of the features of Isabelle/Isar, calculational proofs typically need neither the full complexities of the Isar proof language, nor all the capabilities of the Isabelle tactics language used in the arguments to "by".

The features that are used are limited to a set that can be learned relatively easily, and the way in which these features are used mostly follows rather systematic patterns. Even for some unexpected failures there are recipes, like "if *auto* fails to discharge type correctness conditions, then try *best+* and *(simp_all (no_asm_simp))*".

Even though more understanding of Isabelle and Isar is definitely helpful, and for many basic and auxiliary pre-proven lemmas it is necessary to remember their names for being able to invoke them explicitly, we expect the learning curve for doing calculational Isar proofs based on our theory library to be much smoother than for Isar proofs in predicate logic.

3 Theory Organisation

The kernel of our theory collection (which is still work in progress) is a hierarchy of theories defining the structures of discourse and providing useful facts and derived concepts:

- **Categories:** built on homsets, identities, and composition. Provides concepts like mono-, epi-, and isomorphisms, initial and terminal objects, direct sums and products. Functors will also be included here.

- **Ordered categories:** categories with an ordering $\sqsubseteq$ on every homset. Provides monotonicity and transitivity rules for reasoning with inclusions. Predicates for bounds and residuals, where existing, are also already defined here.
- **Allegories:** extend ordered categories by meets and converse. Provides concepts like univalent, total, ..., symmetric, transitive, ... Relators come in at this level.
- **Distributive allegories:** extend allegories by joins $\sqcup$ and least relations $\perp\!\!\!\perp$.
- **Division allegories:** extend distributive allegories by left and right residuals, and therefore also by symmetric quotients.
- **Dedekind categories**, or **locally complete distributive allegories** allow arbitrary joins, and therefore also greatest relations $\top\!\!\!\top$.
- **Relation algebras** have homsets that are Boolean algebras and therefore also provide complements.
- **Atomic relation algebras satisfying the Tarski rule** are the setting used in a large part of the literature, for example in [SS93].
- **Concrete relation algebras with subobjects, quotients, finite sums and finite products** are the setting of the relational programs of RelView [BBS97, BHL99]. This theory will allow verification of RelView programs in a recognised theorem prover.

Each of these classes of structures is defined in Isabelle/Isar/HOL as a *locale*, which corresponds to the mathematical habit of "assuming an arbitrary, but fixed category/allegory/... throughout".

Besides this "relational hierarchy", we also add support for Kleene algebras and related structures, since reasoning in these structures shares many aspects with relational reasoning. Although Kleene algebra is usually presented as a homogeneous algebra, we axiomatise heterogeneous versions on top of ordered categories since this way their relations with our hierarchy of categories and allegories are more natural.

- Heterogeneous **Kleene algebras**, according to Kozen's axiomatisation [Koz91], share the join structure with distributive allegories, and add the Kleene star, which corresponds to the reflexive transitive closure present in the relational hierarchy in Dedekind categories.
- **Residuated Kleene algebras** add residuals to Kleene algebras, these are shared with division allegories.
- **Action lattices** [Koz94] essentially are residuated Kleene algebras where homsets are lattices.
- **Kleene algebras with domain** (KAD) [DMS03] add to Kleene algebras a domain operator corresponding to that definable in allegories by $\mathsf{dom} R := \mathbb{I} \sqcap R{:}R^{\smile}$.

A simplified overview of our theory dependency graph is shown in Fig. 1.

In order to be able to profit from the shared structures, we provide them separately in small theories. For example, the shared join properties are collected in our "heterogeneous idempotent semirings" (HISR), defined on top of locally ordered categories with join.

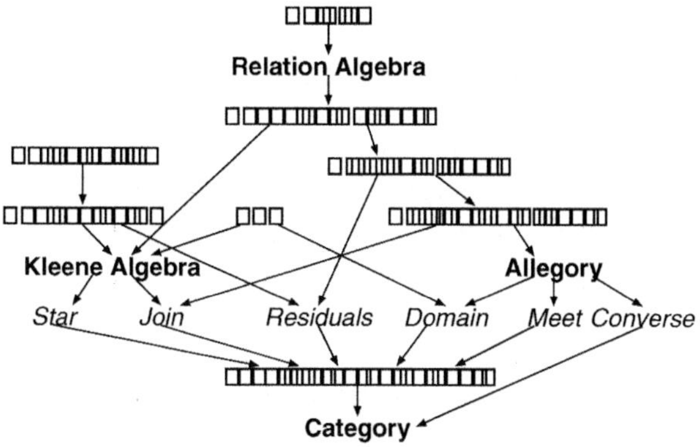

Fig. 1. Theory dependency summary

For the domain elements of Kleene algebras with domains, we do not follow the approach of embedding a Boolean algebra of tests into the subidentities. Instead we observe that the key properties of multiplicatively-idempotent subidentities presented in [DMS03] do not depend on the semiring structure, but already hold in ordered categories. We included a separete theory to collect those properties and use them to axiomatise the domain operator; all the resulting material is then made available to both allegories and Kleene algebras with domain. (See [DG00] for argumentation why the domain operator should be primitive in allegories.)

4 Structure Representation Aspects

For treating functors and relators, we obviously need to be able to deal with at least two categories/allegories in the same context. For structures encapsulated in Isabelle's locales, this implies that each of these locales has to be based on a record type. Such a record is essentially an explicit data structure aggregating all the information defining an algebra; for allegories and relation algebras, a similar organisation can be found in the Haskell library RATH [KS00]. In comparison with Haskell, the extensible records allowed by Isabelle simplify the setup; only the linear hierarchy that is enforced on extensions appears slightly unintuitive in some contexts, but does not lead to theoretical or practical problems. For example, within Kleene algebras, there is always a meet component accessible, but nothing can be proved about it. By organising record extensions and axiomatic theory definitions (i.e., locales) into separate files we achieved a setup which guarantees that reasoning in, for example, Kleene algebras is not burdened in any way by the presence of meets in the underlying records.

For being able to access derived concepts simultaneously in two structures (as, for example, in the statement "relators preserve univalence"), a particular way of defining them with explicit reference to the underlying structure is required; this structure is of type $('o,\ 'm,\ 'r)$ *Allegory-scheme*, the type of all extensions of allegory structure records — these are currently the smallest records containing converse and inclusion. In addition, such derived concepts should, for the sake of demonstrable consistency, be introduced via meta-level equations — for technical reasons, these currently cannot employ user-defined syntax. Therefore, definitions of derived concepts cannot be stated in a simple intuitive shape. For example, ideally we would be able to define univalence by stating something like the following:

$$R : a \leftrightarrow b \Longrightarrow \mathit{univalent}\ R == (R^{\smile} ; R \sqsubseteq \mathit{Id}\ b)$$

Instead, we have to define an internal identifier *isUnivalent* and declare *univalent* with index "ı" for reference to possibly different structures as user-level syntax for this. For the definition, we have to refer to all allegory primitives via the internal identifiers, such as *incl* for $\sqsubseteq$, and have to explicitly supply the structure argument that corresponds to the index position. We also have to impose the well-typedness conditions via a conditional, with the other branch containing the pseudo-value *arbitrary* about which nothing can be proven. For user-friendly access to such a definition, an additional lemma is necessary that contains the well-typedness condition as an assumption:

```
constdefs
  isUnivalent :: ('o, 'm, 'r) Allegory-scheme ⇒ 'm ⇒ bool (univalentı - [1000] 999)
  isUnivalent s R == if isMor s R
                        then incl s (cmp s (conv s R) R) (CId s (Ctrg s R))
                        else arbitrary
```

```
lemma (in ConvOrdCat) univalent-def:
  R : a ↔ b ⟹ univalent R = (R˘ ; R ⊑ Id b)
by (unfold isUnivalent-def, auto)
```

All subsequent reasoning will then use that lemma — its name, *univalent-def*, has been chosen to essentially hide the real definition. For derived predicates, such as univalence, this "definition-lemma" might be almost sufficient; typically, more user-friendly lemmas are also provided, like the one used in the second example proof:

```
lemma (in ConvOrdCat) univalent[elim?]:
  assumes univalent R
  assumes [simp]: R : a ↔ b
  shows R˘ ; R ⊑ Id b
by (rule univalent-def [THEN iffD1])
```

For derived operations, one also needs to state and prove the derived rules that help to automatically discharge well-typedness conditions.

This way of defining derived concepts applies in the same way to user-defined extensions to our theory as to our theory library itself. Although this is of course somewhat inconvenient, there is a systematic procedure to produce the necessary material for user-defined extensions. Adhering to that procedure guarantees that new derived concepts can be used with the same flexibility as those contained in the library, and with the same safety with respect to consistency.

5 Conclusion and Outlook

We have shown that with new features of Isabelle, calculational reasoning in relation algebras can be supported in a way that makes proofs both readable and writable — indeed, in many cases producing a correct proof with the support of the XEmacs interface ProofGeneral [Asp00] is probably slightly easier than producing a good-looking proof directly in LaTeX.

In particular, we have been able to add the flexibility of dealing with several relation algebras at the same time, and of instantiating the abstract arguments in concrete structures, without incurring a prohibitive cost in terms of dealing with well-typedness conditions — the careful arrangement of homset rules allows us to discharge those systematically and automatically.

For the future, we plan to tackle correctness proofs of RelView programs as realistic case studies, and hope to be able to enlist more automated support from Isabelle for "trivial" calculational proof steps.

Another natural application of our framework and its many small component theories will be a study of the recently proposed "Kleene algebras with relations" [Des03].

While the proofs we have shown are by themselves not interesting at all, they demonstrate the style of prover-supported calculational reasoning that is possible already now. As an important step towards more user-friendly support for calculational reasoning we plan to incorporate decision procedures, where feasible, into the corresponding theories, for example for the equational theory of allegories [DG00], or for Kleene algebras. Once such decision procedures are available they can eliminate many tedious steps that are currently still necessary.

Acknowledgements

I am grateful to Millie Rhoss de Guzman and Hitoshi Furusawa for their comments on previous versions of this paper and many related discussions, and also to the anonymous reviewers for their valuable comments and to Kevin Everets for careful proofreading.

References

[Asp00] David Aspinall. Proof General: A generic tool for proof development. In *Tools and Algorithms for the Construction and Analysis of Systems (TACAS)*, volume 1785 of *LNCS*, pages 38-42. Springer, 2000. See also `http://www.proofgeneral.org`. 187

[BBS97] Ralf Behnke, Rudolf Berghammer, and Peter Schneider. Machine support of relational computations: The Kiel RELVIEW system. Technical Report 9711, Institut für Informatik und Praktische Mathematik, Christian-Albrechts- Universität Kiel, June 1997. 184

[BH94] Rudolf Berghammer and Claudia Hattensperger. Computer-aided manipulation of relational expressions and formulae using RALF. In Bettina Buth and Rudolf Berghammer, editors, *Systems for Computer-Aided Specification, Development and Verification*, Bericht Nr. 9416, pages 62-78. Universität Kiel, 1994. 179

[BHL99] Rudolf Berghammer, Thorsten Hoffmann, and Barbara Leoniuk. Rechnergestützte Erstellung von Prototypen für Programme auf relationalen Strukturen. Technical Report 9905, Institut für Informatik und praktische Mathematik, Christian-Albrechts-Universität Kiel, July 1999. 184

[BW01] Gertrud Bauer and Markus Wenzel. Calculational reasoning revisited, an Isabelle/Isar experience. In R. J. Boulton and P. B. Jackson, editors, *Theorem Proving in Higher-Order Logics: TPHOLs 2001*, volume 2152 of *LNCS*, pages 75-90. Springer, 2001. 180

[Des03] Jules Desharnais. Kleene algebras with relations. In Rudolf Berghammer and Bernhard Möller, editors, *Proc. RelMiCS 7, International Seminar on Relational Methods in Computer Science, in combination with the 2nd Intl. Workshop on Applications of Kleene Algebra*, LNCS. Springer, 2003. (Invited Talk). 187

[DG00] Dan Dougherty and Claudio Gutiérrez. Normal forms and reduction for theories of binary relations. In Leo Bachmair, editor, *Rewriting Techniques and Applications, Proc. RTA 2000*, volume 1833 of LNCS, pages 95-109. Springer, 2000. 185, 187

[DM01] Jules Desharnais and Bernhard Möller. Characterizing determinacy in Kleene algebras. *Information Sciences*, 139:253-273, 2001. 178

[DMS03] Jules Desharnais, Bernhard Möller, and Georg Struth. Kleene algebra with domain. Technical Report 2003-7, Universität Augsburg, Institut für Informatik, 2003. 178, 184, 185

[dS02] H.C.M. de Swart, editor. *Proc. RelMiCS 6, International Workshop on Relational Methods in Computer Science, Oisterwijk near Tilburg, Netherlands, 16-21 October 2001*, volume 2561 of LNCS. Springer, 2002.

[FS90] Peter J. Freyd and Andre Scedrov. *Categories, Allegories*, volume 39 of *North-Holland Mathematical Library*. North-Holland, Amsterdam, 1990. 178, 181

[Fur98] Hitoshi Furusawa. *Algebraic Formalisations of Fuzzy Relations and Their Representation Theorems*. PhD thesis, Department of Informatics, Kyushu University, March 1998. 178

[HBS94] Claudia Hattensperger, Rudolf Berghammer, and Gunther Schmidt. RALF – A relation-algebraic formula manipulation system and proof checker. Notes to a system demonstration. InMaurice Nivat, Charles Rattray, Teodore Rus, and Giuseppe Scollo, editors, AMAST '93, Workshops in Computing, pages 405-406. Springer, 1994. 179

[Jip01] Peter Jipsen. Implementing quasi-equational logic on the web. Talk given at the AMS Sectional Meeting, University of South Carolina, March 16-18 2001. `http://www.chapman.edu/~jipsen/PCP/usctalk.html`. 179

[Jip03] Peter Jipsen. PCP: Point and click proofs. Web-based system at URL: `http://www.chapman.edu/~jipsen/PCP/PCPhome.html`, 2003. 179

[Kah01] Wolfram Kahl. A relation-algebraic approach to graph structure transformation, 2001. Habil. Thesis, Fakultät für Informatik, Univ. der Bundeswehr München, Techn. Bericht 2002-03. 178

[Kah02] Wolfram Kahl. A relation-algebraic approach to graph structure transformation. In de Swart [dS02], pages 1-14. (Invited Talk). 178

[KH98] Wolfram Kahl and Claudia Hattensperger. Second-order syntax in HOPS and in RALF. In Bettina Buth, Rudolf Berghammer, and Jan Peleska, editors, *Tools for System Development and Verification*, volume 1 of *BISS Monographs*, pages 140-164, Aachen, 1998. Shaker Verlag. ISBN: 3-8265-3806-4. 179

[Koz91] Dexter Kozen. A completeness theorem for Kleene algebras and the algebra of regular events. *Inform. and Comput.*, 110(2):366-390, 1991. 184

[Koz94] Dexter Kozen. On action algebras. In J. van Eijck and A. Visser, editors, *Logic and Information Flow*, pages 78-88. MIT Press, 1994. 184

[Koz97] Dexter Kozen. Kleene algebra with tests. *ACM Transactions on Programming Languages and Systems*, pages 427-443, May 1997. 178

[Koz98] Dexter Kozen. Typed Kleene algebra. Technical Report 98-1669, Computer Science Department, Cornell University, March 1998. 179

[KS00] Wolfram Kahl and Gunther Schmidt. Exploring (finite) Relation Algebras using Tools written in Haskell. Technical Report 2000-02, Fakultät für Informatik, Universität der Bundeswehr München, October 2000. see also the RATH page `http://ist.unibw-muenchen.de/relmics/tools/RATH/`. 179, 185

[KWP99] Florian Kammüller, Markus Wenzel, and Lawrence C. Paulson. Locales - a sectioning concept for Isabelle. In Y. Bertot, G. Dowek, A. Hirschowitz, C. Paulin, and L. Théry, editors, *Theorem Proving in Higher-Order Logics, 12th International Conference, TPHOLs'99*, volume 1690 of *LNCS*, pages 149-166. Springer, 1999. 180

[Nip03] Tobias Nipkow. Structured proofs in Isar/HOL. In H. Geuvers and F. Wiedijk, editors, *Types for Proofs and Programs, International Workshop TYPES 2002*, LNCS. Springer, 2003. 180

[NPW02] Tobias Nipkow, Lawrence C. Paulson, and Markus Wenzel. *Isabelle/HOL - A Proof Assistant for Higher-Order Logic*, volume 2283 of LNCS. Springer, 2002. 180

[SHW97] Gunther Schmidt, Claudia Hattensperger, and Michael Winter. Heterogeneous relation algebra. In Chris Brink, Wolfram Kahl, and Gunther Schmidt, editors, *Relational Methods in Computer Science*, Advances in Computing Science, chapter 3, pages 39-53. Springer, Wien, New York, 1997. 179

[SS93] Gunther Schmidt and Thomas Ströhlein. *Relations and Graphs, Discrete Mathematics for Computer Scientists.* EATCS-Monographs on Theoretical Computer Science. Springer, 1993. 179, 184

[Str02] Georg Struth. Calculating Church-Rosser proofs in Kleene algebra. In de Swart [dS02], pages 276-290. 180

[VB99] Richard Verhoeven and Roland Backhouse. Towards tool support for program verification *and* construction. In Jeanette Wing, Jim Woodcock, and Jim Davies, editors, *FM '99 - Formal Methods*, volume 1709 of LNCS, pages 1128-1146. Springer, September 1999. 179

[vOG97] David von Oheimb and Thomas F. Gritzner. RALL: Machine-supported proofs for relation algebra. In William McCune, editor, *Conference on Automated Deduction - CADE-14*, volume 1249 of LNCS, pages 380-394. Springer, 1997. 180

[Wen97] Markus Wenzel. Type classes and overloading in higher-order logic. In Elsa L. Gunter and Amy Felty, editors, *Theorem Proving in Higher-Order Logics, TPHOLs '97*, volume 1275 of *LNCS*, pages 307-322. Springer, 1997. 180

[Wen02] Markus M. Wenzel. *Isabelle/Isar - A Versatile Environment for Human-Readable Formal Proof Documents.* PhD thesis, Technische Universität München, Fakultät für Informatik, February 2002. 180

A Calculus of Typed Relations*

Wendy MacCaull[1]** and Ewa Orłowska[2]***

[1] Department of Mathematics, Statistics and Computer Science
St. Francis Xavier University, PO Box 5000, Antigonish, NS, CANADA B2G 2W5
wmaccaul@stfx.ca

[2] National Institute of Telecommunications
Szachowa 1, 04-894, Warsaw, POLAND
orlowska@itl.waw.pl

Abstract. A calculus of typed relations subsuming the classical relational database theory is presented. An associated sound and complete deduction mechanism is developed. The formulation is generalized to deal with nondeterministic databases and information relations in the rough set-style.

Keywords: Relational proof system, typed relations, relational database, nondeterministic databases, information relations

1 Introduction

We present a relational calculus which is intended to incorporate the database relational calculus originated by Codd [3]. We develop a proof system for this calculus in the Rasiowa and Sikorski [16] style. Our conclusion outlines a number of applications including application of this calculus to nondeterministic databases and information relations in the rough set-style of Pawlak (see Pawlak [15]).

Three features of our calculus distinguish it from the calculus of Tarski-style relations.

• First, with each relation there is associated its type. A type is a finite subset of a set whose members are interpreted as attributes. In this way we cope with the fact that database relations are determined by (finite) subsets of a set of attributes. Therefore, the relations of the calculus are relative in the sense of Demri and Orłowska [5].

• Second, as with ordinary relations, each typed relation has an arity, which is the cardinality of its type. However, for any $n \geq 1$, the order of the elements in the n-tuples belonging to a relation does not matter. This reflects the well-known property of database relations that the order of the attributes in the data

* This work was performed within the framework of the COST Action 271. Both authors were supported by a NATO PST Collaborative Linkages Grant.

** Supported by the Natural Sciences and Engineering Research Council of Canada.

*** Support from St. Francis Xavier University, as James Chair Visiting Professor, is gratefully acknowledged.

table is immaterial. Tuples are treated as mappings that assign to an attribute an element of its domain.

• Third, the calculus is comprised of relations of various arities and some operations may act on relations not necessarily of the same arity.

The original motivation for this paper was to construct a Rasiowa and Sikorski style proof system for relational databases. In order to define such a proof system, a precise language for the theory of relational databases was required to allow us to represent database relations and their schemes. We found that no precise logic-style formalism existed in the literature to serve our purpose. Therefore the first step was to define such a language. It seemed reasonable to slightly generalize some of the notions; for example, the notions of scheme and selection operation. Therefore the resulting calculus includes the Codd calculus of relational databases as a special case. This paper fits into the program of giving a general framework for deduction applicable to both classical and non-classical logics (Orłowska [13], MacCaull and Orłowska [11]). It augments relational approaches to dependency theory (see Orłowska [12], Buszkowski and Orłowska [1], MacCaull [9], and Düntsch and Mikulas [6]). MacCaull and Orłowska [10] is the full version of the the present paper, which includes, among other things, properties of the operations on typed relations and full proofs of the theorems presented here.

2 Typed Relations with Implicit Renaming

In this section we propose a notion of typed relation and a calculus of typed relations which is intended to be a formal tool both for representing relational databases and also for reasoning about them.

Let Ω be an infinite set whose elements are referred to as *attributes*. To each $a \in \Omega$ there is associated a nonempty set D_a called the *domain of attribute a*. Types (of relations), usually denoted by capital letters $A, B,\ldots$, etc., are finite subsets of Ω; clearly, if A and B are types, then $A \cup B$, $A \cap B$, and $A - B$ are types. The notation $A \uplus B$ is meant to represent the union of disjoint sets A' and B', obtained from A and B, respectively, by renaming their elements, if necessary. Consequently, $\mathrm{card}(A) = \mathrm{card}(A')$, and $\mathrm{card}(B) = \mathrm{card}(B')$, and $A' \cap B' = \emptyset$. It is a common practise in database systems to rename attributes as needed. This understanding of $\uplus$ allows us to assume that $\uplus$ is commutative and associative, and that $A \uplus \emptyset = A$. To enable renaming, we assume that for every attribute $a \in \Omega$, there are infinitely many attributes a_i such that $D_{a_i} = D_a$. When forming $A \uplus B$, if $a' \in A$ and $b' \in B$ correspond to $a \in A$ and $b \in B$, respectively, it is necessary that $D_{a'} = D_a$ and $D_{b'} = D_b$. The set of all types will be denoted by $\mathcal{T}_\Omega$. Our definition of the disjoint union involves renaming implicitly; we do not need the explicit treatment of types as equivalence classes of an equivalence relation determined by renaming. We refer to this notion of typed relations as *typed relations with implicit renaming*. Let $D_A = \bigcup\{D_a : a \in A\}$; in particular, $D_\emptyset = \emptyset$. A tuple of type A is a map $u : A \rightarrow D_A$ such that for every $a \in A$, $u(a) \in D_a$. The collection of all tuples of type A is called the relation 1^A; for each $a \in \Omega$ the collection of tuples of type $\{a\}$ is denoted by 1^a. Let $1^\emptyset = \{e\}$, where e is the empty tuple. For each $a \in \Omega$, $D_a \neq \emptyset$; therefore $1^a \neq \emptyset$. Consequently $1^A \neq \emptyset$ for all $A \in \mathcal{T}_\Omega$. The above definitions imply that $1^{A \uplus B} = \{t : \exists u \in 1^A, \exists v \in 1^B$ such that if $a \in A$ then $t(a) = u(a)$ and if $b \in B$ then $t(b) = v(b)\}$. Observe that we have defined $\uplus$ so that $1^{A \uplus B} = 1^{B \uplus A}$, $1^{A \uplus (B \uplus C)} = 1^{(A \uplus B) \uplus C}$ and $1^A = 1^{B \uplus (A-B)}$. We often denote tuples $t \in 1^{A \uplus B}$ by uv, and say $t = uv$. Clearly, uv is a mapping, $uv : A \uplus B \to D_{A \uplus B}$, where $D_{A \uplus B} = D_A \cup D_B$. Thus $uv = vu$; similarly, $(uv)w = u(vw)$. Our notation uv is analogous to the relational database notation for unions of sets of attributes: uv is the union of two mappings (where a mapping is a set of pairs). Finally, for any $A \in \mathcal{T}_\Omega$, and for any $u \in 1^A$, $ue = eu = u$.

Definition 1. *A relation R of type A is a subset of* 1^A.

We now describe the basic operations on typed relations. Let $A, B \in \mathcal{T}_\Omega$:

(i) Intersection ($\cap$) Let $R, S \subseteq 1^A$; then
$R \cap^A S = \{u \in 1^A : u \in R \text{ and } u \in S\}$.

(ii) Projection (Π) Let $B \subseteq A$ and let $R \subseteq 1^A$; then
$\Pi^A_B R = \{u \in 1^B : \exists v \in 1^{A-B} \text{ such that } uv \in R\}$.

(iii) Product ($\times$) Let $R \subseteq 1^A$ and $S \subseteq 1^B$; then
$R \times^{A \uplus B} S = \{uv \in 1^{A \uplus B} : u \in 1^A,\ v \in 1^B,\ u \in R \text{ and } v \in S\}$.

(iv) Complement ($-$) Let $R \subseteq 1^A$; then
$-^A R = (1^A - R) = \{u \in 1^A : u \notin R\}$.

We define the constant $0^A = -^A 1^A$; clearly, $0^A = \emptyset$ for all $A \in \mathcal{T}_\Omega$. The reader can easily define union, $R \cup^A S$, and complement of S with respect to R, $R -^A S$, in terms of operations (i)-(iv); we use the notation $R \to^A S$ to denote $-^A R \cup^A S$. Other operations typically used in modern approaches to database theory (as in Simovici and Tenney [17] and Ullman [18]) may be defined in terms of operations (i)-(iv); for example:

Natural join ($\bowtie$) Let $R \subseteq 1^A$ and $S \subseteq 1^B$; then
$R \bowtie^{A \cup B} S = \{uvw \in 1^{A \cup B} : u \in 1^{A-(A\cap B)},\ v \in 1^{A \cap B},\ w \in 1^{B-(A\cap B)},$
$uv \in R \text{ and } vw \in S\}$
$= (R \times^{A \uplus (B-(A\cap B))} 1^{B-(A\cap B)}) \cap^{A \uplus (B-(A\cap B))} (S \times^{B \uplus (A-(B\cap A))} 1^{A-(B\cap A)})$.

Division ($\div$) Let $B \subseteq A$, let $R \subseteq 1^A$ and $S \subseteq 1^B$, $S \neq 0^B$; then
$R \div^A_B S = \{t \in 1^{A-B} : \forall s \in S,\ ts \in R\}$
$= \Pi^A_{A-B} R -^{A-B} (\Pi^A_{A-B}((\Pi^A_{A-B} R \times^{(A-B)\uplus B} S) -^A R))$.

We introduce a more general notion of selection operation, namely *select S in R*, which is defined for any $B \subseteq A$, $S \subseteq 1^B$ and $R \subseteq 1^A$; its application to such S and R yields the tuples $ut \in R$ such that $u \in S$:

Selection (σ) Let $B \subseteq A$, let $R \subseteq 1^A$ and let $S \subseteq 1^B$; then
$\sigma^A_B(S, R) = \{ut \in 1^A : u \in 1^B,\ t \in 1^{A-B},\ u \in S \text{ and } ut \in R\}$
$= (S \times^{B \uplus (A-B)} 1^{A-B}) \cap^A R$.

We define two binary operations, $\odot^A_{a,b}$ and $\supset^A_{a,b}$, which will allow us discuss entailment. To improve readability, we suppress the typing for $\times$.

Definition 2. *Let $R, S \subseteq 1^A$ and let $a \neq b \in A$; then $R \odot^A_{a,b} S = (1^{A-\{a,b\}} \times \Pi^A_a R) \times \Pi^A_b S$.*

Definition 3. *Let $R, S \subseteq 1^A$ and let $a, b \in A$; then $R \supset^A_{a,b} S = ((1^A \odot^A_{a,b} -^A R) \odot^A_{a,b} 1^A) \cup^A S$.*

The result of applying $\supset^A_{a,b}$ does not depend on the choice of $a \neq b$ in A; we use $R \supset^A S$ to denote $R \supset^A_{a,b} S$, for arbitrary $a \neq b \in A$.

The diagonal-free cylindric set algebra of dimension n is the set algebra $\mathbf{A_n} = < A, \cap, -, \emptyset, X^n, C_i >$, where $(A, \cap, -, \emptyset, X^n)$ is a Boolean algebra of (not necessarily all of the) subsets of X^n and for any $R \subseteq X^n$, $C_i = \{y \in X^n : \exists y' \in R \; \forall j \neq i \; (y(j) = y'(j))\}$. In [7], Imielinski and Lipski showed that relational algebra operations may be defined in terms of intersection, complement and cylindrification and conversely. The approach to relational databases via cylindrification has the advantage that all operations are total, since all relations are of the same type. In [4], Cosmadakis, and in [6], Düntsch and Mikulas, exploited the connection between cylindric algebras and database theory to use new methods and to prove new results. The disadvantage of this approach is that all relations are forced to be of the same arity, and in real life databases, query checking is computationally more efficient if we use relations with varying arities. Moreover, there is no completeness theorem for the cylindrical version of relational database theory. For this reason, we choose to develop a typed calculus, with the four basic operations defined above. Other operations definable in terms of (i)-(iv) include the update operations and other joins. Explicit renaming may be effected by defining appropriate unary operations. As suggested by Kozen, a connection between this work and that of Chandra and Harel [2] is investigated in [10]; there the set of types is extended to include types of the form A^2 and tuples of type A^2 correspond to ordered pairs of tuples of type A. The transitive closure operation on relations is defined and its deduction rules are given; this allows us to make queries about transitive closures.

We can easily see that with typed relations we can express all of the fundamental notions of relational database theory: *schema* - a set of attributes, *relation over a schema* - a typed relation, *tuple* and *database* - a set of typed relations.

3 A Language for a Logic of Typed Relations and Its Semantics

Motivated by the above relations and operations on concrete typed relations, we now develop a language (in fact a scheme of languages) of typed relations, whose intended models are databases. Let T_Ω be a set of types for some set Ω of attributes; the expressions of a language L of typed relations over Ω are built from the following disjoint sets of symbols:

$\{e\}$, where e is interpreted as the empty tuple;

$\mathcal{O}_v^a$, an infinite set of object variables of type a, for each $a \in \Omega$;

$\mathcal{O}_c^a$, a set of constants of type a, for each $a \in \Omega$;

$\mathcal{O}^A$, a set of mixed objects of type A, for each $A \in \mathcal{T}_\Omega$, defined as follows: if $A \neq \emptyset$, $u \in \mathcal{O}^A$ iff u is a mapping from A into $\bigcup\{\mathcal{O}_v^a \cup \mathcal{O}_c^a : a \in A\}$ such that for each a, $u(a) \in \mathcal{O}_v^a$ or $u(a) \in \mathcal{O}_c^a$. $\mathcal{O}_v^A$, the variables of type A, and $\mathcal{O}_c^A$, the constants of type A, are subsets of $\mathcal{O}^A$; $u \in \mathcal{O}_v^A$ iff u is a mapping from A into $\bigcup\{\mathcal{O}_v^a : a \in A\}$, and similarly for $\mathcal{O}_c^A$; we presume $\mathcal{O}_c^\emptyset = \{e\}$;

$\mathcal{R}_v^A$, a set of relation variables of type A, for each $A \in \mathcal{T}_\Omega$;

$\mathcal{R}_c^A$, a set of relation constants of type A, for each $A \in \mathcal{T}_\Omega$; $\mathbf{0}^A, \mathbf{1}^A \in \mathcal{R}_c^A$;

$\mathcal{OP}$, a set of operation symbols of varying arities such that: for every k-ary operation $\bigotimes \in \mathcal{OP}$ $(k \geq 1)$, there is associated a sequence $\tau(\bigotimes) = (A_1, ..., A_k, A)$ of $k+1$ elements of $\mathcal{T}_\Omega$; (A_i is the type of the i-th argument of $\bigotimes$, $i = 1, ..., k$, A is the type of the expression obtained by performing the operation $\bigotimes$).

We presume: $\mathcal{OP} \supseteq \{\Pi_B^A, \cap^A, -^A, \times^{A \uplus C}, : A, B, C \in \mathcal{T}_\Omega, B \subseteq A$; where $\tau(\Pi_B^A) = (A, B)$; $\tau(\cap^A) = (A, A, A)$; $\tau(-^A) = (A, A)$ and $\tau(\times^{A \uplus C}) = (A, C, A \uplus C)$. Assumptions concerning the elements of $\mathcal{O}^{A \uplus B}$ and $\mathcal{O}^\emptyset$, analogous to the corresponding assumptions on the set of tuples, are assumed to hold. It follows from the definitions that $\mathcal{O}^{A \uplus B} = \mathcal{O}^{B \uplus A}$ and $\mathcal{O}^{(A \uplus B) \uplus C} = \mathcal{O}^{A \uplus (B \uplus C)}$. As with the notation defined for the tuples, if u denotes a variable from $O_v^{A \uplus B}$ it may be replaced by an expression vw, where $v \in \mathcal{O}_v^A$ and $w \in \mathcal{O}_v^B$. $\mathcal{O}^a$ denotes $\mathcal{O}_v^a \cup \mathcal{O}_c^a$, and $\mathcal{R}^A$ denotes $\mathcal{R}_v^A \cup \mathcal{R}_c^A$. We assume that for all A, $\mathcal{O}^A \neq \emptyset$.

For each $A \in \mathcal{T}_\Omega$, T_A, the set of *terms of type* A, is the smallest set such that: (i) $\mathcal{R}^A \subseteq T_A$; and (ii) if $\bigotimes \in \mathcal{OP}$ such that $\tau(\bigotimes) = (A_1, ...A_m, A)$ and $F_i \in T_{A_i}$, $i = 1, ..., m$, then $\bigotimes(F_1, ..., F_m) \in T_A$. A *formula* in the language L of typed relations over Ω is an expression of the form $F(u)$, where $F \in T_A$ and $u \in \mathcal{O}^A$, for any $A \in \mathcal{T}_\Omega$.

A *model* for the language L of typed relations over Ω is a system $M = \{\{A : A \in \mathcal{T}_\Omega\}, \{U^A : A \in \mathcal{T}_\Omega\}, e, m\}$, where U^A is a nonempty set of tuples of type A, $U^\emptyset = \{e\}$ and m is a meaning function subject to the following conditions:

(1) If $u \in \mathcal{O}_c^A$, then $m(u) \in U^A$ and if $u = vw$, then $m(u) = m(v)m(w)$; $m(e) = e$: we slightly abuse the language here and use e both as a symbol in the language denoting the empty tuple and also as the empty tuple itself.
(2) If $R \in \mathcal{R}^A$, then $m(R) \subseteq U^A$; in particular, $m(\mathbf{1}^A) = U^A$.
(3) If $\bigotimes \in \mathcal{OP}$ and $\tau(\bigotimes) = (A_1, ..., A_k, A)$, then $m(\bigotimes)$ is a k-ary operation acting on relations of types $A_1, ..., A_k$, returning a relation of type A. The operations (i)-(iv) defined in Section 2 receive their usual database theoretic meaning.
(4) If $F = \bigotimes(F_1, ..., F_k)$, then $m(F) = m(\bigotimes)(m(F_1), ..., m(F_k))$.

It follows that $m(F \rightarrow^A G) = -^A m(F) \cup^A m(G)$. It is easy to see that each database over Ω determines a model for the language of typed relations.

A *valuation* in a model $M = \{\{A : A \in \mathcal{T}_\Omega\}, \{U^A : A \in \mathcal{T}_\Omega\}, e, m\}$ is a function *val*: $\bigcup\{\mathcal{O}^A : A \in \mathcal{T}_\Omega\} \rightarrow \bigcup\{U^A : A \in \mathcal{T}_\Omega\}$ such that: if $u \in \mathcal{O}_v^A$, then $val(u) \in U^A$; if $u \in \mathcal{O}^A_c$, then $val(u) = m(u)$; if $u \in \mathcal{O}^A$ and $v \in \mathcal{O}^B$ then $val(vw) = val(v)val(w)$. It follows from the definition of tuple that $val(uv) = val(vu))$, $val((uv)w) = val(u(vw))$ and $val(ue) = val(eu) = val(u)$.

Let $F \in T_A$ and let $u \in \mathcal{O}^A$. We say that the valuation val in the model M *satisfies* the formula $F(u)$, and write $M, val \models F(u)$, iff $val(u) \in m(F)$. We say that the formula $F(u)$ is *true* in the model M iff it is satisfied by every valuation in M. Therefore, if $u \in \mathcal{O}^A_v$, $F(u)$ is true in the model M iff $m(F) = U^A$. Let $\mathcal{C}$ be a class of models. We say that the formula $F(u)$ is $\mathcal{C}$*-valid*, and write $\models_{\mathcal{C}} F(u)$ iff it is true in all models $M \in \mathcal{C}$. We say that $F(u)$ is *valid* if it is valid in the class of all models as defined above and we write $\models F(u)$. We conclude that if $u \in \mathcal{O}^A_v$, then $(F \rightarrow^A G)(u)$ is valid iff for all models defined above $m(F) \subseteq m(G)$.

By a logic of typed relations we mean a system $\mathcal{L} = (L, \mathcal{C})$, where L is a language of typed relations and $\mathcal{C}$ is a class of models for L.

4 Deduction System for a Logic of Typed Relations

A proof system for a logic of typed relations consists of some finite sequences of formulas, called axiomatic sequences, and rules of the form:

$$\frac{H}{J_1 \quad | \quad J_2 \quad \ldots \quad | \quad J_n} \quad (\text{for } 1 \leq n < \omega)$$

where $H, J_1, ..., J_n$ are finite sequences of formulas. A *sequence of formulas*, $H = \alpha_1, \alpha_2, ..., \alpha_k$, *is* $\mathcal{C}$*-valid* iff for every model $M \in \mathcal{C}$, every valuation in the model M satisfies one of the formulas α_i. A rule of the above form is *admissible* for the class $\mathcal{C}$ of models in the case that H is $\mathcal{C}$-valid if and only if each J_i, $i = 1, ..., n$ is $\mathcal{C}$-valid. Axiomatic sequences take the place of axioms: an axiomatic sequence is admissible for a class $\mathcal{C}$ of models iff it is $\mathcal{C}$-valid.

There are two kinds of rules: *decomposition rules*, which enable us to decompose a formula into simpler formulas and *specific rules*, which enable us to modify a sequence of formulas. The intended interpretations of the operation symbols determine decomposition rules; constraints on the intended interpretation of object constants or relation variables or constants determine specific rules. MacCaull and Orłowska [11] outline a strategy for developing relational deduction systems.

If $F(u)$ is the formula whose validity is in question, we generate a tree by placing $F(u)$ at the root and applying deduction rules, until all branches close or there is an open (i.e., non-closed) branch that is complete. Closure of a branch means we have reached an axiomatic sequence on the branch. Completeness of a branch means that all the rules that can be applied have been applied. If all the branches close, the formula $F(u)$ is said to be *provable*. We present the system $\mathcal{D}$, the rules and axiomatic sequences for the class of models satisfying (1)-(4) of Section 3. Below, K and H represent (possibly empty) sequences of formulas, and new $v \in \mathcal{O}^A$ means v is not on the branch when we apply the rule.

Decomposition Rules for $\mathcal{D}$

Let $F, G \in T_A$, let $B \subseteq A$ and let $u \in \mathcal{O}^A$, $w \in \mathcal{O}^B$; then:

$(\cap)\quad \dfrac{K,\ (F\cap^A G)(u),\ H}{K,\ F(u),\ H \quad | \quad K,\ G(u),\ H}$ $\qquad (-\cap)\quad \dfrac{K,\ -^A(F\cap^A G)(u),\ H}{K,\ -^A F(u),\ -^A G(u),\ H}$

$(\Pi)\quad \dfrac{K,\ (\Pi_B^A F)(w),\ H}{K,\ F(wt),\ H,\ (\Pi_B^A F)(w)}$ any $t \in \mathcal{O}^{A-B}$ $\qquad (-\Pi)\quad \dfrac{K,\ -^B(\Pi_B^A F)(wt),\ H}{K,\ -^A F(wt),\ H}$ new $t \in \mathcal{O}_v^{A-B}$

$$(--)\quad \frac{K,\ -^A(-^A F)(u),\ H}{K,\ F(u),\ H}$$

Let $F \in T_A$, $G \in T_B$, $v \in \mathcal{O}^A$, $w \in \mathcal{O}^B$ and $z \in \mathcal{O}^{A-B}$; then:

$(\times)\quad \dfrac{K,\ (F\times^{A\uplus B} G)(vw),\ H}{K,\ F(v),\ H \quad | \quad K,\ G(w),\ H}$ $\qquad (-\times)\quad \dfrac{K,\ -^{A\uplus B}(F\times^{A\uplus B} G)(vw),\ H}{K,\ -^A F(v),\ -^B G(w),\ H}$

Specific Rules for $\mathcal{D}$

Let $F \in T_A$, $u \in \mathcal{O}^A$, $A = A_1 \uplus ... \uplus A_n$, $n \geq 1$, $u_j \in \mathcal{O}_v^{A_j}$, $j = i, ..., n$, $e \in \mathcal{O}_v^{\emptyset}$ and π, a permutation on $\{1, ..., n\}$; then:

$(\pi)\quad \dfrac{K,\ F(u_1...u_n),\ H}{K,\ F(u_{\pi(1)}...u_{\pi(n)}),\ H}$ $\qquad (e)\quad \dfrac{K,\ F(u_1...u_{i-1}(eu_i)...u_n),\ H}{K,\ F(u_1...u_{i-1}u_i...u_n),\ H}$ $n \geq 1,\ 1 \leq i \leq n$

$$(\uplus)\quad \frac{K,\ F(u_1u_2...u_n),\ H}{K,\ F(u_1...u_{i-1}v_1v_2u_{i+1}...u_n),\ H,\ F(u_1...u_n)}$$

where $A = A_1 \uplus ... \uplus A_n$, $u_j \in \mathcal{O}^{A_j}$, $j = 1, ..., n$, for some i, $1 \leq i \leq n$, $u_i \in \mathcal{O}_v^{A_i}$, $A_i = B_1 \uplus B_2$ and $v_1 \in \mathcal{O}_v^{B_1}$, $v_2 \in \mathcal{O}_v^{B_2}$.

The rule (π) shows that objects can be permuted without changing the meaning and the rules (e) and $(\uplus)$ show that any variable can be split into components in such a way that the type is preserved. The following sequences are clearly valid.

Axiomatic Sequences for $\mathcal{D}$

(1) $R(u), (-^A R)(u)$ for any $R \in \mathcal{R}^A$ and any $u \in \mathcal{O}^A$;
(2) $\mathbf{1}^A(u)$ for any $u \in \mathcal{O}^A$ (in particular, if $A = \emptyset$);
(3) $-^A\mathbf{0}^A(u)$ for any $u \in \mathcal{O}^A$;
(4) $(\Pi_\emptyset^A F)(e)$ for any $F \in T_A$, $F \neq 0^A$, $e \in \mathcal{O}^\emptyset$.

We say the L-formula $F(u)$ is *$\mathcal{D}$-provable*, and write $\vdash_{\mathcal{D}} F(u)$, iff it is provable using rules and axiomatic sequences from the system $\mathcal{D}$. The proof of the following may be found in MacCaull and Orłowska [10].

Theorem 1. *(Soundness and completeness) For all L-formulas $F(u)$, $\vdash_{\mathcal{D}} F(u)$ iff $\models F(u)$.*

We may restrict $\mathcal{C}$, the class of models, by adding constraints on constants or relation variables; in this case, the deduction system must be augmented by adding corresponding specific rules or axiomatic sequences (see MacCaull and Orłowska [11]).

5 Applications

I. The usual theorems for Boolean algebras and relational algebras are provable. We detail some examples at the end of the section.

II. Query Checking. For particular databases we define deduction systems with axiomatic sequences of the form $S(c)$ where c is an individual constant of the appropriate type indicating that tuple $c \in S$, and, possibly, with specific rules corresponding to constraints on particular relation constants. To determine what tuples are in a relation R in a particular database, we proceed as follows: develop a proof tree for $R(u)$; suppose all branches close except in the following case: $u = v_1 ... v_n$ where $v_i \in \mathcal{O}_v^{A_i}$, and for each $i = 1, ..., n$, $S(v_i)$ appears on an open branch; if for all $i = 1, ..., n$, $S(c_i)$ is an axiomatic sequence, where $c_i \in \mathcal{O}_c^{A_i}$, then the tuple $c_1 ... c_n \in R$. For example, if $S(c), T(d_1), T(d_2)$ are axiomatic sequences for constants c, d_1, d_2, and if R is the relation $S \times T$, then we can use this procedure to conclude that tuples cd_1 and $cd_2 \in R$.

III. Model Checking. Let R be a typed relation; the constant tuple $c \in R$ if there is a closed proof tree for $R(c)$.

IV. Selections. In classical database theory, the selections are accomplished vis a vis the notion of *comparison formula*; from simple comparison formulas such as $x = y$ or $x \leq y$, compound comparison formulas are built using the connectives of propositional logic. For each comparison formula there is an associated selection operation. Here we opt for a more unified presentation via relations and operations on them. This more general selection captures all the selections encountered in database theory. Moreover, by adding specific rules, we can make selections with respect to relations S with various properties. Since soundness holds, the proof of a formula demonstrates truth in a class of models which includes models where the interpretation of S has the desired properties.

V. Entailment. The proof of the following may be found in MacCaull and Orłowska [10].

Theorem 2. *Let $F \in T_A$, and let $G \in T_B$; if $C = A \cup B$, and $u \in \mathcal{O}^A$, $v \in \mathcal{O}^B$ and $w \in \mathcal{O}^C$, then the following are equivalent:*

(1) for every model M, if $F(u)$ is true in M then $G(v)$ is true in M;
(2) $(F \times 1^{C-A}) \supset^C (G \times 1^{C-B}))(w)$ is $\mathcal{D}$-provable.

VI. Nondeterministic Databases. We use the results of this paper to deal with generalized databases; that is, databases in which a record entry is a set of subsets of attribute domains (Lipski [8]). Let Ω be a set of attributes and let $A \in T_\Omega$.

Definition 4. *A generalized tuple of type A is a map $u : A \rightarrow 2^{D_A}$, such that for all $a \in A$, $u(a) \subseteq D_a$. The collection of all (generalized) tuples of type A is called the relation 1^A. A generalized relation R of type A is a subset of the set 1^A.*

In the relational deduction system presented in previous sections, the tuples may be generalized tuples.

In [14], Orłowska presents numerous information relations for rough set theory, which allow us to reason about nondeterministic databases. Some of these relations have analogues for generalized typed relations. Typed analogues for some information relations of rough set theory are formulated below, where $-u(a)$ is shorthand for $D_a - u(a)$; in what follows, we assume that $a \neq b$.

$$\begin{aligned}
ind^a_b &= \{u \in 1^{\{a,b\}} : u(a) = u(b)\} \text{ (an indiscernibility relation);}\\
sim^a_b &= \{u \in 1^{\{a,b\}} : u(a) \cap u(b) \neq \emptyset\} \text{ (a similarity relation);}\\
nsim^a_b &= \{u \in 1^{\{a,b\}} : -u(a) \cap -u(b) \neq \emptyset\} \text{ (a negative similarity relation);}\\
dis^a_b &= \{u \in 1^{\{a,b\}} : u(a) \cap u(b) = \emptyset\} \text{ (a disjointness relation);}\\
in^a_b &= \{u \in 1^{\{a,b\}} : u(a) \subseteq u(b)\} \text{ (an inclusion relation);}\\
exh^a_b &= \{u \in 1^{\{a,b\}} : u(a) \cup u(b) = D_{\{a,b\}}\} \text{ (an exhaustiveness relation);}\\
com^a_b &= \{u \in 1^{\{a,b\}} : u(a) = -u(b)\} \text{ (a complementarity relation);}\\
pnsim^a_b &= \{u \in 1^{\{a,b\}} : u(a) \cap -u(b) \neq \emptyset\} \text{ (a positive negative similarity relation).}
\end{aligned}$$

Information relations for untyped systems have a number of useful properties (see for example, Orłowska [14]), and it is interesting to consider meaningful properties of information relations for typed systems. For example: $ind^a_b = ind^b_a$; $sim^a_b = sim^b_a$; $nsim^a_b = nsim^b_a$; and $exh^a_b = exh^b_a$. Using the above relations as first arguments when applying the selection operation, we can perform selections in nondeterministic databases analogous to the typical selections of classical databases. For example, $\sigma^A_{a,b}(ind^a_b, R) = \{u \in R : u(a) = u(b)\}$. The following form of transitivity on the indices holds for the collection of relations $\{ind^a_b : a, b \in \Omega\}$:

$$\text{if } a, b, c \in A \text{ then } \sigma^A_{\{a,b\}}(ind^a_b, R) \cap \sigma^A_{\{b,c\}}(ind^b_c, R) \subseteq \sigma^A_{\{a,c\}}(ind^a_c, R).$$

This transitivity on the indices holds for the collection of relations $\{in^a_b : a, b \in \Omega\}$. Other forms of transitivity on the indices hold; for example:

$$\text{if } a, b, c \in A \text{ then } \sigma^A_{\{a,b\}}(sim^a_b, R) \cap \sigma^A_{\{b,c\}}(in^b_c, R) \subseteq \sigma^A_{\{a,c\}}(sim^a_c, R).$$

We may use selections to express an analogue for 3-transitivity on the indices, which holds for the collection $\{com^a_b : a, b \in \Omega\}$; refer to MacCaull and Orłowska [11] for the deduction rules corresponding to the above properties.

We conclude this section with some examples; below, u is a variable of the appropriate type.

1. Let us prove that if R is a relation of type A and $B \subseteq A$, then $R \subseteq (\Pi^A_B R) \times^{B \uplus (A-B)} 1^{A-B}$; it suffices to demonstrate that $\vdash_{\mathcal{D}} (R \rightarrow^A (\Pi^A_B R) \times^{B \uplus (A-B)} 1^{A-B})(u)$.

$$\dfrac{(R \rightarrow^A ((\Pi^A_B R) \times^{B \uplus (A-B)} 1^{A-B})(u)}{\dfrac{(-^A R)(u),\ (\Pi^A_B R) \times^{B \uplus (A-B)} 1^{A-B}(u)}{\dfrac{(-^A R)(vw),\ (\Pi^A_B R) \times^{B \uplus (A-B)} 1^{A-B}(vw)}{\dfrac{(-^A R)(vw),\ (\Pi^A_B R)(v)}{(-^A R)(vw),\ R(vw)} \quad \Big| \quad (-^A R)(vw),\ 1^{A-B}(w)}}}$$

We have applied the rule $(\cup)$, the rule $(\uplus)$ (since $A = B \uplus (A - B)$, we may replace u by vw where $v \in \mathcal{O}_v^B$ and $w \in \mathcal{O}_v^{A-B}$), the rule $(\times)$ and finally on the left branch, the rule (Π). Each branch ends in an axiomatic sequence.

2. Let us prove that if R, S are relations of type A and $B \subseteq A$ then $\Pi_B^A R \subseteq \Pi_B^A(R \cup^A S)$.

$$\cfrac{((\Pi_B^A R) \rightarrow^A (\Pi_B^A(R \cup^A S)))(u)}{\cfrac{(-^A \Pi_B^A R)(u),\ (\Pi_B^A(R \cup^A S))(u)}{\cfrac{(-^A R)(ut),\ (\Pi_B^A(R \cup^A S))(u)}{\cfrac{(-^A R)(ut),\ (R \cup^A S)(ut)}{(-^A R)(ut),\ R(ut),\ S(ut)}}}}$$

We first used the rule $(\cup)$, then the rule $(-\Pi)$ with a new $t \in \mathcal{O}_v^{A-B}$, then the rule (Π) and finally, the rule $(\cup)$. We could choose any $v \in \mathcal{O}^{A-B}$ for the rule (Π), and we chose t, a variable already on the branch. The branch terminates with an axiomatic sequence.

3. A sampling of $\mathcal{D}$-provable properties of typed relations:

(1) $\Pi_B^A 1^A = 1^B$;
(2) $\Pi_\emptyset^A R \rightarrow^\emptyset \mathbf{1}^\emptyset$;
(3) $\mathbf{0}^A \rightarrow^A (R \cap^A S)$;
(4) $\mathbf{1}^\emptyset \times^{\emptyset \uplus A} R \rightarrow^A R$;
(5) $T \times^{A_1 \uplus A_2} \sigma_B^{A_2}(S, R) \rightarrow^{A_1 \uplus A_2} \sigma_B^{A_1 \uplus A_2}(S, T \times R)$;
(6) $(R \times^{A \uplus B} S) \rightarrow^{A \uplus B} (S \times^{B \uplus A} R)$.

If C is a class of models such that the constant relation S of type A is interpreted as a nonempty relation, then the deduction system $\mathcal{D}_\mathcal{C}$ contains the following rule:

$$(\in \mathbf{S}) \qquad \frac{K}{K, -^A \mathbf{S}(t)} \qquad \text{new } t \in \mathcal{O}_v^A$$

and the following is $\mathcal{D}_\mathcal{C}$-provable:

(7) $\Pi_A^{A \uplus B}(R \times^{A \uplus B} S) = R$.

References

[1] Buszkowski, W. and Orłowska, E.: *Indiscernibility-based formalization of dependencies in information systems*, E.Orłowska, editor, Incomplete Information: Rough Set Analysis, Physica Verlag (1997) 293-315.

[2] Chandra, A. K. and Harel, D.: *Computable Queries for Relational Data Bases*, J. of Computer and System Sciences, **21** (1980) 156-178.

[3] Codd, E.: *A relational model for large shared data banks*, Communications of the ACM **16** (1970) 377-387.

[4] Cosmadakis, S.: *Database theory and cylindric lattices* in: A. K. Chondra, editor, Proceedings of the 28th Annual Symposium on Foundations of Computer Science, IEEE Computer Science Press (1987) 411-420.

[5] Demri, S. and Orłowska, E.: Incomplete Information: Structure, Inference, Complexity, EATCS Monographs in Theoretical Computer Science, Springer (2002).

[6] Düntsch, I. and Mikulas, S.: *Cylindric structures and dependencies in relational databases*, J. of Theoretical Computer Science, **269** (2001) 451-468.

[7] Imielinski, T. and Lipski, W.: *The relational model of data and cylindric algebras*, J. of Computer and System Sciences, **28** (1984) 80-102.

[8] Lipski, W.: *On semantic issues connected with incomplete information databases*, ACM Transactions on Database Sytems, **4** (1979) 262-296.

[9] MacCaull, W.: *Proof theory for generalized dependencies for information relations*, Fundamenta Informaticae, **42** (2000) 1-27.

[10] MacCaull, W. and Orłowska, E.: *A logic of typed relations and its applications to relational database theory*, Preprint.

[11] MacCaull, W. and Orłowska, E.: *Correspondence results for relational proof systems with application to the Lambek calculus*, Studia Logica, **71** (2002) 389-414.

[12] Orłowska, E.: *Algebraic approach to database constraints*, Fundamenta Informaticae, **10** (1987) 57-66.

[13] Orłowska, E.: *Relational Formalization of Nonclassical Logics*, in: C. Brink, W. Kahl and G. Schmidt, editors, Relational Methods in Computer Science, Springer Verlag (1996) 90-105.

[14] Orłowska, E.: *Introduction: What You Always Wanted to Know About Rough Sets*, in: E. Orłowska, editor, Incomplete Information: Rough Set Analysis, Physica Verlag (1997) 1-20.

[15] Pawlak, Z.: Rough Sets, Kluwer, Dordrecht (1991).

[16] Rasiowa, H. and Sikorski, R.: The Mathematics of Metamathematics, Polish Science Publishers, Warsaw (1963).

[17] Simovici, D. A. and Tenney, R. L.: Relational Database Systems, Academic Press (1995).

[18] Ullman, J.: Database and Knowledge-Base Systems Volume 1, Computer Science Press (1988).

Greedy-Like Algorithms in Modal Kleene Algebra*

Bernhard Möller and Georg Struth

Institut für Informatik, Universität Augsburg
Universitätsstr. 14, D-86135 Augsburg, Germany
{moeller,struth}@informatik.uni-augsburg.de

Abstract. This study provides an algebraic background for the formal derivation of greedy-like algorithms. We propose Kleene algebra as a particularly simple alternative to previous approaches such as relation algebra. Instead of converse and residuation we use modal operators that are definable in a wide class of algebras, based on domain/codomain or image/pre-image operations. By abstracting from earlier approaches we arrive at a very general theorem about the correctness of loops that covers particular forms of greedy algorithms as special cases.

Keywords: Idempotent semiring, Kleene algebra, image and preimage operation, modal operators, confluence, Geach formula, program development and analysis

1 Introduction

This study is concerned with algebraic derivations and correctness proofs of greedy-like algorithms that use a simple loop to calculate a global optimum. We present a fairly general correctness criterion and give sufficient criteria when iteration in a discrete partially ordered problem domain correctly implements the general algorithm scheme. Proper greedy algorithms are further specializations in which the loop steps are driven by an additional local optimality criterion.

Earlier work (e.g. [1, 2]) has used relation algebra and other algebraic formalisms (e.g. [4]). We show that modal Kleene algebra allows a particularly simple, concise and more general treatment, avoiding the concepts of converse and residual, as used in relation algebra, and exhibiting the derivation structure more clearly.

The central correctness conditions require that certain optimality criteria semi-commute with certain program statements to achieve global optimality. These semi-commutation properties, of the same shape as confluence properties in rewriting systems, can be expressed without converse using modal operators as in dynamic logic; these can be added to Kleene algebra in a simple way.

---

* Research partially sponsored by DFG Project InopSys — Interoperability of Calculi for System Modelling.

R. Berghammer et al. (Eds.): RelMiCS/Kleene-Algebra Ws 2003, LNCS 3051, pp. 202–215, 2004.

2 Looping for Optimality

Greedy algorithms solve certain optimization problems, proceeding in a stepwise fashion without backtracking. At each step there is a set of choices from which a greedy algorithm always takes the one that seems best at the moment, i.e., it works locally without lookahead to the global optimum to be achieved eventually. Instances of this scheme are shortest path and minimum spanning tree problems in graphs, the construction of Huffman codes and scheduling problems. Of course, the greedy approach only works for certain types of problems: as is well-known from hiking in the mountains, always choosing the steepest path will rarely lead to the highest summit of the whole area. The central correctness requirement for the greedy scheme is that

a local choice must not impair reaching the global optimum.

In this and the following section we derive, within the framework of relational algebra, general conditions under which a loop satisfies this principle; later we abstract to Kleene algebra. It turns out that local optimality is inessential; so we obtain a more general class of loops that we call *greedy-like*.

We start with a specification relation T that connects inputs to admissible outputs and a relation C that compares outputs and captures the notion of (global) optimality. The derivation will exhibit our precise requirements on C.

A relation R *improves* T w.r.t. C if it always relates inputs to outputs that are at least as good as those prescribed by T, in formulas

$$\forall\, x, y, z : xTy \wedge xRz \Rightarrow y\; C\; z\;, \qquad \text{or, equivalently,} \qquad T^{\smile}; R \subseteq C\;,$$

where ; denotes relational composition and $T^{\smile}$ is the converse of T. Since $\emptyset$ trivially improves T, we are interested in the greatest improvement and define it by the Galois connection

$$X \subseteq \mathrm{GIMP}(T, C) \overset{\mathrm{def}}{\Leftrightarrow} T^{\smile}; X \subseteq C\;. \tag{1}$$

Using a residual, this could be expressed as $\mathrm{GIMP}(T, C) = T^{\smile}\backslash C$. However, we will not need any special properties of residuals.

An implementation of specification T that always produces optimal solutions then is a relation that refines and improves T. So we define

$$\mathrm{OPT}(T, C) \overset{\mathrm{def}}{=} T \cap \mathrm{GIMP}(T, C)\;.$$

We now consider a loop program

$$W \overset{\mathrm{def}}{=} \textsf{while } P \textsf{ do } S \overset{\mathrm{def}}{=} (P\,;S)^*\,;\neg P\;.$$

Here the loop condition P is represented by a subidentity $P \subseteq I$, where I is the identity relation, and $\neg P \overset{\mathrm{def}}{=} I - P$ is its relative complement. We want to calculate a sufficient criterion for W to be a refinement of $\mathrm{OPT}(T, C)$, i.e., for

$$W \subseteq T\,, \qquad (2) \qquad\qquad W \subseteq \mathrm{GIMP}(T,C)\,, \qquad (3)$$

where we defer the treatment of (2) to the next section.

Spelling out the definitions in (3) results in $T^{\smile} ; (P ; S)^* ; \neg P \subseteq C$. We abstract a bit and try to answer the question when, for $Q \subseteq I$, we have

$$U ; V^* ; Q \subseteq C\,. \qquad (4)$$

A standard result from regular algebra (see (15) in Section 5) is the semi-commutation property $W ; X \subseteq Y ; Z \Rightarrow W ; X^* \subseteq Y^* ; Z$. Hence (4) can be established given the conditions

$$U ; V \subseteq C ; U\,, \qquad (5) \qquad\qquad U ; Q \subseteq C\,, \qquad (6)$$

since then

$$U ; V^* ; Q \subseteq C^* ; U ; Q \subseteq C^* ; C = C^+\,.$$

If we now assume C to be transitive, which is reasonable for a comparison relation, we have $C^+ \subseteq C$ and can draw the desired conclusion.

How can we, in turn, establish (5) and (6), at least in our special case? Translating back we get the proof obligations

$$T^{\smile} ; P ; S \subseteq C ; T^{\smile}, \qquad (7) \qquad\qquad T^{\smile} ; \neg P \subseteq C\,. \qquad (8)$$

Let us interpret these conditions. (7) means that every pass through the loop body preserves the possibility of obtaining a solution that is at least as good as all possible solutions before. (8) means that upon loop termination no possible solution is better than the termination value.

3 Iterating Through the Problem Domain

We now decompose the specification relation T into the iteration of a set E of elementary steps between elements of the problem domain. We admit, as initial approximations, arbitrary inputs but as outputs only terminal elements from which no further elementary steps are possible. Therefore we assume now that T has the special shape

$$T = \mathsf{exhaust}\, E \stackrel{\text{def}}{=} E^* ; \neg\ulcorner E = \mathsf{while}\ \ulcorner E\ \mathsf{do}\ E\,, \qquad (9)$$

where the *domain* of a relation R is, as usual, defined by

$$\ulcorner R \stackrel{\text{def}}{=} R ; R^{\smile} \cap I\,. \qquad (10)$$

Such a problem structure is found, e.g., in matroids and greedoids [6, 7] where it is additionally assumed that T is a discrete strict-order and that all terminal (or maximal) elements, the *bases*, have the same height (also known as *rank* or *dimension*) in the associated Hasse diagram.

We try to calculate an implementation that traverses the problem domain without backtracking, i.e., using elementary steps only forward. This suggests trying $P ; S \subseteq E$. Now, by monotonicity of the star operation, proof obligation (2) can be fulfilled if additionally we can achieve $\neg P \subseteq \neg\ulcorner E$ or, equivalently, $\ulcorner E \subseteq P$. Sufficient conditions for these properties are

$$P ; S \subseteq E \qquad\qquad \ulcorner(P ; S) \supseteq \ulcorner E \,. \tag{11}$$

These are reasonable requirements, since they prevent that the iteration blocks at a non-terminal element. They even imply $\ulcorner(P ; S) = \ulcorner E$.

Next, we deal with proof obligation (8), assuming (9). We calculate

$$T^{\smile} ; \neg\ulcorner E \subseteq C \Leftrightarrow \neg\ulcorner E ; T \subseteq C^{\smile} \Leftrightarrow \neg\ulcorner E ; E^* ; \neg\ulcorner E \subseteq C^{\smile} \Leftrightarrow$$
$$\neg\ulcorner E ; (I \cup E ; E^*) ; \neg\ulcorner E \subseteq C^{\smile} \Leftrightarrow \neg\ulcorner E \subseteq C \,.$$

Step one employs properties of converse. Step two uses (9). Step three unfolds the star. Step four uses distributivity, $\neg\ulcorner E ; E = \emptyset$ and idempotence of $\neg\ulcorner E$.

So (8) is established if C is reflexive on terminal elements. This holds, in particular, if C is fully reflexive, i.e., a pre-order. But in some applications one may choose to leave C partially reflexive. E.g., when constructing a Huffman code, the non-terminal elements are proper forests, for which a comparison relation is not given as easily as for the terminal elements, which are single code trees.

As for proof obligation (7), it is a generic condition that has to be considered individually in each case. Our derivation can be summed up as follows.

Theorem 1. *Suppose that* $T = \mathsf{exhaust}\, E$ *and that* C *is reflexive on* $\neg\ulcorner E$ *and transitive. Then* $(7) \wedge (11) \Rightarrow \mathsf{while}\ \ulcorner E\ \mathsf{do}\ S \subseteq \mathrm{OPT}(T, C)$.

So far we still have a general scheme that does not specifically mention greediness. But we can refine S further to choose in every step a locally optimal element. To this end we need another pre-order L and stipulate

$$S \subseteq \mathrm{GIMP}(E, L) \,. \tag{12}$$

This now provides a truly greedy algorithm, the correctness of which is already shown by Theorem 1. It corresponds to Curtis's "Best-Global" algorithm [2].

4 From Converses to Modalities

The central step in the above derivation, viz. exhibiting conditions (7) and (8), uses only regular algebra. It is an interesting question whether the whole derivation can be ported to Kleene algebra by eliminating the converse operation in some way. This would generalize the result to a much wider class of models.

In the above formulas converse is used only in a very restricted way reminiscent of the relational formulation of a general diamond (or confluence) property:

$$R^{\smile} ; S \subseteq T ; U^{\smile} \,. \tag{13}$$

To bring (8) into this form, just compose the right hand side with $I^{\smile}$.

This observation is the key to success if one also remembers modal correspondence theory (see e.g. [12]), according to which the above formula is equivalent to each of the modal Geach formulas

$$\langle R\rangle[T]P \Rightarrow [S]\langle U\rangle P\ , \qquad \langle S\rangle[U]P \Rightarrow [R]\langle T\rangle P\ . \tag{14}$$

Now we are in good shape, since the modal operators $\langle X\rangle$ and $[X]$ can be defined as predicate transformers in Kleene algebra, cf. [11].

We shall use many instances of these formulas. Since one can easily confuse the rôles of the relations involved in the modal formulation, we shall illustrate these formulas by the type of diagram shown on the right hand side. When read as a confluence-type diagram, the solid arrows and their end points symbolize given elements and relations between them, whereas the dotted ones stand for a quantifier stipulating existence of a further element and appropriate relations to given elements. If one of the arrows is an identity, the diagram shrinks to a triangle.

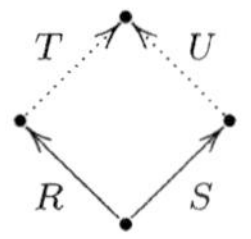

5 Abstracting to Modal Kleene Algebra

We now abstract from the relational setting to the more general formalism of modal Kleene algebra.

First, a *Kleene algebra* (KA) [8] is a structure $(K, +, \cdot, ^*, 0, 1)$ such that $(K, +, \cdot, 0, 1)$ is an idempotent semiring and * is a unary operation axiomatized by the (Horn) identities

$$1 + aa^* \leq a^*\ , \quad (*\text{-}1) \qquad b + ac \leq c \Rightarrow a^*b \leq c\ , \quad (*\text{-}3)$$

$$1 + a^*a \leq a^*\ , \quad (*\text{-}2) \qquad b + ca \leq c \Rightarrow ba^* \leq c\ , \quad (*\text{-}4)$$

for all $a, b, c \in K$. Here, $\leq$ denotes the natural ordering on K defined by $a \leq b$ iff $a + b = b$. An important property that follows from these axioms is the semi-commutation law

$$a\,b \leq ca \ \Rightarrow a\,b^* \leq c^*a\ . \tag{15}$$

A special case of this establishes (4). A KA is *-*continuous* if for all a, b, c we have $ab^*c = \sum_{i \in \mathbb{N}} a\,b^i c$.

A *Kleene algebra with tests* (KAT) [9] is a KA with a Boolean subalgebra $\mathsf{test}(K) \subseteq K$ of *tests* or *predicates*, in which 1 is the greatest element, 0 is the least element and $\cdot$ coincides with the meet operation. In a KAT K one again defines $\mathsf{while}\ p\ \mathsf{do}\ a \stackrel{\text{def}}{=} (pa)^*\neg p$ for $p \in \mathsf{test}(K)$ and $a \in K$.

Finally, a *Kleene algebra with domain* (KAD) [3] is a KAT with an additional operation $\ulcorner\_ : K \to \mathsf{test}(K)$ such that for all $a, b \in K$ and $p, q \in \mathsf{test}(K)$,

$$a \leq \ulcorner a\, a\ , \qquad \ulcorner(pa) \leq p\ , \qquad \ulcorner(ab) = \ulcorner(a\ulcorner b)\ . \tag{dom}$$

Let us explain these axioms. As in the algebra of relations, multiplication with a test from the left or right means domain or range restriction, resp. Now first, since $\ulcorner a \leq 1$ by $\ulcorner a \in \mathsf{test}(K)$, monotonicity of multiplication shows that the first axiom can be strengthened to an equality expressing that restriction to the full domain is no restriction at all. The second axiom means that after restriction the remaining domain must satisfy the restricting test. The third axiom serves to make the modal operators below well-behaved w.r.t. composition. An important consequence of the axioms is that $\ulcorner\_$ preserves arbitrary existing suprema [11].

Examples of KADs are the algebra of concrete relations, where $\ulcorner\_$ coincides with the operation defined in (10), the algebra of path sets in a directed graph (see e.g. [10]) and Kleene's original algebra of formal languages.

In a KAD, the (forward) modal operators *diamond* and *box* are defined by

$$\langle a\rangle p \stackrel{\text{def}}{=} \ulcorner(ap) , \qquad [a]p \stackrel{\text{def}}{=} \neg\langle a\rangle\neg p ,$$

resulting in a *modal Kleene algebra*. This is adequate, since it makes the diamond coincide with the inverse image operator. The third axiom in (dom) implies

$$\langle ab\rangle p = \langle a\rangle\langle b\rangle p , \qquad [ab]p = [a][b]p . \tag{16}$$

The modal operators for a test q are given by

$$\langle q\rangle p = qp , \qquad [q]p = q \rightarrow p \stackrel{\text{def}}{=} \neg q + p . \tag{17}$$

For the above-mentioned connection between confluence-type formulas and the Geach formula we introduce the following notion. A KAD *with converse* is a KAD K with an additional operation $\_^{\smile} : K \rightarrow K$ that is an involution, distributes over $+$, is the identity on tests and is contravariant over $\cdot$, i.e., satisfies $(ab)^{\smile} = b^{\smile}a^{\smile}$. One can show (see again [3]) that over a KAD with converse the first two axioms of (dom) imply the Galois connection

$$\langle a^{\smile}\rangle p \leq q \Leftrightarrow p \leq [a]q . \tag{18}$$

It follows that in a KAD with converse all predicate transformers $\lambda p \,.\, \langle a\rangle p$ are universally disjunctive, i.e., preserve all existing suprema (and that all predicate transformers $\lambda p \,.\, [a]p$ are universally anti-disjunctive, i.e., transform all existing suprema into corresponding infima). This generalizes to KADs that also provide a codomain operation, since there one can also define the backward modal operators and replace $\langle a^{\smile}\rangle$ by the backward diamond of a.

Moreover, using (17) and the shunting rule from Boolean algebra we get, even in KADs without converse, for tests p, q, r

$$p \leq [r]q \Leftrightarrow \langle r\rangle p \leq q . \tag{19}$$

Finally, using the star induction axioms, one can show the following induction principle for the diamond operator (cf. [3]):

$$\langle a\rangle p + q \leq p \Rightarrow \langle a^*\rangle q \leq p . \tag{20}$$

6 Properties of Modal Operators

Our previous relation-algebraic derivation can now be mimicked more abstractly at the level of predicate transformers over an arbitrary KAD. In particular, we do not need the carrier of the algebra to be a complete Boolean algebra as in the case of relation algebra. Although the particular case of greedy algorithms can be treated in the weaker system of power allegories (see [1, 2]), we can even avoid residuals and converse.

Assume a KAT $(K, \mathsf{test}(K), +, \cdot, 0, 1, ^*)$. By a *predicate transformer* we mean a function $f : \mathsf{test}(K) \to \mathsf{test}(K)$. It is *disjunctive* if $f(p+q) = f(p) + f(q)$ and *conjunctive* if $f(p \cdot q) = f(p) \cdot f(q)$.

Let P be the set of *all* predicate transformers and D the set of strict and disjunctive (and hence monotonic) ones. Under the pointwise ordering

$$f \leq g \overset{\text{def}}{\Leftrightarrow} \forall\, p \,.\, f(p) \leq g(p)$$

P forms a lattice in which the supremum $f \oplus g$ and infimum $f \odot g$ of f and g are the pointwise liftings $\oplus$ and $\odot$ of $+$ and $\cdot$, resp.

We now concentrate on the modal predicate transformers $\langle \_ \rangle$ and $[\_]$ from KAD. For the rest of the paper we will work as much as possible at the point-free level. To smoothen the notation, we will denote composition of predicate transformers by mere juxtaposition.

In a KAD with converse the test-level Galois connection (18) implies the predicate-transformer-level cancellation laws and Galois connection

$$\langle a^{\smile} \rangle [a] \leq \langle 1 \rangle \leq [a] \langle a^{\smile} \rangle \;, \tag{21}$$

$$\langle a^{\smile} \rangle f \leq g \Leftrightarrow f \leq [a] g \;. \tag{22}$$

We will now give an abstract proof of equivalence of the Geach formula (14) and the confluence property (13). We do this in KADs that are *extensional* (or *separable* as they are called in dynamic algebra), i.e., satisfy

$$a \leq b \Leftrightarrow \langle a \rangle \leq \langle b \rangle \;. \tag{23}$$

Note that only the direction from right to left must be required, the other one holds in KAD by monotonicity.

Theorem 2. *In an extensional* KAD *with converse,*

$$a^{\smile} b \leq c d^{\smile} \Leftrightarrow \langle b \rangle [d] \leq [a] \langle c \rangle \;.$$

Proof. ($\Rightarrow$) We calculate
$a^{\smile} b \leq c d^{\smile} \;\Leftrightarrow\; \langle a^{\smile} b \rangle \leq \langle c d^{\smile} \rangle \;\Leftrightarrow\; \langle a^{\smile} \rangle \langle b \rangle \leq \langle c \rangle \langle d^{\smile} \rangle \;\Leftrightarrow$
$\langle b \rangle \leq [a] \langle c \rangle \langle d^{\smile} \rangle \;\Rightarrow\; \langle b \rangle [d] \leq [a] \langle c \rangle \langle d^{\smile} \rangle [d] \;\Rightarrow\; \langle b \rangle [d] \leq [a] \langle c \rangle$.
The first step uses (23). The second step employs (16). The third step applies (18). The fourth step uses monotonicity. The fifth step follows by (21).
($\Leftarrow$) Let $\langle b \rangle [d] \leq [a] \langle c \rangle$. Then $\langle b \rangle \leq \langle b \rangle [d] \langle d^{\smile} \rangle \leq [a] \langle c \rangle \langle d^{\smile} \rangle$. Now the proof continues like for ($\Rightarrow$), read upside down. □

The Geach formula has the desired interpretation even in KAs without converse. To understand this, think of KA elements as describing sets of computation paths leading from initial to final states. The modal operators forget the intermediate states and hence induce a relation-like view of the KA elements, and so the Geach formula carries over its meaning to the more general setting.

It should also be noted that the dual formula $ab^{\smile} \leq c^{\smile}d$ cannot be treated in the same way; it requires a KA with codomain rather than domain and the corresponding backward modal operators. So greatest flexibility is achieved in KAs with both domain and codomain.

A special case of the Geach formula deals with the domain operator. Its relational definition (10) implies $\ulcorner R \subseteq R ; R^{\smile}$, admitting a certain relaxation of domain constraints. Using the Geach formula, this translates into the modal Kleene formula $\langle \ulcorner a\rangle [a] \leq \langle a\rangle$, which holds in *all* KADs, not only in extensional KADs with converse. An easily verified consequence is

$$ab \leq b \Rightarrow \langle \ulcorner a\rangle [b] \leq \langle a\rangle [b] \ . \tag{24}$$

7 Looping for Optimality in Kleene Algebra

We can now replay our previous derivation in the more abstract setting of KAD. Specifications and implementations are now simply elements of a KAD K with complete test algebra $\mathsf{test}(K)$.

A difference to the relational setting is that we cannot carry over GIMP directly, since in general KADs residuals need not exist. But for our derivation there is no need to internalize the concept of greatest improvement; rather we use a characterizing predicate in which the right hand side of (1) is replaced by the corresponding modal formula and c now plays the rôle of C:

$$\mathrm{IMP}(x, t, c) \stackrel{\mathrm{def}}{\Leftrightarrow} \langle x\rangle \leq [t]\langle c\rangle \ . \tag{25}$$

Now we need to find a sufficient criterion for IMP($\mathsf{while}\, p\ \mathsf{do}\, s$, t, c), which spells out to $\langle (ps)^* \neg p\rangle \leq [t]\langle c\rangle$.

As in Section 2, we abstract and want to achieve, for $v \in K$ and $q \in \mathsf{test}(K)$, that $\langle v^*\, q\rangle \leq [t]\langle c\rangle$, which by (20) is implied by $\langle v\rangle[t]\langle c\rangle p + qp \leq [t]\langle c\rangle p$, in equivalent point-free notation

$$\langle q\rangle \leq [t]\langle c\rangle \ \wedge \ \langle v\rangle [t]\langle c\rangle \leq [t]\langle c\rangle \ . \tag{26}$$

The second conjunct, in turn, follows from

$$\langle v\rangle [t] \leq [t]\langle c\rangle \ , \tag{27}$$

provided $\langle c\rangle\langle c\rangle \leq \langle c\rangle$. In this case we call c *weakly transitive*. This is a much weaker requirement than transitivity $cc \leq c$. To see this, view the Kleene elements again as sets of computation paths. If c consists of paths with exactly two states each (i.e., is isomorphic to a binary relation on states) then cc consists of paths with exactly three states, and so $cc \leq c$ holds only if $cc = 0$. But c is still weakly transitivity if it is transitive considered as a binary relation.

The full KA specification reads $\mathrm{KAOPT}(x,t,c) \stackrel{\mathrm{def}}{=} x \leq t \wedge \mathrm{IMP}(x,t,c)$, analogously to Section 2 and yields the proof obligation, abstracted from (2),

$$\mathsf{while}\ p\ \mathsf{do}\ s \leq t\ . \tag{28}$$

Assume now, as in (9), that $t = \mathsf{exhaust}\, e = e^* \,;\, \neg\ulcorner e$. The same derivation as in Section 3 yields that the following abstraction of (11) is sufficient for (28):

$$ps \leq e \ \wedge\ \ulcorner(ps) \geq \ulcorner e\ . \tag{29}$$

Next we note that (27) and the first conjunct of (26) spell out to

$$\langle ps\rangle[t] \leq [t]\langle c\rangle\ , \tag{30}$$ $$\langle \neg p\rangle \leq [t]\langle c\rangle\ . \tag{31}$$

Again, (29) implies (31) if $\langle \neg\ulcorner e\rangle \leq \langle c\rangle$. Then we call c *weakly reflexive* on $\neg\ulcorner e$.

Summing up, we have the following KA variant of Theorem 1:

Theorem 3. *Suppose that c is weakly reflexive on $\neg\ulcorner e$ and weakly transitive and that $t =$* exhaust *e. Then (29) $\wedge$ (30) $\Rightarrow$* $\mathrm{KAOPT}(\mathsf{while}\ \ulcorner e\ \mathsf{do}\ s, t, c)$.

In [1, 2] it is pointed out that under certain circumstances the above proof obligations will not succeed for the full c. Rather one has to add "context information", restricting c by intersecting it with an element of the form $d\,d^{\smallsmile}$. E.g., the relation *perm*, holding between lists that are permutations of each other, can be expressed as *bagify bagify*$^{\smallsmile}$, where *bagify* transforms lists into bags. Although an element of the shape $d\,d^{\smallsmile}$ can be mimicked by a predicate transformer when the backward modal operators are available, we offer an alternative here. Using the Geach formula again, we calculate, in an extensional KAD,

$$x \leq dd^{\smallsmile} \ \Leftrightarrow\ 1^{\smallsmile}x \leq dd^{\smallsmile} \ \Leftrightarrow\ \langle x\rangle[d] \leq [1]\langle d\rangle \ \Leftrightarrow\ \langle x\rangle[d] \leq \langle d\rangle\ .$$

So the context restriction yields another conjunct in the proof obligations.

8 Classifying Greedy Algorithms

We now demonstrate that modal Kleene algebra is indeed convenient for further applications. We demonstrate this in an abstract reconstruction of Curtis's classification of Greedy algorithms in [2] to which we also refer the reader for concrete examples of the various types of algorithms. The modal operators again lead to considerably more concise proofs than the original relational ones.

Throughout this section we assume the following. First, c and l are pre-orders that model global and local comparison, respectively. Second, $t = \mathsf{exhaust}\, e$ is the construction operation that completes initial approximative solutions to terminal ones using elementary steps e. Third, $g \leq e$ is supposed to be a greedy step that satisfies, analogously to (11) and (12),

$$\ulcorner g = \ulcorner e\ , \tag{32}$$

$$\mathrm{IMP}(g,e,l)\ ,\ \text{ i.e., } \langle g\rangle \leq [e]\langle l\rangle \quad (\Leftrightarrow e^{\smallsmile} g \leq l)\ . \tag{33}$$

In the following theorems we will always list the conditions both in modal notation as obtained by the Geach formula from Theorem 2 and in the one of KADs with converse and illustrate them by diagrams.

Immediately from Theorem 3 we obtain the following description of the first class of greedy algorithms:

Theorem 4. (Best-Global) KAOPT(exhaust g, t, c) *follows from*

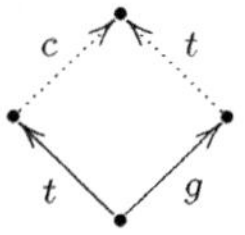

$$\langle g\rangle[t] \le [t]\langle c\rangle \qquad (\Leftrightarrow t^{\smile} g \le ct^{\smile}) \ . \tag{34}$$

The next class is characterized by

Theorem 5. (Better-Global) KAOPT(exhaust g, t, c) *follows from*

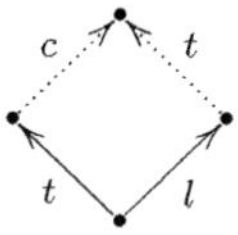

$$\langle l\rangle[t] \le [t]\langle c\rangle \qquad (\Leftrightarrow t^{\smile} l \le ct^{\smile}) \ . \tag{35}$$

This condition says that for any pair of local choices the locally better one has a completion at least as good as any completion of the locally worse one.

Proof. We show that the assumptions imply condition (34) of Theorem 4.

$$\langle g\rangle[t] \le [e]\langle l\rangle[t] \le [e][t]\langle c\rangle = [\ulcorner e][t]\langle c\rangle \ .$$

The first step uses (33), the second one (35) and the third one the definition of $t =$ exhaust e. But by (19) and (32) this is equivalent to the claim. □

The third class of greedy algorithms has a more elaborate set of preconditions.

Theorem 6. (Best-Local) *If we assume* **-continuity,* KAOPT(exhaust g, t, c) *follows from*

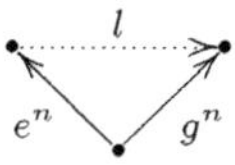

$$\forall\, n \in \mathbb{N} : \langle g^n\rangle \le [e^n]\langle l\rangle \ (\Leftrightarrow \forall\, n \in \mathbb{N} : (e^n)^{\smile} g^n \le l) \ , \tag{36}$$

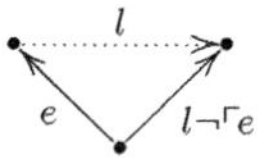

$$\langle l\neg\ulcorner e\rangle \le [e]\langle l\rangle \ (\Leftrightarrow e^{\smile} l\neg\ulcorner e \le l) \ , \tag{37}$$

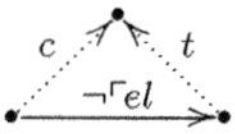

$$\langle l\rangle[t] \le [\neg\ulcorner e]\langle c\rangle \ (\Leftrightarrow \neg\ulcorner el \le ct^{\smile}) \ . \tag{38}$$

Here the local choice is made depending on the history of choices before. The first of these conditions says that each step produces an approximation to the final optimum that is optimal among the approximations that can be obtained with the same number of elementary steps. The other two conditions state that once the sequence of greedy steps finishes, completions of other approximations cannot improve the result any more.

Proof. First we note that, by an easy induction using idempotence of tests, in particular $\langle\neg\ulcorner e\rangle = \langle\neg\ulcorner e\rangle\langle\neg\ulcorner e\rangle$, condition (37) generalizes to

$$n \geq 0 \Rightarrow \langle l\rangle\langle\neg\ulcorner e\rangle \leq [e^n]\langle l\rangle\ . \tag{39}$$

The proof proper is performed by showing that the assumptions imply condition (34) of Theorem 4, i.e., $\langle g\rangle[t] \leq [t]\langle c\rangle$.

Using *-continuity we get $[\sum_{n\in\mathbb{N}} e^n\neg\ulcorner e] = \bigodot_{n\in\mathbb{N}}[e^n\neg\ulcorner e]$, so that the claim reduces to

$$\forall\, n \in \mathbb{N} : \langle g\rangle[t] \leq [e^n][\neg\ulcorner e]\langle c\rangle\ .$$

For $n = 0$, we use idempotence of tests and (19) to see that

$$\langle g\rangle[t] \leq [\neg\ulcorner e]\langle c\rangle \Leftrightarrow \langle\neg\ulcorner e\rangle\langle g\rangle[t] \leq [\neg\ulcorner e]\langle c\rangle\ .$$

Now we calculate, using (32),

$$\langle\neg\ulcorner e\rangle\langle g\rangle[t] = \langle\neg\ulcorner g\rangle\langle g\rangle[t] = \langle\neg\ulcorner g g\rangle[t] = \langle 0\rangle[t] = \langle 0\rangle\ ,$$

and the claim is shown.

For fixed $n > 0$ we split the greedy step g into the part $g\ulcorner(g^{n-1})$ that admits at least $n-1$ further greedy steps, and its relative complement $g\neg\ulcorner(g^{n-1})$, and show separately $\langle g\ulcorner(g^{n-1})\rangle[t] \leq r$ and $\langle g\neg\ulcorner(g^{n-1})\rangle[t] \leq r$, where $r \stackrel{\mathrm{def}}{=} [e^n][\neg\ulcorner e]\langle c\rangle$. For the first part we calculate

$$\langle g\rangle\langle\ulcorner(g^{n-1})\rangle[t] \leq \langle g\rangle\langle g^{n-1}\rangle[t] \leq [e^n]\langle l\rangle[t] \leq [e^n][\neg\ulcorner e]\langle c\rangle\ .$$

The first step uses $g^{n-1}t \leq t$ and (24). The second step joins powers and uses (36). The final step employs (38).

For the second part we want to use (24) again and so have to replace $\neg\ulcorner(g^{n-1})$ by a positive domain expression. We calculate, for arbitrary i,

$$\neg\ulcorner(g^i) = \neg(\ulcorner(g^i\ulcorner g) + \ulcorner(g^i\neg\ulcorner g)) = \neg\ulcorner(g^i\ulcorner g)\neg\ulcorner(g^i\neg\ulcorner g) = \neg\ulcorner(g^{i+1})\neg\ulcorner(g^i\neg\ulcorner g)\ .$$

Using only the $\geq$ half of this equality and shunting we obtain

$$\neg\ulcorner(g^{i+1}) \leq \neg\ulcorner(g^i) + \ulcorner(g^i\neg\ulcorner g)\ ,$$

and an easy induction shows $\neg\ulcorner(g^{n-1}) \leq \sum_{i<n-1}\ulcorner(g^i\neg\ulcorner g)$. By disjunctivity of $\langle\_\rangle$ our claim is thus established if $\langle g\rangle\langle\ulcorner(g^i\neg\ulcorner g)\rangle[t] \leq r$ for all $i < n-1$. We calculate

$$\begin{aligned}&\langle g\rangle\langle\ulcorner(g^i\neg\ulcorner g)\rangle[t] \leq \langle g\rangle\langle g^i\neg\ulcorner g\rangle[t] = \langle g^{i+1}\rangle\langle\neg\ulcorner e\rangle[t] \leq \\ &[e^{i+1}]\langle l\rangle\langle\neg\ulcorner e\rangle[t] \leq [e^{i+1}][e^{n-i-1}]\langle l\rangle[t] \leq [e^n][\neg\ulcorner e]\langle c\rangle\ .\end{aligned}$$

The first step uses $g^i\neg\ulcorner gt \leq t$ and (24). The second step joins powers and uses (32). The third step employs condition (36). The fourth step uses (39) and $n > i+1$. The final step joins powers and employs condition (38). □

The final class of algorithms is given by

Theorem 7. (Better-Local) *Under* **-continuity,* KAOPT(exhaust g, t, c) *follows from*

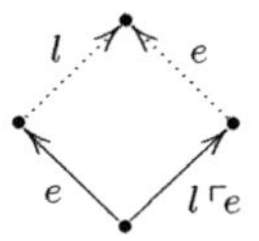

$$\langle l\rangle\langle\ulcorner e\rangle[e] \le [e]\langle l\rangle\ (\Leftrightarrow e^{\smile}\, l\, \ulcorner e \le l e^{\smile})\ , \tag{40}$$

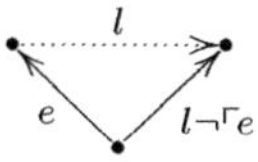

$$\langle l\rangle\langle\neg\ulcorner e\rangle \le [e]\langle l\rangle\ (\Leftrightarrow e^{\smile}\, l\, \neg\ulcorner e \le l)\ , \tag{41}$$

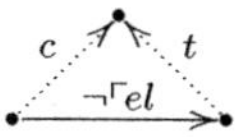

$$\langle l\rangle[t] \le [\neg\ulcorner e]\langle c\rangle\ (\Leftrightarrow \neg\ulcorner e\, l \le c t^{\smile})\ . \tag{42}$$

This essentially says that for any two local choices and any one-step extension of the locally worse one there is a locally better one-step extension of the locally better one.

Proof. We show that condition (35) of Theorem 5 is satisfied. First, using antitonicity of [_] and distributivity, we obtain

$$\begin{array}{l}(40) \wedge (41) \;\Leftrightarrow\; \langle l\rangle(\langle\ulcorner e\rangle[e] \oplus \langle\neg\ulcorner e\rangle[1]) \le [e]\langle l\rangle \;\Rightarrow \\ \langle l\rangle(\langle\ulcorner e\rangle[e+1] \oplus \langle\neg\ulcorner e\rangle[e+1]) \le [e]\langle l\rangle \;\Leftrightarrow\; \langle l\rangle[e+1] \le [e]\langle l\rangle\ .\end{array}$$

Now dualization of (15) together with the equivalence in (14) allows us to infer $\langle l\rangle[e^*] = \langle l\rangle[(e+1)^*] \le [e^*]\langle l\rangle$, from which by condition (42) we get $\langle l\rangle[t] = \langle l\rangle[e^*][t] \le [e^*]\langle l\rangle[t] \le [e^*][\neg\ulcorner e]\langle c\rangle = [t]\langle c\rangle$. □

Curtis's classification is completed by showing the following relationship between the algorithm classes:

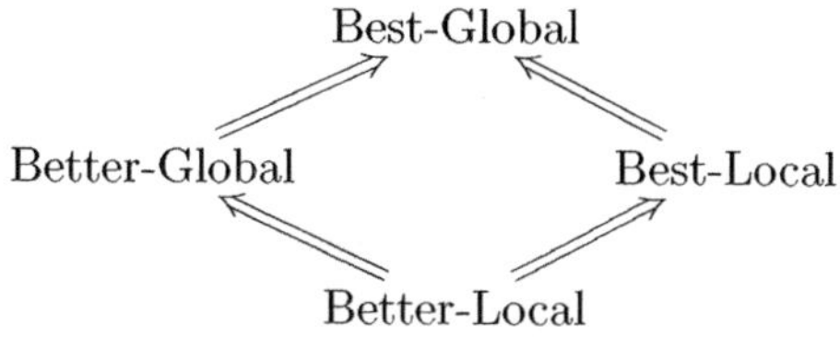

Except for the relation between Better-Local and Best-Local this was was established by the proofs of the previous theorems.

Theorem 8. *The Better-Local conditions imply the Best-Local conditions.*

Proof. It suffices to show (36). This is done by induction on n. For $n = 0$ the claim follows from $1 \le l$. For the induction step we calculate

$$\begin{array}{l}\langle g^{n+1}\rangle \le [e^n]\langle l\rangle\langle g\rangle \le [e^n]\langle l\rangle\langle\ulcorner g\rangle\langle g\rangle \le \\ \qquad [e^n]\langle l\rangle\langle\ulcorner g\rangle[e]\langle l\rangle \le [e^n][e]\langle l\rangle\langle l\rangle \le [e^{n+1}]\langle l\rangle\ .\end{array}$$

The first step splits a power and uses the induction hypothesis. The second step uses a domain law. The third step employs (33). The fourth step uses (32) and (40). The last step joins powers and uses transitivity of l. □

9 Conclusion

We have shown that a concise algebraic derivation of a general greedy-like algorithm can be obtained in the framework of Kleene algebra. The more pristine framework avoids detours through residuals and leads to a simpler correctness proof than in [2, 4].

The treatment has exhibited an interesting relation with semi-commutation properties as known from rewriting and allegories [5]. The connection to KA has already been explored in [13].

Omitting converse has led us into the interesting and very well-behaved algebra of predicate transformers. In it we can prove properties such as $\langle a^* \rangle = \langle a \rangle^*$ that cannot even be expressed in dynamic logic. We are therefore convinced that this algebra will have many further applications; for an example see [11].

Acknowledgement

We are grateful to S. Curtis for an enlightening discussion on greedy algorithms. Helpful remarks were provided by R. Backhouse, H. Bherer, J. Desharnais, T. Ehm, O. de Moor and M. Winter.

References

[1] R. S. Bird, O. de Moor: Algebra of programming. Prentice Hall 1997.

[2] S. A. Curtis: A relational approach to optimization problems. D.Phil. Thesis. Technical Monograph PRG-122, Oxford University Computing Laboratory 1996.

[3] J. Desharnais, B. Möller, and G. Struth: Kleene algebra with domain. Technical Report 2003-07, Universität Augsburg, Institut für Informatik, 2003.

[4] J. E. Durán: Transformational derivation of greedy network algorithms from descriptive specifications. In: E. A. Boiten, B. Möller (eds.): Mathematics of program construction. Lecture Notes in Computer Science **2386**. Springer 2002, 40–67.

[5] P. Freyd, A. Scedrov: Categories, allegories. North-Holland 1990.

[6] P. Helman, B. M. E. Moret, H. D. Shapiro: An exact characterization of greedy structures. SIAM Journal on Discrete Mathematics **6**, 274-283 (1993).

[7] B. Korte, L. Lovász, R. Schrader: Greedoids. Heidelberg: Springer 1991.

[8] D. Kozen:A completeness theorem for Kleene algebras and the algebra of regular events. Information and Computation **110**, 366–390 (1994).

[9] D. Kozen: Kleene algebra with tests. Transactions on Programming Languages and Systems **19**, 427–443 (1997).

[10] B. Möller: Derivation of graph and pointer algorithms. In: B. Möller, H. A. Partsch, S. A. Schuman (eds.): Formal program development. Lecture Notes in computer science **755**. Springer 1993, 123–160.

[11] B. Möller and G. Struth: Modal Kleene algebra and partial correctness. Technical Report 2003-08, Universität Augsburg, Institut für Informatik, 2003.

[12] S. Popkorn: First steps in modal logic. Cambridge University Press 1994.

[13] G. Struth: Calculating Church-Rosser proofs in Kleene algebra. In: H. C. M. de Swart (ed.): Relational Methods in Computer Science, 6th International Conference. Lecture Notes in Computer Science **2561**. Springer 2002, 276–290.

Rasiowa-Sikorski Style Relational Elementary Set Theory*

Eugenio Omodeo[1], Ewa Orłowska[2], and Alberto Policriti[3]

[1] Dipartimento di Informatica, Università di L'Aquila, Italy
[2] Institute of Telecommunications, Warsaw, Poland
[3] Dipartimento di Matematica e Informatica, Università di Udine, Italy

Abstract. A Rasiowa-Sikorski proof system is presented for an elementary set theory which can act as a target language for translating propositional modal logics. The proposed system permits a modular analysis of (modal) axioms in terms of deductive rules for the relational apparatus. Such an analysis is possible even in the case when the starting modal logic does not possess a first-order correspondent. Moreover, the formalism enables a fine-tunable and uniform analysis of modal deductions in a simple and purely set-theoretic language.

Keywords: Modal logic, relational systems, translation methods

1 Introduction

The subject of this paper lies at the intersection of two lines of investigation on systems of non-classical logic. Both lines rely on Kripkean semantics in order to translate such systems into a classical first-order setting, which regards relation algebras under one approach, weak set theory under the other. For a significant collection of cases including mono-modal propositional logics, algebras of *dyadic* relations provide an adequate background; since set theory is concerned with a very special dyadic relation, *membership*, it is not surprising that the two approaches can be reconciled.

Accessibility, which relates worlds in the domain of a Kripke frame, must be taken into account, of course: in order to view this as a sub-relation of membership, we must on the one hand

- renounce some of the beloved features of standard set theory, namely extensionality (reflected by the assumption that 'sets whose elements are the same are identical') and regularity (stating that membership is a well-founded, and therefore an acyclic, relation); on the other hand, we must

- amalgamate the domains of all Kripke frames into a single structure which can play, relative to our non-standard set theory, the role of domain of discourse,

* Work partially supported by MURST/MIUR project *Aggregate- and number-reasoning for computing: from decision algorithms to constraint programming with multisets, sets, and maps*. This research benefited from collaborations fostered by the European action COST n. 274 (TARSKI, see `www.tarski.org`).

R. Berghammer et al. (Eds.): RelMiCS/Kleene-Algebra Ws 2003, LNCS 3051, pp. 215–226, 2004.

$$\begin{array}{ll} x \in y \cup z \leftrightarrow x \in y \ \vee \ x \in z & \text{(union)} \\ x \in y \setminus z \leftrightarrow x \in y \ \wedge \ x \notin z & \text{(difference)} \\ x \subseteq y \leftrightarrow \forall v \, (v \in x \rightarrow v \in y) & \text{(inclusion)} \\ x \in \mathscr{P}(y) \leftrightarrow x \subseteq y & \text{(powerset)} \end{array}$$

Fig. 1. Axioms of the minimal aggregate theory Ω

much like the von Neumann cumulative hierarchy does relative to the Zermelo-Fraenkel theory.

Rather than resorting to a non-well-founded set theory (such as Aczel's one), we prefer to analyze modal logics from the standpoint of an absolutely minimal aggregate theory Ω (cf. Figure 1), more readily amenable to computational reasoning methods. In Ω, the operations $\cup, \setminus$ are counterparts of the extensional propositional connectives, and its construct $\mathscr{P}$ corresponds to the modal necessity-operator $\Box$.

Notice that an enhancement to this axiomatic system Ω would result from the addition, not very engaging indeed, of a *pair axiom* such as $x \in \{y, z\} \leftrightarrow x = y \vee x = z$: after this addition, one can easily translate the whole resulting theory into a ground relation calculus, by following the Tarski-Givant recipe for 'set theory without variables'. Such a translation can already be carried out with the ingredients of Ω alone (cf. [FOP03]), but for the time being we have made no deep investigations in this entirely equational direction; instead, we have designed a proof-system *à la* Rasiowa-Sikorski [RS63] whose specific rules, tailored for the axioms of Ω, are detailed in Section 3.

Previous examples of uses of Rasiowa-Sikorski systems in various contexts relevant for Computer Science and Logic include [SO96, FO95, FO98] and [DO00].

As we will recall in Section 4 from [DMP95], [BDMP97], and [COP01, Chapter 12], there is a known translation $\phi \mapsto \phi^*$ of modal propositional sentences into set-terms of Ω which enjoys the following properties:

- A sentence schema $\phi \equiv \phi[p_1, \ldots, p_n]$ built from n distinct propositional meta-variables p_i becomes a term $\phi^* \equiv \phi^*[f, x_1, \ldots, x_n]$ involving $n+1$ distinct set-variables, one of which, f, is meant to represent a generic frame.
- If $\phi^* \equiv \phi^*[f, \boldsymbol{x}]$ and $\psi^* \equiv \psi^*[f, \boldsymbol{y}]$ result from propositional schemata ϕ, ψ, then the biimplication

$$\psi \models_{\mathsf{K}} \phi \Leftrightarrow \Omega \vdash \forall f \big(f \subseteq \mathscr{P}(f) \wedge \forall \boldsymbol{y} (f \subseteq \psi^*) \rightarrow \forall \boldsymbol{x} (f \subseteq \phi^*) \big)$$

  holds, where K is the minimal modal logic.

Hence, by combining this translation with a proof system for Ω, e.g. the one based on the rules of Section 3 (which presupposes no explicit axioms), we achieve a proof system which can be exploited to semi-decide any finitely axiomatized mono-modal propositional logic. From now on we will call REST (*Relational Elementary Set Theory*) the Rasiowa-Sikorski system variant of Ω, target of the translation.

2 The System REST of Set Theory

2.1 Syntax

The symbols of our language are the following:
- a denumerable set **Var** of individual variables;
- one unary predicate $\texttt{Trans}(\cdot)$ and two binary predicates $\in$ and $\subseteq$, to be used in infix notation, which should not be confused with the corresponding set-theoretic symbols used at the meta-level;
- one unary function symbol $\mathscr{P}$ and two binary function symbols $\cup$ and $\setminus$, again to be used in infix notation;
- propositional connectives and quantifiers.

The set of terms **T** is inductively defined as follows:
- $\mathbf{Var} \subseteq \mathbf{T}$;
- if $t, t' \in \mathbf{T}$, then $t \cup t', t \setminus t', \mathscr{P}(t) \in \mathbf{T}$.

The set of formulas **F** is inductively defined as follows:
- if $t, t' \in \mathbf{T}$, then $t \in t', t \subseteq t', \texttt{Trans}(t) \in \mathbf{F}$ (atomic formulas);
- **F** is closed with respect to uses of propositional connectives and quantifiers.

2.2 Semantics

This section focusses on special models of Ω—which will, accordingly, be models of REST—to act as our set-theoretic counterparts of Kripke frames. More specifically, we associate with each Kripke frame a cumulative hierarchy whose structure amalgamates the frame accessibility relation with the recursively constructed membership relation. The resulting universe will encompass all Kripke models sharing the initially given frame (W, R). Our set universes are built starting with $W = U_0$ which, insofar as the domain of a relational structure, cannot be empty. (On the opposite, in more classical approaches to set theory, where the first level U_0 is taken to be $\emptyset$, the resulting structure turns out to be an initial segment of the well-known *von Neumann's cumulative hierarchy.*)

Definition 1. *Let W be a collection of not-set elements*[1] *and R be a binary relation on W. The* model $\mathbf{U} = (U, \in^{\mathbf{U}})$ generated by (W, R) *consists of the* (set) universe

$$U = \bigcup_{n \geqslant 0} U_n,$$

where $U_0 = W$ and $U_n = \mathscr{P}(U_{n-1}) \cup U_{n-1}$, for $n > 0$; and by the membership

$$v \in^{\mathbf{U}} u \Leftrightarrow_{\texttt{def}} \begin{cases} vR^{-1}u & \textit{if } u, v \in W; \\ v \in u & \textit{otherwise.} \end{cases}$$

[1] That is, atomic elements of some unspecified sort.

Hence, the lower levels of a set universe such as the one defined above contain the given relational structure represented using $\in$ only.[2] The higher levels, being defined using the powerset operator, are populated enough to contain all possible interpretations of (modal) formulas.

The notions of truth in a model and validity in REST are defined as usual.

The above definition is the adaptation to the case of K-derivability of the model construction presented in [BDMP97] which referred to *general* frames (see Definition 3 therein). A more stringent definition was given in the proofs of Theorem 2 in [DMP95] and Theorem 12.2 in [COP01].

The set-theoretic $\Box$-*as*-$\mathscr{P}$ *translation* represents any Kripke frame as a set, with the accessibility relation modeled using the membership relation $\in$. The theory in which the translation is carried out, when specified *à la* Hilbert, is a very weak, *finitely* axiomatizable, first-order set theory called Ω (cf. [DMP95]). The axioms, in the language with relational symbols $\in$ and $\subseteq$ and functional symbols $\cup, \setminus$, and $\mathscr{P}$, are the ones in Figure 1. Notice that, since the corresponding restrictions cannot be imposed on generic accessibiity relations, neither the extensionality axiom nor the axiom of foundation are in Ω.

Lemma 1. *The model* $\mathbf{U} = (U, \in^{\mathbf{U}})$, *generated by a frame* (W, R) *is an* Ω-*model.*

Proof. See [BDMP97] where a more general result is provided (cf. Lemma 4 therein). The key point in that proof, that applies directly also to our current context, is the introduction of the following function

$$F(x) = \begin{cases} \{v \in W : xRv\} & \text{if } x \in W, \\ x & \text{otherwise,} \end{cases}$$

on the ground of which, assuming all elements of W to be pairwise distinct (sets of the same rank), we can prove that

$$x \in^{\mathbf{U}} y \text{ if and only if } x \in F(y),$$

and

$$y \in W \Rightarrow x \in^{\mathbf{U}} y \text{ if and only if } x \in W \wedge yRx.$$

The range of F can be proved to be the set $U \setminus W$ and to be closed under $\cup$ and $\setminus$, with F being the identity function on $U \setminus W$. Hence, we can define:

$$x \cup^{\mathbf{U}} y = F(x) \cup F(y), \qquad x \setminus^{\mathbf{U}} y = F(x) \setminus F(y);$$

and since $\{y \in U : F(y) \subseteq F(x)\}$ is in $U \setminus W$, we can define

$$\mathscr{P}^{\mathbf{U}}(x) = \{y \in U : F(y) \subseteq F(x)\}.$$

[2] The reader can easily check that $|W|$ levels suffice to supply representatives of all elements in the frame.

3 Rasiowa-Sikorski Deduction System for REST

Our system will consist of only one kind of axioms, namely any sequence of formulas containing a formula α and its negation $\neg\alpha$, plus the following rules, where K and H denote finite (possibly empty) sequences of formulas. By a new variable we mean a variable that does not appear in the upper sequence of the respective rule. A rule is admissible whenever the upper sequence is valid if and only if the lower sequences are valid, where validity of a sequence of formulas means first order validity of the disjunction of its elements. The proofs have the form of trees obtained by application of the rules; a tree is closed whenever all of its leaves contain an axiom.

Propositional Rules.

$$\frac{K,\ \alpha \vee \beta\ ,H}{K, \alpha, \beta\ , H}\ (\vee) \qquad \frac{K,\ \neg(\alpha \vee \beta)\ ,H}{K, \neg\alpha, K \mid K, \neg\beta, H}\ (\neg\vee)$$

$$\frac{K,\ \alpha \wedge \beta\ ,H}{K, \alpha, H \mid K, \beta, H}\ (\wedge) \qquad \frac{K,\ \neg(\alpha \wedge \beta)\ ,H}{K, \neg\alpha, \neg\beta\ , H}\ (\neg\wedge)$$

$$\frac{K,\ \alpha \to \beta,\ H}{K, \neg\alpha, \beta\ , H}\ (\to) \qquad \frac{K,\ \neg(\alpha \to \beta),\ H}{K, \alpha, K \mid K, \neg\beta, H}\ (\neg \to)$$

$$\frac{K,\ \neg\neg\alpha,\ H}{K, \alpha\ , H}\ (\neg\neg)$$

First-order Rules. (z new variable and t arbitrary term.)

$$\frac{K,\ (\forall x)\alpha(x),\ H}{K, \alpha(z)\ , H}\ (\forall) \qquad \frac{K,\ \neg(\forall x)\alpha(x),\ H}{K, \neg\alpha(t), \neg(\forall x)\alpha(x)\ , H}\ (\neg\forall)$$

$$\frac{K,\ (\exists x)\alpha(x),\ H}{K, \alpha(t), (\exists x)\alpha(x)\ , H}\ (\exists) \qquad \frac{K,\ \neg(\exists x)\alpha(x),\ H}{K, \neg\alpha(z)\ , H}\ (\neg\exists)$$

Specific Rules for Set Theory.

$$\frac{K,\ x \in y \cup z,\ H}{K, x \in y, x \in z, H}\ (\in \cup) \qquad \frac{K,\ x \notin y \cup z,\ H}{K, x \notin y, H \mid K, x \notin z, H}\ (\notin \cup)$$

$$\frac{K,\ x \in y \setminus z,\ H}{K, x \in y, H \mid K, x \notin z, H}\ (\in \setminus) \qquad \frac{K,\ x \notin y \setminus z,\ H}{K, x \notin y, x \in z, H}\ (\notin \setminus)$$

$$\frac{K, \quad x \subseteq y, \quad H}{K, z \notin x, z \in y, H} (\subseteq) \qquad \frac{K, \quad x \not\subseteq y, \quad H}{K, t \in x, x \not\subseteq y, H \mid K, t \notin y, x \not\subseteq y, H} (\not\subseteq)$$

z new variable, t arbitrary term,

$$\frac{K, \quad x \in \mathscr{P}(y), \quad H}{K, x \subseteq y, H} (\in \mathscr{P}) \qquad \frac{K, \quad x \notin \mathscr{P}(y), \quad H}{K, x \not\subseteq y, H} (\notin \mathscr{P})$$

$$\frac{K, \quad \mathtt{Trans}(x), \quad H}{K, x \subseteq \mathscr{P}(x), H} (\mathtt{Trans}) \qquad \frac{K, \quad \neg\mathtt{Trans}(x), \quad H}{K, x \not\subseteq \mathscr{P}(x), H} (\neg\mathtt{Trans})$$

Cut Rule.

$$\frac{K}{K, \alpha \mid K, \neg\alpha} (\mathtt{cut})$$

Notice that there is only one specific rule that introduces new variables (namely $(\subseteq)$), and that there is only one rule that instantiates over arbitrary terms and requires repetition (namely $(\not\subseteq)$). The infinite trees to be produced in the following examples are generated through an interplay between these two rules.

Lemma 2. *For every model* $\mathbf{U} = (U, \in^{\mathbf{U}})$*:*
rule $(\in \cup)$ *is admissible* $\Leftrightarrow$ *if* $x \in^{\mathbf{U}} y$ *or* $x \in^{\mathbf{U}} z$, *then* $x \in^{\mathbf{U}} y \cup^{\mathbf{U}} z$;
rule $(\notin \cup)$ *is admissible* $\Leftrightarrow$ *if* $x \in^{\mathbf{U}} y \cup^{\mathbf{U}} z$, *then* $x \in^{\mathbf{U}} y$ *or* $x \in^{\mathbf{U}} z$;
rule $(\in \setminus)$ *is admissible* $\Leftrightarrow$ *if* $x \in^{\mathbf{U}} y$ *or* $x \notin^{\mathbf{U}} z$, *then* $x \in^{\mathbf{U}} y \setminus^{\mathbf{U}} z$;
rule $(\notin \setminus)$ *is admissible* $\Leftrightarrow$ *if* $x \in^{\mathbf{U}} y \setminus^{\mathbf{U}} z$, *then* $x \in^{\mathbf{U}} y$ *or* $x \notin^{\mathbf{U}} z$;
rule $(\subseteq)$ *is admissible* $\Leftrightarrow$ *if for all* z $(z \in^{\mathbf{U}} x \rightarrow z \in^{\mathbf{U}} y)$, *then* $x \subseteq^{\mathbf{U}} y$;
rule $(\not\subseteq)$ *is admissible* $\Leftrightarrow$ *if* $x \subseteq^{\mathbf{U}} y$, *then for all* z $(z \in^{\mathbf{U}} x \rightarrow z \in^{\mathbf{U}} y)$;
rule $(\in \mathscr{P})$ *is admissible* $\Leftrightarrow$ *if* $x \subseteq^{\mathbf{U}} y$, *then* $x \in \mathscr{P}^{\mathbf{U}}(y)$;
rule $(\notin \mathscr{P})$ *is admissible* $\Leftrightarrow$ *if* $x \in \mathscr{P}^{\mathbf{U}}(y)$, *then* $x \subseteq^{\mathbf{U}} y$;
rule $(\mathtt{Trans})$ *is admissible* $\Leftrightarrow$ *if* $x \subseteq^{\mathbf{U}} \mathscr{P}^{\mathbf{U}}(y)$, *then* $\mathtt{Trans}^{\mathbf{U}}(x)$;
rule $(\neg\mathtt{Trans})$ *is admissible* $\Leftrightarrow$ *if* $\mathtt{Trans}^{\mathbf{U}}(x)$, *then* $x \subseteq^{\mathbf{U}} \mathscr{P}^{\mathbf{U}}(y)$.

4 The Translation from Mono-modal Logic to REST

Given a modal formula $\phi(P_1, ..., P_n)$, with propositional variables $P_1, ..., P_n$, we define its translation to be the set-theoretic *term* $\phi^*(f, x_1, ..., x_n)$, with set variables $f, x_1, ..., x_n$, built using $\cup, \setminus$, and $\mathscr{P}$. Intuitively speaking, the term $\phi^*(f, x_1, ..., x_n)$ represents the set of those worlds (in the frame f) in which the formula ϕ holds. The inductive definition of $\phi^*(f, x_1, ..., x_n)$ is the following:

- $P_i^* = x_i$;
- $(\phi \vee \psi)^* = \phi^* \cup \psi^*$;
- $(\neg\phi)^* = f \setminus \phi^*$;
- $(\Box\phi)^* = \mathscr{P}(\phi^*)$.

For all modal formulas ϕ, ψ, let us put

$$(\phi,\psi)^* \Leftrightarrow_{\mathtt{def}} \forall f\Big(\big(\mathtt{Trans}(f) \wedge \forall \boldsymbol{x}(f \subseteq \phi^*(f,\boldsymbol{x}))\big) \to \forall \boldsymbol{y}(f \subseteq \psi^*(f,\boldsymbol{y}))\Big),$$

where $\mathtt{Trans}(f)$ stands for $\forall z\ (z \in f \to z \subseteq f)$ and expresses the fact that f is a *transitive* set (cf. [Jec78]).

Then the following results showing the adequacy of the translation hold [DMP95]:

$$\phi \vdash_K \psi \ \Rightarrow \Omega \vdash (\phi,\psi)^* \text{ (Completeness)}$$
$$\Omega \vdash (\phi,\psi)^* \Rightarrow \phi \models_f \psi \quad \text{(Soundness)}$$

Completeness is proved (proof-theoretically) by induction on the length of the derivation in K, while soundness is proved (semantically) by constructing a model of Ω from a Kripke frame using an argument relying on non-well-founded set theory. It is worth noting that, for frame-complete theories, the above translation captures exactly the notion of K-derivability.

The following completeness theorem holds for the presented Rasiowa-Sikorski system REST:

Theorem 1. *For every formula ϕ specifying a modal logic extending* K *and every ψ, the following are equivalent:*

- *the REST formula $(\phi,\psi)^*$ is valid in REST,*
- *there is a closed decomposition tree for $(\phi,\psi)^*$.*

We conclude this section recalling that representing the accessibility relation R by the (unique) membership relation $\in$ does not restrict the field of applicability of the translation to mono-modal logics. This is essentially because the set-theoretic structure of worlds can be used to represent *multiple* accessibility relations (in a completely *symmetric* way, as a matter of fact) as was proved in [DMP95].

5 Indirect Method and Direct Method for Modal Derivability

Modal derivability can be tackled (indirectly) by translating Hilbert's axioms together with the modal formula to be proved, and using REST.

To obtain a more direct translation of a modal logic specified by Hilbert's axioms, it is instructive to compare what happens when we treat by our REST rules the K-axiom $\Box(y \to z) \to (\Box y \to \Box z)$ and other sentence schemata which are often adopted as axioms in specific modal logics. In the case of K, we will get a finite tree all of whose branches are closed (i.e., each disjunction labelling a leaf contains complementary literals, and hence is valid): this means that the counterpart of this sentence schema is a theorem of Ω. On the other hand, when, for example, we elaborate the schema $\Box y \to \Box\Box y$, a tree all of whose branches are closed save one will result (see Section 5.1). This approach will allow us to infer specific rules from REST trees and will provide a more direct derivability technique tailored for the modal logic under study. In the following we illustrate our ideas with two case studies: one regarding a modal logic which admits a first-order correspondent and the other one which does not.

5.1 Rule Generation: First-Order Rules

As an example of (first-order) rule generation, consider the following treatment of the axiom expressing transitivity: $\Box y \rightarrow \Box\Box y$.

We underline the literal to which the rule is applied. When two rules are applied in sequence, we underline the corresponding literal twice.

$$\underline{f \subseteq (f \setminus \mathscr{P}(y)) \cup (\mathscr{P}\mathscr{P}(y))}$$

$$\downarrow (\subseteq)\ z \text{ new variable}$$

$$z \notin f, \underline{z \in (f \setminus \mathscr{P}(y)) \cup (\mathscr{P}\mathscr{P}(y))}$$

$$\downarrow (\in \cup)$$

$$z \notin f, \underline{z \in f \setminus \mathscr{P}(y)}, z \in \mathscr{P}\mathscr{P}(y)$$

$$\swarrow (\in \setminus) \searrow$$

$$z \notin f, z \in f, \ldots \textbf{closed} \qquad z \notin f, z \notin \mathscr{P}(y), \underline{z \in \mathscr{P}\mathscr{P}(y)}$$

$$\downarrow (\in \mathscr{P})$$

$$z \notin f, \underline{z \notin \mathscr{P}(y)}, z \subseteq \mathscr{P}(y)$$

$$\downarrow (\notin \mathscr{P})$$

$$z \notin f, z \not\subseteq y, \underline{\underline{z \subseteq \mathscr{P}(y)}}$$

$$\downarrow (\subseteq)\ w \text{ new variable}$$
$$\downarrow (\in \mathscr{P})$$

$$z \notin f, z \not\subseteq y, w \notin z, \underline{w \subseteq y}$$

$$\downarrow (\subseteq)\ u \text{ new variable}$$

$$z \notin f, \underline{z \not\subseteq y}, w \notin z, u \notin w, u \in y$$

At this point an application of rule ($\not\subseteq$) with u as arbitrary term (variable) produces one closed branch (with $u \in y$ and $u \notin y$ as matching formulas) and the following sequence:

$$z \notin f, u \in z, w \notin z, u \notin w, u \in y, z \not\subseteq y.$$

Now notice that since all the variables can be considered to be universally quantified and since the commas correspond to disjunctions, the above sequence can be rewritten as:

$$(\forall z, f, u, w)(z \in f \rightarrow ((w \in z \wedge u \in w) \rightarrow u \in z) \vee (\forall y)(u \in y \vee z \not\subseteq y)).$$

Denoting as μ the matrix of the above formula (that is, $(z \in f \rightarrow ((w \in z \wedge u \in w) \rightarrow u \in z) \vee (\forall y)(u \in y \vee z \not\subseteq y)))$, we have that:

$$(\forall z, f, u, w)((u \in z \rightarrow \mu) \vee (u \notin z \rightarrow \mu)). \tag{1}$$

On the other hand, since $\Omega \vdash (\forall z, u)(u \notin z \rightarrow (\exists y)(u \notin y \wedge z \subseteq y))$, and $\Omega \vdash (\forall z, u)(u \in z \rightarrow (\forall y)(z \subseteq y \rightarrow u \in y))$, we have that (1) simplifies into

$$(\forall z, f, u, w)(z \in f \rightarrow ((w \in z \wedge u \in w) \rightarrow u \in z)).$$

From the above formula we can deduce the (first-order) rule for transitivity (u, w arbitrary terms):

$$\frac{K,\ z \notin f,\ H}{K,\ u \notin z, z \notin f,\ H \mid K,\ w \in z, z \notin f,\ H \mid K,\ u \in w, z \notin f,\ H}$$

The above result is in complete accordance with the one presented in [MO02].

An inspection of the above technique shows how the first-order rule is generated by eliminating the part of μ containing the variable y which represents the (modal) second-order condition in the translation. Such an elimination was possible since the literal $u \in z$ in μ was proved to be equivalent in Ω to the part of μ involving y (namely $(\forall y)(z \subseteq y \rightarrow u \in y)$). The reader can easily check that to this purpose *any* disjunction of literals in μ (clearly not involving y-variables) would do.

Remark 1. Whenever a term has to be chosen in order to instantiate the applications of rule ($\not\subseteq$), the choice is not really free. It should be clear that, at least, the following two heuristics can always be applied: 1) it makes sense to instantiate only with (variables representing) worlds; 2) whenever a witness for a non-inclusion literal in chosen, we should try to produce as many matching literals (i.e. closed branches) as possible (see the above case with variable u and the next one with variable z).

Let us consider another example in some detail, namely the axiom for reflexivity $\Box y \rightarrow y$:

$$\underline{\underline{f \subseteq (f \setminus \mathscr{P}(y)) \cup y}}$$

$\downarrow(\subseteq)$ z new variable

$\downarrow (\in \cup)$

$$z \notin f, \underline{z \in (f \setminus \mathscr{P}(y))}, z \in y$$

$\downarrow (\in \setminus)$

$$z \notin f, \underline{z \notin \mathscr{P}(y)}, z \in y$$

$$\downarrow (\in \mathscr{P})$$

$$z \notin f, \underline{z \not\subseteq y}, z \in y$$

At this point the only literal to be processed is the underlined $\not\subseteq$-literal in the above formula and the only variable that allows some matching is z. Hence, applying rule ($\not\subseteq$) with z as chosen term, we obtain a closed branch and the sequence

$$z \notin f, z \in z, z \in y, z \not\subseteq y,$$

which can be rewritten as

$$(\forall z, f)(z \notin f \lor z \in z \lor (\forall y)(z \in y \lor z \not\subseteq y)). \tag{2}$$

Now, observing that

$$\Omega \vdash (\forall z)(z \notin z \rightarrow (\exists y)(z \notin y \land z \subseteq y)),$$

and

$$\Omega \vdash (\forall z)(z \in z \rightarrow (\forall y)(z \subseteq y \rightarrow z \in y)),$$

we have that (2) simplifies into

$$(\forall z, f)(z \in f \rightarrow z \in z),$$

from which we get the (first-order) rule for reflexivity (z a chosen variable):

$$\frac{K,\ z \notin f,\ H}{K,\ z \notin z, z \notin f,\ H.}$$

5.2 Rule Generation: Second-Order Rules

Let us now consider a case in which the first-order correspondent does not exist: Löb's axiom $\Box(\Box y \rightarrow y) \rightarrow \Box y$ that we will study together with the above axiom for transitivity.

Remark 2. Since the formulas we are analyzing in our examples have always the frame f as left sub-term in formulas whose principal constructor is $\setminus$, it can easily be checked that we can denote $f \setminus t$ as $\bar{t}$ and that we can modify the rule ($\in \setminus$) into

$$\frac{K,\ x \in \bar{t},\ H}{K,\ x \notin t,\ H}$$

Here below, we do not spell out all the rules used in simple steps:

$$f \subseteq \bar{\mathscr{P}}(\bar{\mathscr{P}}(y) \cup y) \cup \mathscr{P}(y)$$

$$\downarrow \; v_0 \text{ new variable}$$

$$v_0 \notin f, v_0 \not\subseteq (\bar{\mathscr{P}}(y) \cup y), v_0 \subseteq y$$

$$\downarrow \; v_1 \text{ new variable}$$

$$v_0 \notin f, v_0 \not\subseteq (\bar{\mathscr{P}}(y) \cup y), v_1 \notin v_0, v_1 \in y$$

Since, as we observed above, our goal is always to close some branch, we reduce the $\not\subseteq$-literal using v_1 as arbitrary term and we obtain a closed branch and the following sequence:

$$v_0 \notin f, v_1 \notin (\bar{\mathscr{P}}(y) \cup y), v_1 \notin v_0, v_1 \in y, v_0 \not\subseteq (\bar{\mathscr{P}}(y) \cup y);$$

at this point we apply rule ($\notin \cup$) and we obtain a closed branch and the sequence:

$$v_0 \notin f, v_1 \notin \bar{\mathscr{P}}(y), v_1 \notin v_0, v_1 \in y, v_0 \not\subseteq (\bar{\mathscr{P}}(y) \cup y),$$

from which we obtain

$$v_0 \notin f, v_1 \subseteq y, v_1 \notin v_0, v_1 \in y, v_0 \not\subseteq (\bar{\mathscr{P}}(y) \cup y).$$

The above formula calls for the introduction of a new variable v_2 entering the game when we reduce the inclusion $v_1 \subseteq y$; and thus we obtain

$$v_0 \notin f, v_2 \notin v_1, v_2 \in y, v_1 \notin v_0, v_1 \in y, v_0 \not\subseteq (\bar{\mathscr{P}}(y) \cup y).$$

The derivation to be produced from the above sequence will produce a class of branches to be closed using transitivity, and the following *infinitary* sequence:

$$v_0 \notin f, v_{i+1} \notin v_i, v_{i+1} \in y, v_0 \not\subseteq (\bar{\mathscr{P}}(y) \cup y) \text{ for all } i \geqslant 0.$$

A close inspection of the sequence of applications of rules for reducing $\subseteq$ (introducing new variables) and $\not\subseteq$ (using the newly introduced variables), shows that the path is somehow "forced" and that the situation we are now, apart from the infinitary sequence, is very similar to the previously treated cases. In fact, even the elimination of the y variable is now possible on the ground of the following observation:

$$\Omega \vdash \forall y\Big(\big(v_0 \in f \wedge \bigwedge_{i \geqslant 0} v_{i+1} \in v_i\big) \rightarrow \bigvee_{i \geqslant 0} v_{i+1} \in y\Big) \leftrightarrow \neg\Big(v_0 \in f \wedge \bigwedge_{i \geqslant 0} v_{i+1} \in v_i\Big).$$

From the above formula we can deduce the following infinitary rule expressing the well-known second-order condition equivalent to Löb's axiom:

$$\frac{K,\ v_0 \notin f,\ H}{K,\ v_1 \in v_0, v_0 \notin f,\ H \mid \cdots \mid K,\ v_{i+1} \in v_i, v_0 \notin f,\ H \mid \cdots}$$

6 Conclusion and Future Developments

The problem of establishing whether a modal axiom admits a first-order correspondent is notoriously undecidable; notwithstanding, syntactic manipulations of the kind we can carry out, even manually, with our proposed Rasiowa-Sikorski system of rules seem to lead us to the essence of the problem as quickly and as close as possible by analytic means.

Potential optimizations to be studied include: a) the fact that rules involving arbitrary instantiations never use either f or compound terms and, moreover, seem to use terms already appearing on the branch only; b) attempts to design heuristics aiming at apply cut to atomic formulas only.

References

[BDMP97] J. F. A. K. van Benthem, G. D'Agostino, A. Montanari, and A. Policriti, *Modal deduction in second-order logic and set theory-I*, Journal of Logic and Computation **7** (1997), no. 2, 251–265. 216, 218

[COP01] D. Cantone, E. G. Omodeo, and A. Policriti, *Set theory for computing. from decision procedures to declarative programming with sets*, Monographs in Computer Science, Springer-Verlag, 2001. 216, 218

[DMP95] G. D'Agostino, A. Montanari, and A. Policriti, *A set-theoretic translation method for polymodal logics*, Journal of Automated Reasoning **3** (1995), no. 15, 317–337. 216, 218, 221

[DO00] I. Düntsch and E. Orłowska, *A proof system for contact relation algebras*, Journal of Philosophical Logic (2000), no. 29, 241–262. 216

[FO95] M. Frias and E. Orłowska, *A proof system for fork algebras and its applications to reasoning in logics based on intuitionism*, Logique et Analyse (1995), no. 150-151-152, 239–284. 216

[FO98] ______, *Equational reasoning in nonclassical logics*, Journal of Applied Non-Classical Logics **8** (1998), no. 1-2, 27–66. 216

[FOP03] A. Formisano, E. G. Omodeo, and A. Policriti, *Three-variable statements of set-pairing*, submitted (2003). 216

[Jec78] T. J. Jech, *Set theory*, Springer-Verlag, New York, 1978. 221

[MO02] W. MacCaull and E. Orłowska, *Correspondence results for relational proof systems with application to the Lambek calculus*, Studia Logica **71** (2002), 279–304. 223

[RS63] H. Rasiowa and R. Sikorski, *The mathematics of metamathematics*, Polish Scientific Publishers, Warsaw, 1963. 216

[SO96] S.Demri and E. Orłowska, *Logical analysis of demonic nondeterministic programs*, Theoretical Computer Science (1996), no. 166, 173–202. 216

Relational Data Analysis

Gunther Schmidt*

Institute for Software Technology, Department of Computing Science
Federal Armed Forces University Munich
Schmidt@Informatik.UniBw-Muenchen.DE

Abstract. Given a binary relation R, we look for partitions of its row set and its column set, respectively, that behave well with respect to selected algebraic properties, i.e., correspond to congruences related to R. Permutations are derived from these congruences that allow to rearrange R visualizing the decomposition.

1 Introduction

Known and new methods of decomposing a relation are presented together with methods of making the decomposition visible. Such aspects as difunctionality, Moore-Penrose inverses, independence and line covering, chainability, game decompositions, matchings, Hall conditions, term rank, and others are handled under one common roof.

One area of application is multicriteria decision making. There, relations are given beforehand and one asks for dichotomies generated by the relation [1, 3] following certain rational ideas. One may call this *theorem extraction* or theorem formulation — as opposed to *theorem proving*. Once formulated, theorem provers would certainly establish the theorem. But in practical situations it is more important *to find* the theorem on which one is willing to base decisions.

Much of the basic approach can be demonstrated by the initial example of a sparse boolean matrix A, which looks rather randomly distributed. It is a big problem in marketing, e.g., to analyze such raw data so as to finally arrive at $A_{\text{rearranged}}$. We have visualized that certain theorems are valid, namely that elements of $\{1, 9, 11, 15, 16, 17\}$ are only related to elements from $\{3, 6, 7, 13\}$, etc.

Having knowledge of this type may obviously be considered a major marketing advantage. Our basic question is, therefore: Are there methods to generate such theorems? One will certainly not blindly generate all theorems and scan them in some sort of a Laputa method [7] for interesting ones. Rather, there should be some concept as to how these theorems might look like. In the environment of RelMiCS, one will concentrate on "relation-algebraic theorem patterns". Several of these have been investigated and formulated as ontologies.

* Cooperation and communication around this research was partly sponsored by the European COST Action 274: TARSKI (Theory and Application of Relational Structures as Knowledge Instruments), which is gratefully acknowledged.

R. Berghammer et al. (Eds.): RelMiCS/Kleene-Algebra Ws 2003, LNCS 3051, pp. 227–237, 2004.

$A =$

	1	2	3	4	5	6	7	8	9	10	11	12	13
1	0	0	0	0	0	1	0	0	0	0	0	0	0
2	0	0	0	1	0	0	0	0	0	0	0	0	0
3	1	0	0	0	0	0	0	1	0	0	0	0	0
4	0	0	0	0	1	0	0	0	0	1	0	0	0
5	0	0	0	0	0	0	0	0	0	0	0	0	0
6	1	0	0	0	0	0	0	1	0	0	0	0	0
7	0	0	0	1	0	0	0	1	0	0	0	0	0
8	0	0	0	0	0	0	0	0	0	0	0	0	0
9	0	0	1	0	0	1	0	0	0	0	0	0	0
10	0	0	0	0	0	0	0	0	0	0	0	0	0
11	0	0	0	0	0	1	1	0	0	0	0	0	0
12	0	0	0	0	0	0	0	0	0	0	0	0	0
13	0	0	0	0	1	0	0	0	1	0	0	0	0
14	0	1	0	0	0	0	0	0	0	1	0	0	0
15	0	0	0	0	0	0	1	0	0	0	0	0	1
16	0	0	1	0	0	0	0	0	0	0	0	0	1
17	0	0	0	0	0	1	0	0	0	0	0	0	0

Fig. 1. An initial example of a sparse relation

$A_{\text{rearranged}} =$

	3	6	7	13	1	4	8	2	5	9	10	11	12
1	0	1	0	0	0	0	0	0	0	0	0	0	0
9	1	1	0	0	0	0	0	0	0	0	0	0	0
11	0	1	1	0	0	0	0	0	0	0	0	0	0
15	0	0	1	1	0	0	0	0	0	0	0	0	0
16	1	0	0	1	0	0	0	0	0	0	0	0	0
17	0	1	0	0	0	0	0	0	0	0	0	0	0
2	0	0	0	0	0	1	0	0	0	0	0	0	0
3	0	0	0	0	1	0	1	0	0	0	0	0	0
6	0	0	0	0	1	0	1	0	0	0	0	0	0
7	0	0	0	0	0	1	1	0	0	0	0	0	0
4	0	0	0	0	0	0	0	0	1	0	1	0	0
13	0	0	0	0	0	0	0	0	1	1	0	0	0
14	0	0	0	0	0	0	0	1	0	0	1	0	0
5	0	0	0	0	0	0	0	0	0	0	0	0	0
8	0	0	0	0	0	0	0	0	0	0	0	0	0
10	0	0	0	0	0	0	0	0	0	0	0	0	0
12	0	0	0	0	0	0	0	0	0	0	0	0	0

Fig. 2. The initial example partitioned and rearranged

By ontology we mean relational games, irreducibility, line covering, or matching concepts, e.g. With one of these ontologies, the relation in question is analyzed.

The full report underlying this article, obtainable via

`http://ist.unibw-muenchen.de/People/schmidt/DecompoHomePage.html`,

strives to provide the methodological basis together with a toolkit for the task of decomposing relations in various application disciplines. This report is not just

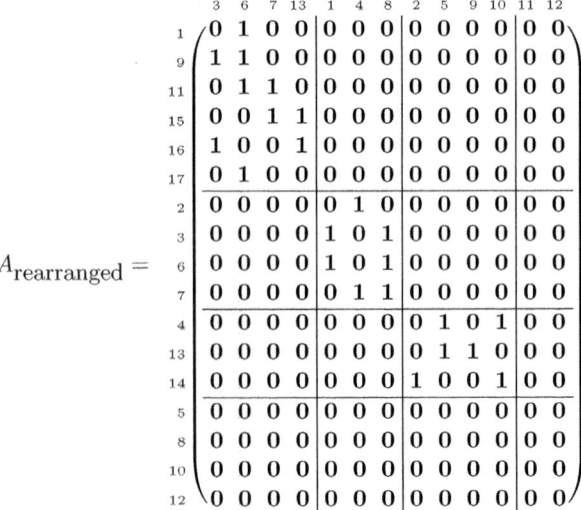

a research report but also a Haskell program in literate style sufficient to run the programs.

2 Prerequisites and Tools

We have used Haskell [2] as the underlying programming language. It is a purely functional language, and it is currently widely accepted in research and university teaching. For more information see the Haskell WWW site at

`URL: http://www.haskell.org/.`

In the example presented in the introduction, the resulting permutation has only been shown via the permuted row and column numbers. The permutations, however, should be fully available in the program. There, they may be given as a function, decomposed into cycles, or as a permutation matrix. Either form has its specific merits. Sometimes also the inverted permutation is useful. Therefore, types and functions to switch from one form to the other, and to apply a permutation to some list are provided for. Also a basic technique is introduced determining permutations from partitions. Let, for example, a list of partitioning subsets be given as

```
[[False,True, False,True, False,False],
 [False,False,True, False,False,True ],
 [True, False,False,False,True, False]]
```

We are interested to obtain a permutation like `[5,1,3,2,6,4]`, which directs `True` entries of the first row to the front, followed by `True` entries of the second row, etc.

Relations are throughout handled as rectangular Boolean matrices. Often we represent their entries `True` by **1** and `False` by **0** when showing matrices in the text. The basic relational operators `&&&`, `|||`, `***`, `<==` for intersection, union, composition, containment, etc., of relations are all formulated in Haskell.

3 Congruences

Whenever some equivalence behaves well with regard to some other structure, we are accustomed to call it a congruence. This is well-known for algebraic structures, i.e., those defined by mappings on some set. We define it correspondingly for the non-algebraic case, including heterogeneous relations. Let R be a relation and Ξ, Θ be equivalences on the source, resp. the target set of that relation. Then the pair (Ξ, Θ) is called a R-**congruence** if $\Xi; R \subseteq R; \Theta$.

We have formulated a generalisation of Birkhoffs famous theorem on the lattice of congruences for algebraic structures, extending it to relational structures.

Proposition 1. *Let some finite heterogeneous relation R be given. Then all R-congruences (P, Q) which satisfy both, $R; R^\mathsf{T} \subseteq P$ and $R^\mathsf{T}; R \subseteq Q$, form a complete lattice, the least element of which is $(\Xi, \Theta) := ((R; R^\mathsf{T})^*, (R^\mathsf{T}; R)^*)$, the pair of natural congruences wrt. R.*

Proof. Among other items, the proof has to show that the set
$\{(P,Q) \mid (P,Q) \text{ is an } R\text{-congruence satisfying } R;R^\mathsf{T} \subseteq P \text{ and } R^\mathsf{T};R \subseteq Q\}$
is $\cap$-hereditary:

$$\begin{aligned}(P_1 \cap P_2);R &\subseteq P_1;R \cap P_2;R \subseteq R;Q_1 \cap R;Q_2\\ &\subseteq (R \cap R;Q_2;Q_1^\mathsf{T});(Q_1 \cap R^\mathsf{T};R;Q_2) \text{ using Dedekind's rule}\\ &\subseteq R;(Q_1 \cap Q_2;Q_2) = R;(Q_1 \cap Q_2)\end{aligned}$$ □

Every congruence leads to a partition in equivalence classes, and these partitions in turn give rise to the permutations we compute for presentation purposes.

4 Difunctional Decomposition

Two different cases of decomposition will be distinguished: For *heterogeneous* relations between two sets — be it that they have equal cardinality —, we will permute rows and columns *independently* and for *homogeneous* relations rows and columns will be permuted *simultaneously.*

In the introductory example, difunctionality leads to an important decomposition. It groups rows as well as columns according to the "natural congruence".

A relation R is called **difunctional** if $R;R^\mathsf{T};R = R$. For every relation R, the least difunctional relation containing it is well defined and we define (according to J. Riguet, [4]) the difunctional closure as

$$h_{\mathsf{difu}}(R) := \mathtt{inf}\,\{H \mid R \subseteq H \text{ with } H \text{ difunctional}\}.$$

We also ask for the practical aspects of this definition, which has long been discussed and is known among numerical analysts. To describe the main property of any diagonal block of the rearranged initial example, let a relation R be given that is conceived as a chessboard with dark squares or white according to whether R_{ik} is `True` or `False`. A rook shall operate on the chessboard in horizontal or vertical direction; however, it is only allowed to change direction on dark squares. Using this interpretation, a relation R is called **chainable** if the non-vanishing entries (i.e., the dark squares) can all be reached from one another by a sequence of "rook moves", or else if $h_{\mathsf{difu}}(R) = \mathbb{T}$. The diagonal blocks mentioned are chainable when considered as a separate matrix. It is easy to prove that a total and surjective relation R is chainable precisely when its so-called edge-adjacency $K := \overline{\mathbb{I}} \cap R;R^\mathsf{T}$ is strongly connected.

The concept of being difunctional is related to linear algebra for numerical problems. A relation G is called a **Moore-Penrose inverse** of A if the following four conditions hold $A;G;A = A$, $G;A;G = G$, $(A;G)^\mathsf{T} = A;G$, $(G;A)^\mathsf{T} = G;A$. Moore-Penrose inverses are uniquely determined provided they exist. The Moore-Penrose inverse of a difunctional matrix always exists and turns out to be its converse.

5 Line Covering and Independence

Some relations may be decomposed in such a way, that there is a subset of row entries that is completely unrelated to a subset of column entries. In this context, a relation A may admit vectors x and y (with $\perp\!\!\perp \neq x \neq \top\!\!\top$ or $\perp\!\!\perp \neq y \neq \top\!\!\top$ to avoid degeneration), such that $Ay \subseteq x$ or, equivalently, $A \subseteq \overline{\overline{x};y^{\mathsf{T}}}$. Given appropriate permutations P of the rows, and Q of the columns, respectively, we then have

$$P;A;Q^{\mathsf{T}} = \begin{pmatrix} * & \perp\!\!\perp \\ * & * \end{pmatrix} \qquad P;x = \begin{pmatrix} \perp\!\!\perp \\ \top\!\!\top \end{pmatrix} \qquad Q;y = \begin{pmatrix} \perp\!\!\perp \\ \top\!\!\top \end{pmatrix}.$$

Given $A;y \subseteq x$, to enlarge the $\perp\!\!\perp$-zone is not so easy a task, which may be seen in the case of the identity relation $\mathbb{I}$: All shapes from $1 \times (n-1), 2 \times (n-2), \dots (n-1) \times 1$ may be chosen. To avoid this counter-running, one usually studies this effect with one of the sets x and y negated. So we consider pairs of subsets (s,t) taken from the domain and from the range side and define

$$\begin{array}{lll} (s,t) \text{ is a } \textbf{line covering} & :\Longleftrightarrow & A;\overline{t} \subseteq s. \\ (s,t) \text{ is a } \textbf{pair of independent sets} & :\Longleftrightarrow & A;t \subseteq \overline{s}. \end{array}$$

For the moment, call rows and columns lines. Then we are able to cover all entries **1** by $|\overline{y}|$ vertical plus $|x|$ horizontal lines. Given a relation A, the **term rank** is defined as the minimum number of lines necessary to cover all entries **1** in A, i.e. $\mathtt{min}\{|s| + |t| \mid A;\overline{t} \subseteq s\}$.

Consider

$$\begin{pmatrix} A_{11} & \perp\!\!\perp \\ A_{21} & A_{22} \end{pmatrix} ; \begin{pmatrix} \perp\!\!\perp \\ \top\!\!\top \end{pmatrix} \subseteq \begin{pmatrix} \perp\!\!\perp \\ \top\!\!\top \end{pmatrix}.$$

Hoping to arrive at fewer lines than the columns of A_{11} and the rows of A_{22} to cover, one might start a first naive attempt and try to cover with s and t but with row i, e.g., omitted. If (s,t) is already minimal, there will be an entry in row i of A_{22} containing a **1**. Therefore, A_{22} is a total relation. In the same way, A_{11} turns out to be surjective. But we may also try to get rid of a set $x \subseteq s$ of rows and accept that a set of columns be added instead. It follows from minimality that regardless of how we choose $x \subseteq s$, there will be at least as many columns necessary to cover what has been left out. For a relation A and a set x, we therefore say that x satisfies the **Hall condition** if $|z| \leq |A^{\mathsf{T}};z|$ holds for every subset $z \subseteq x$. If we have a line covering with $|s| + |t|$ minimal, then A_{11}^{T} as well as A_{22} will satisfy the Hall-condition.

We will later find minimum line coverings and maximum independent sets without just checking them all exhaustively. We postpone this, until further prerequisites are at hand and concentrate on the following aspect.

Proposition 2. *Let a finite relation A be given. Then A is either chainable or it admits a pair (s,t) which is nontrivial, such that both s,t as well as $\overline{s},\overline{t}$, constitute at the same time a pair of independent sets and a line covering.*

Difunctionality and line coverings are related in the following way.

Proposition 3. *A relation A admits a pair (x,y) such that (x,y) and $(\overline{x},\overline{y})$ are line coverings if and only if its difunctional closure admits these line coverings.*

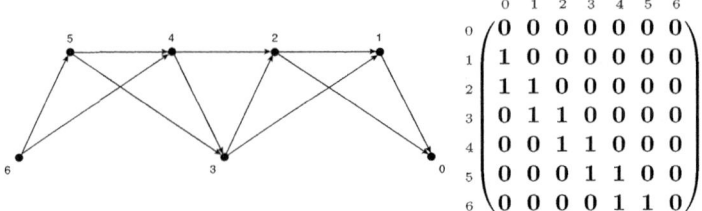

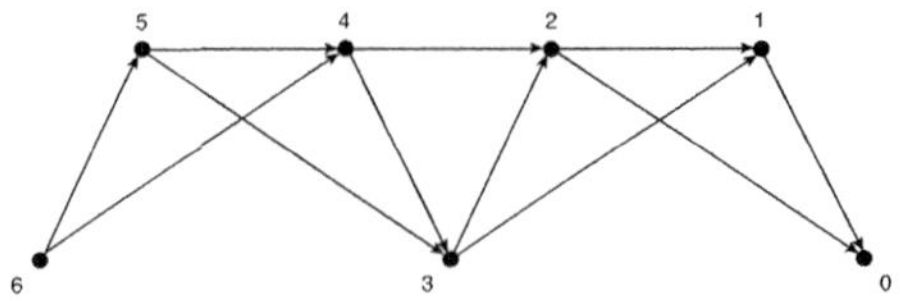

$$
\begin{array}{c} \\ 0\\1\\2\\3\\4\\5\\6 \end{array}
\begin{array}{c} \begin{array}{ccccccc} 0&1&2&3&4&5&6 \end{array} \\
\begin{pmatrix}
0&0&0&0&0&0&0\\
1&0&0&0&0&0&0\\
1&1&0&0&0&0&0\\
0&1&1&0&0&0&0\\
0&0&1&1&0&0&0\\
0&0&0&1&1&0&0\\
0&0&0&0&1&1&0
\end{pmatrix}\end{array}
$$

Fig. 3. Situations in a game starting with a pile of 6 matches

6 Game Decomposition

While so far heterogeneous relations have been treated by permuting their rows and columns independently, we now specialize to the homogeneous case and apply permutations simultaneously. There is a well-developed theory of standard iterations for Boolean matrices in order to solve a diversity of application problems such as matching, line covering, assignment, games, etc. We will present a general framework for executing these iterations.

In all of these cases, we need two antitone mappings between power sets, which we call $\sigma : \mathcal{P}(V) \rightarrow \mathcal{P}(W)$ and $\pi : \mathcal{P}(W) \rightarrow \mathcal{P}(V)$, according to [5, 6]. In case $V = W$, let an arbitrary homogeneous relation $B : V \leftrightarrow V$ be given with $\pi(x) = \sigma(x) = \overline{B;x}$. Two players are supposed to make moves alternatingly according to B in choosing a consecutive arrow to follow. The player who has no further move, i.e., who is about to move and finds an empty row in the relation B, or a terminal vertex in the graph, has lost.
Such a game is easily visualized by taking a relation B represented by a graph, on which players have to determine a path in an alternating way. We study the Nim game starting with 6 matches from which we are allowed to take 1 or 2.

We have a homogeneous relation, and we easily observe how with equalities in an alternating pattern an iteration evolves

$$\bot \subseteq \overline{B;\top} = \overline{B;\overline{B;\bot}} \subseteq \overline{B;\overline{B;\overline{B;\top}}} = \ldots \subseteq \ldots \subseteq \overline{B;\overline{B;\overline{B;\bot}}} = \overline{B;\overline{B;\top}} \subseteq \overline{B;\bot} = \top.$$

The limit of the iteration is characterised by the formulae $a = \pi(b)$ and $\sigma(a) = b$, which this time turn out to be $a = \overline{B;b}$ and $\overline{B;a} = b$. In addition, we will always have $a \subseteq b$. The smaller set a gives loss positions, while the larger one then indicates win positions as $\overline{b}$ and draw positions as $b \cap \overline{a}$. This is visualized by the following diagram for sets of win, loss, and draw, the arrows of which indicate moves that must exist (the thick black arrows), may exist (the thin arrows), or are not allowed to exist (the crossed out grey arrows).

Proposition 4. *Any finite homogeneous relation may by simultaneously permuting rows and columns be transformed into a matrix satisfying the following*

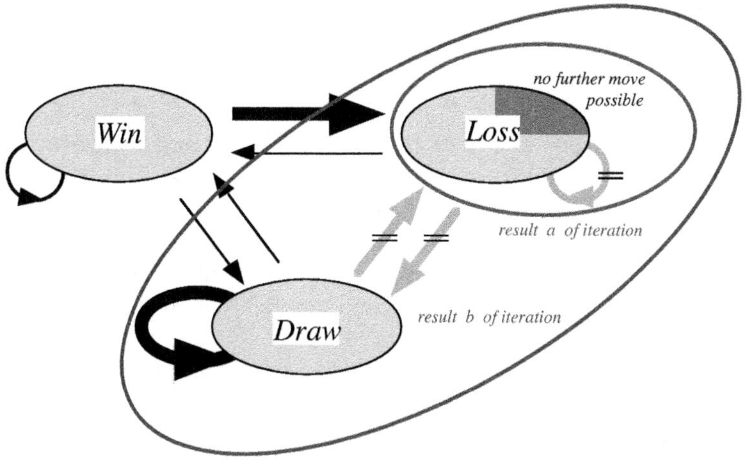

Fig. 4. Decompositions into sets of winning, draw, and losing positions

basic structure with square diagonal entries:

$$\begin{pmatrix} \bot\!\!\!\bot & \bot\!\!\!\bot & * \\ \bot\!\!\!\bot & \mathit{total} & * \\ \mathit{total} & * & * \end{pmatrix}$$

□

Of course, the win or the loss zone may contain no rows. If the win zone is nonempty, so is the loss zone. Nevertheless, the subdivision into loss/draw/win groups is uniquely determined, and indeed

$$a = \begin{pmatrix} \top \\ \bot\!\!\!\bot \\ \bot\!\!\!\bot \end{pmatrix} = \overline{\begin{pmatrix} \bot\!\!\!\bot & \bot\!\!\!\bot & * \\ \bot\!\!\!\bot & \text{total} & * \\ \text{total} & * & * \end{pmatrix} ; \begin{pmatrix} \top \\ \top \\ \bot\!\!\!\bot \end{pmatrix}}$$

$$b = \begin{pmatrix} \top \\ \top \\ \bot\!\!\!\bot \end{pmatrix} = \overline{\begin{pmatrix} \bot\!\!\!\bot & \bot\!\!\!\bot & * \\ \bot\!\!\!\bot & \text{total} & * \\ \text{total} & * & * \end{pmatrix} ; \begin{pmatrix} \top \\ \bot\!\!\!\bot \\ \bot\!\!\!\bot \end{pmatrix}}$$

7 Matching and Assignment

An additional antimorphism situation is known to exist in connection with matchings and assignments. Let two matrices $Q, \lambda : V \leftrightarrow W$ be given, where $\lambda \subseteq Q$ is univalent and injective, i.e. a matching — possibly not yet of maximum cardinality.

We consider Q to be a relation of sympathy between a set of boys and a set of girls and λ the set of current dating assignments, assumed only to be established if sympathy holds. We now try to maximize the number of dating assignments.

Definition 5. *Given the (possibly heterogeneous) relation Q, we call the relation λ a* **Q-matching** *provided it is a univalent and injective relation contained in Q, i.e., if*

$$\lambda \subseteq Q \qquad \lambda;\lambda^{\mathsf{T}} \subseteq \mathbb{I}, \qquad \lambda^{\mathsf{T}};\lambda \subseteq \mathbb{I}.$$

An iteration with $\pi(w) := \overline{\lambda;w}$ and $\sigma(v) := \overline{Q^{\mathsf{T}};v}$ will end with two vectors (a,b) satisfying $a = \pi(b)$ and $\sigma(a) = b$ as before. Here, this means $\overline{a} = \lambda;b$ and $\overline{b} = Q^{\mathsf{T}};a$.
In addition $\overline{a} = Q;b$. This follows from the chain $\overline{a} = \lambda;b \subseteq Q;b \subseteq \overline{a}$, which implies equality at every intermediate state. Only the resulting equalities for a,b have been used together with monotony and the Schröder rule to obtain this decomposition.

We find out that the pair a,b is an inclusion-maximal pair of independent sets for Q, or else $\overline{a},\overline{b}$ is an inclusion-minimal line covering. As of yet, a,b need not be an inclusion-maximal pair of independent sets for λ, nor need $\overline{a},\overline{b}$ be an inclusion-minimal line covering for λ! This will only be the case, when in addition $\overline{b} = \lambda^{\mathsf{T}};a$.

It is thus not uninteresting to concentrate on the condition $\overline{b} = \lambda^{\mathsf{T}};a$. After having found some matching relation and applying the iteration, it may not yet be satisfied. So let us assume $\overline{b} = \lambda^{\mathsf{T}};a$ *not* to hold, which means that

$$\overline{b} = Q^{\mathsf{T}};a \underset{\neq}{\supseteq} \lambda^{\mathsf{T}};a.$$

We make use of the formula $\lambda\overline{S} = \lambda\mathbb{T} \cap \overline{\lambda;S}$, which holds since λ is a univalent relation. The iteration finally ends with equations $\overline{b} = Q^{\mathsf{T}};a$ and $\overline{a} = \lambda;b$. This easily expands to

$$\overline{b} = Q^{\mathsf{T}};a = Q^{\mathsf{T}};\overline{\lambda;b} = Q^{\mathsf{T}};\overline{\lambda;\overline{Q^{\mathsf{T}};a}} = Q^{\mathsf{T}};\overline{\lambda;\overline{Q^{\mathsf{T}};\overline{\lambda;\overline{Q^{\mathsf{T}};a}}}} \ \ldots$$

from which the last but one becomes

$$\begin{aligned}\overline{b} &= Q^{\mathsf{T}};a = Q^{\mathsf{T}};\overline{\lambda;b} = Q^{\mathsf{T}};\overline{\lambda;\mathbb{T} \cap \overline{\lambda;Q^{\mathsf{T}};a}} = Q^{\mathsf{T}};(\overline{\lambda;\mathbb{T}} \cup \lambda;Q^{\mathsf{T}};a)\\ &= Q^{\mathsf{T}};(\overline{\lambda;\mathbb{T}} \cup \lambda;Q^{\mathsf{T}};(\overline{\lambda;\mathbb{T}} \cup \lambda;Q^{\mathsf{T}};a))\end{aligned}$$

indicating how to prove that

$$\overline{b} = (Q^{\mathsf{T}} \cup Q^{\mathsf{T}};\lambda;Q^{\mathsf{T}} \cup Q^{\mathsf{T}};\lambda;Q^{\mathsf{T}};\lambda;Q^{\mathsf{T}} \cup \ \ldots\ );\overline{\lambda;\mathbb{T}}$$

If we have $\lambda^{\mathsf{T}};a \underset{\neq}{\subseteq} \overline{b}$, we may thus find a point in

$$(Q^{\mathsf{T}} \cup Q^{\mathsf{T}};\lambda;Q^{\mathsf{T}} \cup Q^{\mathsf{T}};\lambda;Q^{\mathsf{T}};\lambda;Q^{\mathsf{T}} \cup \ \ldots\ );\overline{\lambda;\mathbb{T}} \cap \overline{\lambda^{\mathsf{T}};a}$$

which leads to the well-known alternating chain algorithm.

When showing the result, some additional care will be taken concerning empty rows or columns in Q showing them at the first or at the last position, respectively. The Hall condition is made visible by arranging λ in a diagonal shape and by introducing an additional subdividing line. In principle, this gives a 4 by 4 pattern. Either or all of the first 3 row zones, or the last 3 column zones may be empty.

	3	15	13	17	4	9	14	1	5	7	12	8	16	2	11	6	10
7	0	0	0	0	0	0	0	0	0	0	0	0	0	0	0	0	0
16	0	0	0	0	0	0	0	0	0	0	0	0	0	0	0	0	0
6	1	0	1	0	0	0	0	0	0	0	0	0	0	0	0	0	0
12	0	0	0	0	1	0	0	0	0	0	0	0	0	0	0	0	0
14	1	1	0	0	0	0	0	0	0	0	0	0	0	0	0	0	0
1	1	0	0	1	0	0	0	0	0	0	0	0	0	0	0	0	0
2	1	1	0	0	0	0	0	0	0	0	0	0	0	0	0	0	0
4	0	0	1	1	0	0	0	0	0	0	0	0	0	0	0	0	0
5	0	0	0	1	0	0	0	0	0	0	0	0	0	0	0	0	0
10	0	0	0	0	1	0	0	0	0	0	0	0	0	0	0	0	0
3	1	0	0	0	0	1	0	0	0	0	1	0	0	0	1	0	0
8	0	0	0	0	0	0	1	0	0	0	0	0	0	0	0	0	0
9	1	0	0	0	0	0	0	1	0	0	0	0	0	0	0	0	0
11	0	0	0	0	0	0	0	0	1	0	0	0	0	0	0	0	0
13	0	0	0	0	0	0	0	0	0	1	0	0	0	0	0	0	0
15	0	0	0	0	1	0	0	1	0	0	1	0	0	0	0	0	0
17	1	0	0	0	1	0	0	1	0	0	0	1	0	0	0	0	0
18	0	0	0	0	0	0	0	0	0	0	0	0	1	0	0	0	0
19	0	1	0	0	1	0	0	0	1	0	1	0	0	1	1	0	0

Fig. 5. Arbitrary heterogeneous relation with a rearrangement according to a cardinality-maximum matching — the diagonals

Proposition 6. *Any given heterogeneous relation Q admits a cardinality-maximum matching $\lambda \subseteq Q$. By independently permuting rows and columns they can jointly be transformed into matrices of a 2 by 2 pattern with not necessarily square diagonal blocks of the following form:*

$$Q = \begin{pmatrix} \mathit{Hall}^{\mathsf{T}} & ⫫ \\ * & \mathit{Hall} \end{pmatrix} \qquad \lambda = \begin{pmatrix} \mathit{univ.} + \mathit{surject.} + \mathit{inject.} & ⫫ \\ ⫫ & \mathit{univ.} + \mathit{total} + \mathit{inject.} \end{pmatrix}$$

Looking back, we see that we have not just decomposed according to a pair of independent sets, but in addition ordered the rest of the relation so as to obtain matchings.

8 Conclusion and Outlook

Based on such investigations, it is possible to automatically develop *theories* combined with some relations, starting from a given ontology (difunctionality, game, e.g.). When applying a relational decomposition, this theory will change as it afterwards fits into the given ontology and, thus, includes the necessary predicates and theorems.

The following diagram shows the idea. There are concrete relations given and the sparse theory with which one may work on these. By this we mean hardly more than asking whether two elements are related. There is, however, also an ontology concerning games, irreducibility, or difunctionality, e.g. What we are constructing is in a sense the pushout. Afterwards, the given model may be viewed with the win-loss-draw structure, e.g., which the game ontology has provided. Then the theory will also contain certain closed formulae describing what holds between the new items.

$$\begin{array}{ccc} \text{Sparse theory} & \models & \text{Given model} \\ \downarrow & & \downarrow \\ \text{Ontology-enhanced theory} & \models & \text{Result model} \end{array}$$

With the methods presented it is possible to analyze a given relation with regard to different concepts and to visualize the results. Some programs are more efficient, others less. This was not a matter of concern in this paper, though. We had in mind data of a size not too big to be handled and even visualized for presentation purposes.

Another point of possible criticism matters more. Should the relations handled stem from rather fuzzy sources, taken by some α-cut, e.g., the results will heavily depend on single entries of the relation. On the other hand, such a single entry may be a rather "weak" one as to its origin from the α-cut as it has hardly passed the threshold. In the approach chosen, therefore, we have extremely high sensitivity depending on the initial data. This is a feature, one is usually not interested in. Rather, one would highly estimate results which stay more or less the same as data are changed only moderately. This would give more "meaning" to the results and would make it easier to base decisions on them.

Nonetheless, we hope that our exposition will lead to future research. We have scanned a diversity of topics for their algebraic properties. On several occasions, we have replaced counting arguments by algebraic ones. Our hope is that these algebraic properties will be of value in the following regard: Only recently, fuzzy relations have been investigated with more intensity. There, the entry of the matrix is not just $\mathbf{1}$ / $\mathbf{0}$ or yes/no. Instead, coefficients from some suitable lattice are taken. In this way it is possible to express given situations in more detail. Astonishingly, much of the algebraic structure of relation algebra still remains valid. In this modified context an investigation should be done using fuzzy relations. Reduced sensitivity of results with respect to given data may be hoped for.

Acknowledgements

The author gratefully acknowledges fruitful discussions with Michael Ebert as well as detailed comments of the referees.

References

[1] Sašo Džeroski and Nada Lavrač, editors. *Relational Data Mining*. Springer-Verlag, 2001. 227

[2] Paul Hudak, Simon L. Peyton Jones, Philip Wadler, et al. Report on the programming language Haskell, a non-strict purely functional language, version 1.2. *ACM SIGPLAN Notices*, 27(5), 1992. See also `http://haskell.org/`. 229

[3] Leonid Kitainik. *Fuzzy Decision Procedures With Binary Relations — Towards a unified theory*, volume 13 of *Theory and Decision Library, Series D: System Theory, Knowledge Engineering and Problem Solving*. Kluwer Academic Publishers, 1993. 227

[4] Jacques Riguet. Quelques propriétés des relations difonctionelles. C. R. Acad. Sci. Paris, 230:1999–2000, 1950. 230

[5] Gunther Schmidt and Thomas Ströhlein. *Relationen und Graphen*. Mathematik für Informatiker. Springer-Verlag, 1989. ISBN 3-540-50304-8, ISBN 0-387-50304-8. 232

[6] Gunther Schmidt and Thomas Ströhlein. *Relations and Graphs - Discrete Mathematics for Computer Scientists*. EATCS Monographs on Theoretical Computer Science. Springer-Verlag, 1993. ISBN 3-540-56254-0, ISBN 0-387-56254-0. 232

[7] Jonathan Swift. Gulliver's Travels. 1726. 227

Two Proof Systems for Peirce Algebras*

Renate A. Schmidt[1], Ewa Orłowska[2], and Ullrich Hustadt[3]

[1] University of Manchester, UK
schmidt@cs.man.ac.uk
[2] National Institute of Telecommunications, Warsaw
orlowska@itl.waw.pl
[3] University of Liverpool, UK
U.Hustadt@csc.liv.ac.uk

Abstract. This paper develops and compares two tableaux-style proof systems for Peirce algebras. One is a tableau refutation proof system, the other is a proof system in the style of Rasiowa-Sikorski.

1 Introduction

The purpose of this paper is twofold. First, we develop two proof systems for the class of Peirce algebras, namely a Rasiowa-Sikorski-style system and tableau system. Second, we present in a formal way a principle of duality between these proof systems.

Procedurally, the two systems are very similar. They both use a top-down approach. Their rules are of the form

$$\frac{X}{X_1 \mid \ldots \mid X_n} \tag{1}$$

where both the numerator X and the denominators $X_1, \ldots, X_n$ ($n \geq 1$) are finite sets of formulae. Given a formula of a logic, the decomposition rules of the systems enable us to decompose it into simpler formulae, or the specific rules enable us to modify a set of formulae. Some sets of formulae have the status of axioms and are used as closure rules. Applying the rules to a given formula we form a tree (or tableau) whose nodes consist of finite sets of formulae. We stop applying the rules to a node of the tree (i.e., we close the corresponding branch) whenever we eventually obtain an axiomatic set of formulae.

The main difference between the two systems is in their underlying semantics. Rasiowa-Sikorski systems are validity checkers. The rules preserve and reflect validity of the sets of formulae which are their premises and conclusions (i.e. branching is interpreted as conjunction), and the axiomatic sets are valid. Validity of a set means first-order validity of the disjunction of its formulae (i.e.,

* This work was supported by EU COST Action 274, and research grants GR/M88761 and GR/R92035 from the UK Engineering and Physical Sciences Research Council. Part of the work by the first author was done while on sabbatical leave at the Max-Planck-Institut für Informatik, Germany, in 2002.

R. Berghammer et al. (Eds.): RelMiCS/Kleene-Algebra Ws 2003, LNCS 3051, pp. 238–251, 2004.

comma is interpreted as disjunction). In order to verify validity of a formula, we place it at the root of a tree and we apply the rules until all the branches close (i.e. we reached a valid set of formulae) or there is an open branch that is complete. Completeness of a branch means that all the rules that can be applied have been applied. A formula is valid if there is a proof tree for it, that is, a tree where all the branches close.

Tableau systems are unsatisfiability checkers. In tableau systems the rules preserve and reflect unsatisfiability of sets of formulae which are their premises and conclusions, axiomatic sets are unsatisfiable, and satisfiability of a set means first order satisfiability of the conjunction of its formulae (i.e., comma is interpreted as conjunction). Equivalently, a rule of the form as above is admissible whenever X is satisfiable iff either of X_i, $(1 \leq i \leq n)$, is satisfiable (i.e. branching is interpreted as disjunction). In order to verify unsatisfiability of a formula, we place it at the root of a tree and we apply the rules until all the branches of the tree close (i.e. we reached an unsatisfiable set of formulae) or there is an open branch that is complete. As before, completeness of a branch means that all the rules that can be applied have been applied. A formula is unsatisfiable if there is a proof tree for it, where all the branches close. Clearly, applying the tableau proof procedure to the negation of a formula we can check validity of the formula itself.

We formalise a duality between Rasiowa-Sikorski and tableau systems in terms of a classification of the rules. It follows that the duality can be observed at the syntactic level. Clearly, its justification is based on semantic features of the two systems.

Since Peirce algebras are, in a sense, a join of a Boolean algebra and a relation algebra, the proof systems presented in the paper inherit many features, on the one hand, from the proof systems for first-order logic [18, 23, 7] and, on the other hand, from the proof systems for relation algebras [25, 10, 22, 14, 9, 4, 17, 6, 13].

2 Peirce Algebra

Peirce algebra, introduced in Britz [3] and refined in Brink, Britz and Schmidt [2], formalises the properties of binary relations and sets, and their interactions. Although these can also be formalised in relation algbera (see remarks at the end of the section), our interest in Peirce algebra is motivated by practical considerations. Peirce algebra provides a natural algebraic framework for various branches of computer science. Applications include the modelling of programming constructs [2], natural language analysis [19, 20], and the interpretation of description logics which are used for knowledge representation and reasoning [2, 19]. Also, the algebraic semantics of modal logics and extended modal logics, such as Boolean modal logic [8] and dynamic modal logic [5] can be studied in the framework of Peirce algebra. It is not difficult to see that all these applications can be modelled in relation algebra. Of particular interest to these applications is the fact that Peirce algebra has more interesting and well-behaved reducts than relation algebra. For instance, many decidable description logics and extended modal logics (see [21] for an overview of the latter) correspond directly to reducts of Peirce algebra. By following the ideas of [17] and exploiting results in [2], problems in these description or modal logics can of course be translated into relation algebra statements. However, experiments with the first-order logic theorem prover MSPASS [11] show that the performance is superior on the first-order encodings of Peirce algebra statements than relation algebra statements corresponding to problems belonging to decidable description and modal logics. This kind of behaviour is not particular to MSPASS; it is also expected from other first-order logic theorem provers. A simple explanation is that generally, in first-order logic provers, unary literals, which correspond to the Boolean elements, can be handled much more effectively than binary predicates.

Formally, a *Peirce algebra* (as defined in Brink et al [2]) is a two-sorted algebra $(\mathfrak{B}, \mathfrak{R}, :, ^c)$ defined equationally by:

1. $\mathfrak{B} = (B, +, \cdot, -, 0, 1)$ is a Boolean algebra,
2. $\mathfrak{R} = (R, +, \cdot, -, 0, 1, ;, ^\smile, e)$ is a relation algebra [24],
3. $:$ is a mapping $R \times B \longrightarrow B$ satisfying ($r, s \in R$ and $a, b \in B$):
   M1 $r:(a+b) = r:a + r:b$
   M2 $(r+s):a = r:a + s:a$
   M3 $r:(s:a) = (r;s):a$
   M4 $e:a = a$
   M5 $0:a = 0$
   M6 $r^\smile : -(r:a) \leq -a$
4. ^{c} is a mapping $B \longrightarrow R$ satisfying ($a \in B$ and $r \in R$):
   P1 $a^c : 1 = a$
   P2 $(r:1)^c = r;1$

The operation $:$ is the *Peirce product* of Brink's Boolean modules [1]; in fact, M1–M6 are the axioms of Boolean modules which capture the interrelationship of the operators, except ^{c} with the Peirce product. A Peirce algebra is therefore an extension of a Boolean module $(\mathfrak{B}, \mathfrak{R}, :)$ with an operation ^{c} defined by 4. The operation ^{c} is called the *left cylindrification* operation.

On the set-theoretic level the Peirce product multiplies a binary relation R with a set A to give the set $R:A = \{x \mid \exists y\ (x,y) \in R \text{ and } y \in A\}$. This gives an algebraic interpretation of the multi-modal diamond operator. The intuitive definition of the left cylindrification of a set A is $A^c = \{(x,y) \mid x \in A\}$, i.e. the relation with domain A and the range consisting of all the elements of a corresponding universe. There are other ways to formalise the relationship from sets to relations. The test operator of propositional dynamic logic (*PDL*), domain restriction and the cross product would be alternatives to the left cylindrification operator [2].

Peirce algebras are expressively equivalent to relation algebras. This follows from the observation in Brink et al [2] that the Boolean elements in a Peirce algebra can be modelled as either right ideal elements or identity elements in the underlying relation algebra. In a Peirce algebra $(\mathfrak{B}, \mathfrak{R}, :, ^c)$, the Boolean algebra of right ideal elements in the underlying relation algebra $\mathfrak{R}$ and the Boolean algebra of identity elements is isomorphic to the Boolean algebra $\mathfrak{B}$ underlying the Peirce algebra [2]. Peirce algebras therefore inherit various properties of relation algebras. For example, it follows that the class of Peirce algebras is not representable. This is a consequence of a well-known result for relation algebras by Lyndon [12]. Consequently, there are properties of binary relations that cannot be proved in the framework of Peirce algebra. Monk [15] has proved that the class of representable relation algebras is not finitely axiomatisable by a set of equational axioms. This means there is no finite equational proof system for reasoning about the (equational) properties of relations. From an applications perspective this is a disadvantage. Therefore, in order to express and derive every property of sets and binary relations we restrict our attention to the reasoning problem of the elementary theory of Peirce algebra as formalised in Peirce logic which is defined next. That is, Peirce logic is intended to be the logic of the class of representable Peirce algebras.

3 Peirce Logic

There are different ways of defining Peirce logic. One is just as first-order logic over one-place or two-place predicates augmented with the operations of Peirce algebra, cf. Nellas [16]. Our definition is similar in style to the logical formalisation of Peirce algebras given by de Rijke [5].

The language $\mathcal{L}$ of Peirce logic consists of two syntactic types: (i) countably many Boolean symbols, called *atomic Boolean formulae* and denoted by A_i, and (ii) countably many relational symbols, called *atomic relational formulae* and denoted by R_i. The logical connectives are the classical connectives, negation, intersection and falsum (or bottom), here denoted by $-$, $\cap$ and 0, respectively, the standard connectives of relational logics, $;$ (composition), $^{\smile}$ (converse), Id (identity), the logical version of Peirce product $:$ and left cylindrification ^{c}.

An *atomic formula* defined over the language $\mathcal{L}$ is any atomic Boolean or relational formula in $\mathcal{L}$. The set of *Boolean formulae* over $\mathcal{L}$ is very similar to Boolean terms in Peirce algebras, similarly for *relational formulae*. In particular, the Boolean and relational formulae are defined inductively by the following BNF production rules.

Boolean formulae:	A, B	$\longrightarrow$	$A_i \mid 0 \mid -A \mid A \cap B \mid R\!:\!A$
Relational formulae:	R, S	$\longrightarrow$	$R_i \mid 0 \mid -R \mid R \cap S \mid R;S \mid R^{\smile} \mid \mathit{Id} \mid A^{c}$

The symbols 0 and Id are nullary connectives which are interpreted as the empty set (or relation) and the identity relation, respectively. The set of *formulae* of Peirce logic is the smallest set of Boolean and relational formulae defined over $\mathcal{L}$. We assume two defined connectives: *Peirce sum* $R \ddagger A = -((-R):(-A))$ and *relational sum* $R \dagger S = -((-R);(-S))$.

We now define the semantics of Peirce logic. A *model* for Peirce logic is a system of the form $M = (U, m)$, where U is a non-empty set, and m is a meaning function subject to the following conditions:

1. If A is a Boolean symbol then $m(A) \subseteq U$ and m extends to all the Boolean formulae as follows, where A and B are arbitrary Boolean formulae over $\mathcal{L}$.

$$m(0) = \emptyset \qquad m(-A) = U \setminus m(A) \qquad m(A \cap B) = m(A) \cap m(B)$$

2. If R is a relational symbol then $m(R) \subseteq U \times U$ and m extends to all the relational formulae as follows, for all relational formulae R and S over $\mathcal{L}$.

$$\begin{array}{lll} m(0) = \emptyset & m(Id) = Id_U & m(R \cap S) = m(R) \cap m(S) \\ m(-R) = U \times U \setminus m(R) & m(R^{\smile}) = m(R)^{\smile} & m(R\,;S) = m(R)\,;m(S) \end{array}$$

3. If A is a Boolean formula and R a relational formula then

$$\begin{aligned} m(R\colon A) &= \{x \in U \mid \text{there is a } y \in m(A) \text{ such that } (a,b) \in m(R)\} \\ m(A^{c}) &= \{(x,y) \in U \times U \mid x \in m(A)\}. \end{aligned}$$

A formula F in $\mathcal{L}$ is said to be *satisfiable* in a model M iff one of the following is true: (i) there is an element s in U such that $s \in m(F)$, if F is a Boolean formula, and (ii) there is a pair of elements s and t in U such that $(s,t) \in m(F)$, if F is a relational formula. In these cases we write $M, s \models F$ or $M, (s,t) \models F$, respectively. A formula F in $\mathcal{L}$ is *valid* in a model M iff F is satisfiable with respect to arbitrary elements in U or pairs of elements in U. In this case we write $M \models F$. Observe that if F is valid in M then $m(F) = U$, if F is a Boolean formula, and $m(F) = U \times U$, if F is a relational formula. A Peirce logic formula is said to be *satisfiable* if there is a model M in which F is satisfiable. A Peirce logic formula F is said to be *valid*, and we write $\models F$, if it is valid in all models of Peirce logic. Let Γ be a set of Peirce logic formulae. A formula F is *semantically entailed* by Γ, written $\Gamma \models F$, iff for all models M, whenever $M \models G$ for all $G \in \Gamma$ then $M \models F$. In Peirce logic semantic entailment can be reduced to validity:

Lemma 1. *Let $\Gamma \cup \{F\}$ be a set of Peirce logic formulae. Suppose Γ is partitioned into two sets Γ_b and Γ_r of the Boolean and relational formulae in Γ, respectively. Then*

$$\begin{aligned} \Gamma \models F \quad &\textit{iff} \quad -((1 \ddagger \cap \Gamma_b) \cap (1 \dagger \cap \Gamma_r \dagger 1) \cap -F) \textit{ is valid} \\ &\textit{iff} \quad (1 \ddagger \cap \Gamma_b) \cap (1 \dagger \cap \Gamma_r \dagger 1) \cap -F \textit{ is unsatisfiable.} \end{aligned}$$

The proof systems we are going to describe for proving or refuting formulae of Peirce logic manipulate labelled formulae defined over two extended languages. One is the language tailored for refutation proofs using tableau and the other is tailored for proofs of validity in the style Rasiowa-Sikorski. In the tableau system the labels are constants and in the Rasiowa-Sikorski system the labels are variables. Therefore, let the tableau language $\mathcal{L}^{\mathsf{T}}$ be an extension of the language $\mathcal{L}$ with a countable set Con of individual constants, denoted by a_i, and the connective $\perp$. Further, let the Rasiowa-Sikorski language $\mathcal{L}^{\mathsf{RS}}$ be an extension of $\mathcal{L}$ with a countable set Var of individual variables, denoted by x_i, and the connective $\top$. The set of formulae over $\mathcal{L}^{\mathsf{T}}$, respectively over $\mathcal{L}^{\mathsf{RS}}$, is defined by

Formulae over $\mathcal{L}^{\mathsf{T}}$:	ψ	$\longrightarrow$	$\bot \mid a\ \ A \mid a\ \ Rb$
Formulae over $\mathcal{L}^{\mathsf{RS}}$:	ψ	$\longrightarrow$	$\top \mid x\,A\ \mid x\,R\,y$

where a and b denote individual constants in Con, and x and y denote individual variables in Var. Subsequently, we use the notation s and t for either both constants or both variables, which will be clear from the context. An *atomic (labelled) formula* over $\mathcal{L}^{\mathsf{T}}$ or $\mathcal{L}^{\mathsf{RS}}$ is a formula of the form $s\,A$ or $s\,R\,t$ are either both individual constants of variables, in which the formula A or R is primitive, i.e. is a Boolean or relational symbols or a nullary connective (0 or Id).

Now we define the semantics of formulae over the extended languages $\mathcal{L}^{\mathsf{T}}$ and $\mathcal{L}^{\mathsf{RS}}$. Assume $M = (U, m)$ is a model defined as above with the meaning function m extended so that it provides also an interpretation of individual constants, that is, for each a in Con, $m(a) \in U$. A *valuation* in M is a mapping v from the set of variables and constants of the language to U such that if a is a constant then $v(a) = m(a)$. The *satisfiability* of formulae in a model $M = (U, m)$ by a valuation v in M is defined by (s and t either both denote constants or variables):

$M, v \models \top$	$M, v \not\models \bot$
$M, v \models s\,A$	iff $v(s) \in m(A)$ for any Boolean formula A
$M, v \models s\,R\,t$	iff $(v(s), v(t)) \in m(R)$ for any relational formula R.

Let ψ be any formula over $\mathcal{L}^{\mathsf{T}}$ (or $\mathcal{L}^{\mathsf{RS}}$). ψ is *satisfiable* whenever there is a model M and a valuation v in M such that $M, v \models \psi$. ψ is said to be *unsatisfiable* whenever it is not satisfiable. Validity of a formula in a model is defined by: A formula ψ is *valid* in a model M whenever $M, v \models \psi$ for every v in M. A formula is *valid* whenever it is true in all the models defined over the extended languages.

Lemma 2. *Let x and y be variables in* Var, *and let a and b be constants in* Con. *(i) If F is a Boolean formula of Peirce logic then: F is valid iff $x\,F$ is valid iff $\neg F$ is unsatisfiable iff $a\ \ F$ is unsatisfiable. (ii) If F is a relational formula of Peirce logic then: F is valid iff $x\,F\,y$ is valid iff $\neg F$ is unsatisfiable iff $a\ \ Fb$ is unsatisfiable.*

In the definition of the inference rules we will be using the symbol $\sim$ which is defined as follows: If F denotes a Boolean or relational formula then $\sim\!F$ denotes G, if $F = \neg G$, and $\neg F$ otherwise.

4 A Tableau Refutation System

A tableau is a finitely branching tree whose nodes are sets of formulae. Given a formula F of Peirce logic to be tested for satisfiability the root node is the

Decomposition rules:

$$(\cap)\ \frac{a\ \ A\cap B}{a\ \ A,\ a\ \ B} \qquad (-\cap)\ \frac{a\ \ -(A\cap B)}{a\ \ {\sim}A \mid a\ \ {\sim}B} \qquad (--)\ \frac{a\ \ --A}{a\ \ A}$$

$$(:)\ \frac{a\ \ R\colon A}{a\ \ Rc,\ c\,A} \quad \text{where } c \text{ is a new constant}$$

$$(-:)\ \frac{a\ \ -(R\colon A)}{a\ \ {\sim}Rc \mid c{\sim}A} \quad \text{where } c \text{ is any constant}$$

$$(\cap)\ \frac{a\ \ R\cap S b}{a\ \ Rb,\ a\ \ Sb} \qquad (-\cap)\ \frac{a\ \ -(R\cap S)b}{a\ \ {\sim}Rb \mid a\ \ {\sim}Sb} \qquad (--)\ \frac{a\ \ --Rb}{a\ \ Rb}$$

$$(\smile)\ \frac{a\ \ R^{\smile}b}{b\,R\,a} \qquad (-\smile)\ \frac{a\ \ -(R^{\smile})b}{b{\sim}R\,a}$$

$$({}^{c})\ \frac{a\ \ A^{c}b}{a\ \ A} \qquad (-{}^{c})\ \frac{a\ \ -(A^{c})b}{a\ \ {\sim}A}$$

$$(;)\ \frac{a\ \ R\,;Sb}{a\ \ Rc,\ c\,S\,b} \quad \text{where } c \text{ is a new constant}$$

$$(-;)\ \frac{a\ \ -(R;S)b}{a\ \ {\sim}Rc \mid c{\sim}S\,b} \quad \text{where } c \text{ is any constant}$$

Specific rules:

$$(\mathrm{sym})\ \frac{a\ \ Idb}{b\,Id\,a}$$

$$(\mathrm{id}_1)\ \frac{b\,A\ ,\ a\ \ Idb}{a\ \ A} \qquad (\mathrm{id}_2)\ \frac{b\,R\,c,\ a\ \ Idb}{a\ \ Rc} \qquad (\mathrm{id}_3)\ \frac{c\,R\,a\ ,\ a\ \ Idb}{c\,R\,b}$$

Closure rules:

$$(\mathrm{cl}_1)\ \frac{a\ \ A,\ a\ \ -A}{\bot} \qquad (\mathrm{cl}_2)\ \frac{a\ \ 0}{\bot}$$

$$(\mathrm{cl}_3)\ \frac{a\ \ Rb,\ a\ \ -Rb}{\bot} \qquad (\mathrm{cl}_4)\ \frac{a\ \ 0b}{\bot} \qquad (\mathrm{refl})\ \frac{a\ \ -Ida}{\bot}$$

Fig. 1. Tableau rules for Peirce logic

set $\{a\ \ F\}$, when F is a Boolean formula and $\{a\ \ Fb\}$, when F is a relational formula. Successor nodes are constructed in accordance with a set of expansion rules. An expansion rule has the form (1), where X, X_i are sets of formulae over $\mathcal{L}^{\mathsf{T}}$ $(1 \leq i \leq n)$. The formulae in X are called premises and the formulae in X_i are called conclusions.

Let T be the *calculus for Peirce logic* defined by the rules of Figure 1. The specific rules express properties of the identity relation. (sym) expresses the symmetry of Id and (id_1)–(id_3) express that $Id\colon A \subseteq A$, $R\,;Id \subseteq R$ and $Id\,;R \subseteq R$. Reflexivity of Id is ensured by the reflexivity rule (refl), classified here as a closure rule. The other closure rules reduce elementary contradictions to $\bot$.

(Observe that the transitivity rule for identity is redundant in T because it is an instance of both identity rules (id_2) and (id_3).)

Concerning the rules for negated main connectives, consider for example the rule ($-\cap$). By the use of $\sim$ (defined above), we have chosen to eliminate immediately the double negations normally introduced for a formula like $a\ -(-A\ \cap B)$, where one of the conjuncts is a negated formula, had we used the rule $a\ -(A \cap B)/a\ -A|\,a\ -B$ instead. This does not make the ($--$) rules superfluous however.

A *tableau derivation* from a set N of formulae over $\mathcal{L}^{\mathsf{T}}$ is a finitely branching, ordered tree T with root N and nodes which are sets of $\mathcal{L}^{\mathsf{T}}$-formulae. The tree is constructed by applications of the expansion rules to the leaves. Let N be a leaf node in a (partially constructed) tableau derivation. A rule (1) is *applicable* to N, if N contains formulae in the form X. Then an application of the rule creates a new tree T' which is the same as T except that the node N has n successor nodes N_i which are extensions of N with the formulae in X_i. That is, $N_i = N \cup X_i$ ($1 \leq i \leq n$). It is assumed that on a branch in any tableau derivation no instance of a rule is applied twice to the same instance of the numerator.

For each rule application to a node N if the following is true, then the rule is said to be *(satisfiability) admissible.*

$$\exists \overline{a}\ (\bigwedge X \wedge \bigwedge N) \text{ is satisfiable} \quad \text{iff} \quad \bigvee_i \exists \overline{a}\ (\bigwedge X_i \wedge \bigwedge N) \text{ is satisfiable,}$$

where $\overline{a}$ denotes the sequence of constants occurring in the corresponding matrix.

Lemma 3. *Each rule in* T *is (satisfiability) admissible.*

That is, in the tableau system (as usual), sets of formulae are interpreted conjunctively and the vertical bar is interpreted disjunctively.

Any path $N_0, N_1, \ldots$ in a derivation T, where N_0 denotes the root node of T, is called a *closed branch* in T iff the set $\bigcup_{j \geq 0} N_j$ contains $\perp$ (a contradiction has occurred), otherwise it is called an *open branch.* We call a branch B in a derivation tree *complete* (with respect to T) iff no new successor nodes can be added to the endpoint of B by T, otherwise it is called an *incomplete branch.* A derivation T is *closed* iff every path $N(= N_0), N_1, \ldots$ in it is a closed branch, otherwise it is called an *open derivation.* A closed derivation tree is also called a *refutation (tree).*

A derivation T from N is called *fair* iff for any path $N(= N_0), N_1, \ldots$ in T, with *limit* $N_\infty = \bigcup_{j \geq 0} N_j$, it is the case that each formula ψ which can be deduced from premises in N_∞ is contained in some N_j. Intuitively, fairness means that no possible application of an inference rule is delayed indefinitely. It also means that the γ rules, i.e. the rules ($-:$) and ($-\,;$), are applied infinitely often. For a finite complete branch $N(= N_0), N_1, \ldots N_n$, the limit N_∞ is equal to N_n.

Theorem 1 (Soundness and completeness of tableau). *Let T be a fair* T *derivation from a set N of formulae in $\mathcal{L}^{\mathsf{T}}$. Then: (i) If $N(= N_0), N_1, \ldots$ is a path with limit N_∞, then N_∞ is closed under the rules of* T. *(ii) N is satisfiable iff there exists a path in T with limit N_∞ such that N_∞ is satisfiable. (iii) N is unsatisfiable iff for every path $N(= N_0), N_1, \ldots$ the limit N_∞ contains $\perp$.*

This result follows immediately from the corresponding result for ground tableau of first-order logic, cf. Fitting [7] for a cut-free tableau calculus and a completeness proof. The reason is that the rules of T mirror the rules of the first-order logic ground tableau calculus (cf. Nellas [16]). By ground first-order logic tableau we mean a Smullyan-style tableau calculus [23], as opposed to free-variable tableau.

Corollary 1. *A Peirce logic formula is unsatisfiable iff the rules of* T *can be used to construct a closed tableau.*

The decomposition rules very clearly reflect the semantics of the top most connective in the premises. Because the decomposition rules are based on semantic equivalences, the following is immediate.

Lemma 4. *The decomposition rules of* T *are invertible.*

Recall, a rule of the form (1) is *invertible*, if the following is satisfied: there is a closed derivation for X iff there are closed derivations for each X_i $(1 \leq i \leq n)$.

There are alternative ways of capturing the properties of the identity relation in the calculus. In the presence of (sym), the following rule combines the rules (id$_1$)–(id$_3$) and can be used instead.

$$(Id)\ \frac{\psi,\ a\ \ Idb}{\psi[b]_\lambda}\quad \text{if } \psi|_\lambda = a$$

This rule corresponds to the familiar substitution axiom of equality in sentence tableau for first-order logic [7]. If the formula $a\ \ Idb$ is in a leaf node then the substitution rule generates the formula ψ, in which the occurrence of the constant a at position λ is replaced by b.

5 A Rasiowa-Sikorski Proof System

Now we turn to a different style of proof system. Rasiowa-Sikorski proof systems aim to prove validity. Given a candidate formula F they aim to prove its validity or, if it is not valid, the aim is to construct a counter-model (i.e. a model for the complement of the candidate formula). Starting with $\{x\,F\}$ (or $\{x\,F\,y\}$), this is done by systematic case analysis until fundamental validities are found. Rasiowa-Sikorski expansion rules have the same form (1) as for tableau and are also applied top-down. The definition of a Rasiowa-Sikorski derivations, and its construction by application of rules, is the same as a tableau derivation with the difference that the language is $\mathcal{L}^{\mathsf{RS}}$ instead of $\mathcal{L}^{\mathsf{T}}$. Crucially the interpretation of the rules is different. As above, X, X_i denote sets of formulae, but different from above sets of formulae are interpreted as disjunctions of formulae, whereas branching is interpreted conjunctively. A rule is *(validity) admissible*, if for any application of the rule to a node N,

$$\forall \overline{x}\,(\bigvee X \vee \bigvee N) \text{ is valid} \quad \text{iff} \quad \bigwedge_i \forall \overline{x}\,(\bigvee X_i \vee \bigvee N) \text{ is valid},$$

where $\overline{x}$ is the sequence of variables occurring in the corresponding matrix.

Any path $N_0, N_1, \ldots$ in a Rasiowa-Sikorski derivation T, where N_0 denotes the root node of T, is called a *closed branch* in T iff the set $\bigcup_{j\geq 0} N_j$ contains $\top$ (an axiomatic set was found), otherwise it is called an *open branch*. A Rasiowa-Sikorski derivation T is *closed* iff every path from the root in it is a closed branch, otherwise it is called an *open derivation*. A closed Rasiowa-Sikorski derivation is also called a *proof (tree)*. The concepts of *(in)complete branches*, *fairness* and *invertible rules* are the same as for tableau.

Let RS be the Rasiowa-Sikorski calculus for Peirce logic defined by the rules of Figure 2. As for tableau we distinguish between three kinds of deduction rules: decomposition rules, specific rules for identity and closure rules. The premises of the closure rules are commonly referred to as *axiomatic sets*.

Lemma 5. *Each rule in* RS *is (validity) admissible.*

Lemma 6. *Each decomposition rule in* RS *is invertible.*

Theorem 2 (Soundness and completeness of Rasiowa-Sikorski). *Let T be a fair* RS *derivation from a set N of formulae in $\mathcal{L}^{\mathsf{RS}}$. Then: (i) If $N(= N_0), N_1, \ldots$ is a path with limit N_∞, then N_∞ is closed under the rules of* RS*. (ii) N is valid iff there exists a path in T with limit N_∞ such that N_∞ is valid. (iii) N is valid iff for every path $N(= N_0), N_1, \ldots$ the limit N_∞ contains $\top$.*

Corollary 2. *A Peirce logic formula is valid iff the rules of* RS *can be used to construct a closed derivation tree.*

We conclude this section with remarks relating our presentation to presentations of Rasiowa-Sikorski systems usually found in the literature. We assume the rules are extension rules similar as for tableau which ignore the issue of repetition by assuming all main premises are retained during an inference step. Another difference is that we use sets instead of sequences of formulae. These differences are logically insignificant, however, and largely a matter of taste, although when developing an implementation of the calculus, the differences will need to be taken into account. Our presentation was chosen for reasons of uniformity.

6 Duality

The two systems presented are clearly dual to each other. This section is a formal discussion of this relationship between T and RS.

Suppose $\mathcal{R}_1$ and $\mathcal{R}_2$ are two expansion rules of the form (1). If $\mathcal{R}_2$ is obtained from $\mathcal{R}_1$ (or vice versa) by interchanging the logical connectives and symbols in accordance with the tables in Figure 3, then $\mathcal{R}_2$ is the *dual rule* to $\mathcal{R}_1$. I.e. all occurrences of $F_1 \cap F_2$ are replaced with $-(F_1 \cap F_2)$, all occurrences of $-(F_1 \cap F_2)$ are replaced with $F_1 \cap F_2$, etc. Notice we assume that $\wedge$ and $\vee$ refer to meta-level conjunction and disjunction, i.e. ',' and '|' for tableau and '|' and ',' for Rasiowa-Sikorski. Thus although the form of dual rules is the same the meta-level interpretation is interchanged.

Decomposition rules:

$$(\cap)\ \frac{x\,A\cap B}{x\,A\ \mid x\,B} \qquad (-\cap)\ \frac{x-(A\cap B)}{x\sim A,\ x\sim B} \qquad (--)\ \frac{x--A}{x\,A}$$

$$(:)\ \frac{x\,R\!:\!A}{x\,R\,z\mid z\,A}\ \text{ where } z \text{ is any variable}$$

$$(-:)\ \frac{x-(R\!:\!A)}{x\sim R\,z,\ z\sim A}\ \text{ where } z \text{ is a new variable}$$

$$(\cap)\ \frac{x\,R\cap S\,y}{x\,R\,y\mid x\,S\,y} \qquad (-\cap)\ \frac{x-(R\cap S)\,y}{x\sim R\,y,\ x\sim S\,y} \qquad (--)\ \frac{x--R\,y}{x\,R\,y}$$

$$(\smile)\ \frac{x\,R^{\smile}\,y}{y\,R\,x} \qquad (-\smile)\ \frac{x-(R^{\smile})\,y}{y\sim R\,x}$$

$$(^{c})\ \frac{x\,A^{c}\,y}{x\,A} \qquad (-^{c})\ \frac{x-(A^{c})\,y}{x\sim A}$$

$$(;)\ \frac{x\,R;S\,y}{x\,R\,z\mid z\,S\,y}\ \text{ where } z \text{ is any variable}$$

$$(-;)\ \frac{x-(R;S)\,y}{x\sim R\,z,\ z\sim S\,y}\ \text{ where } z \text{ is new variable}$$

Specific rules:

$$(\mathrm{sym})\ \frac{x-Id\,y}{y-Id\,x}$$

$$(\mathrm{id}_1)\ \frac{y-A,\ x-Id\,y}{x-A} \qquad (\mathrm{id}_2)\ \frac{y-R\,z,\ x-Id\,y}{x-R\,z} \qquad (\mathrm{id}_3)\ \frac{z-R\,x,\ x-Id\,y}{z-R\,y}$$

Closure rules:

$$(\mathrm{cl}_1)\ \frac{x\,A,\ x-A}{\top} \qquad (\mathrm{cl}_2)\ \frac{x-0}{\top}$$

$$(\mathrm{cl}_3)\ \frac{x\,R\,y,\ x-R\,y}{\top} \qquad (\mathrm{cl}_4)\ \frac{x-0\,y}{\top} \qquad (\mathrm{refl})\ \frac{x\,Id\,x}{\top}$$

Fig. 2. Rasiowa-Sikorski rules for Peirce logic

$\cap$	$-\cap$	$:$	$-:$	$\smile$	$-\smile$	c	$-c$	$;$	$-;$	Id	$-Id$	A	$-A$	R	$-R$	0	-0
$-\cap$	$\cap$	$-:$	$:$	$-\smile$	$\smile$	$-c$	c	$-;$	$;$	$-Id$	Id	$-A$	A	$-R$	R	-0	0

$\bot$	$\top$	x,y,z	a,b,c
$\top$	$\bot$	a,b,c	x,y,z

$\wedge$	$\vee$
$\vee$	$\wedge$

Fig. 3. Dual connectives and symbols

T	$\cap$	$-\cap$	$:$	$-:$	$\smile$	$-\smile$	^{c}	$-^c$	$;$	$-;$	sym	id_i	cl_j	refl
RS	$-\cap$	$\cap$	$-:$	$:$	$-\smile$	$\smile$	$-^c$	^{c}	$-;$	$;$	sym	id_i	cl_j	refl

Fig. 4. Dual rules ($1 \leq i \leq 3$, $1 \leq j \leq 4$)

Lemma 7. *The pair of rules in* T *and* RS *in each column of the table in Figure 4 are dual rules.*

Note the double negation rules are the only rules which do not appear in Figure 4.

Lemma 8. *Let $\mathcal{R}$ be any satisfiability admissible rule in* T*. Then the dual of $\mathcal{R}$ is a validity admissible rule in* RS*. Let $\mathcal{R}$ be any validity admissible rule in* RS*. Then the dual of $\mathcal{R}$ is a satisfiability admissible rule in* T*.*

Theorem 3. *Let F be a Peirce logic formula. Then, starting with $x \sim F$ (or $x \sim F\, y$), every inference step $\mathcal{I}$ (i.e. every rule application) in a* T *derivation for a F (or a Fb) can be mimicked in* RS *by $\mathcal{I}$ itself, when $\mathcal{I}$ involves the application of ($--$), or it can be mimicked by the application of the dual rule. Similarly, every RS inference step from $x\, F$ (or $x\, F\, y$) can be mimicked by a corresponding inference step in* T *starting from a $\sim F$ (or a $\sim Fb$).*

It follows that the systems T and RS step-wise simulate each other in a dual sense. They also p-simulate each other both with respect to derivations and search in a dual sense. See Schmidt and Hustadt [21, §8] for definitions of the notions of step-wise simulation and p-simulation.

It also follows that any prover for one of the systems can be used as a prover for the other system; users only need to keep in mind the dual interpretation of the formulae and rules. Clearly, optimisations compatible with one system will also be compatible in the dual form with the other system. For example, the tableau system admits that the γ rules can be restricted to constants occurring on the current branch (and means that γ rules are not necessarily applied infinitely often on a branch). This property carries over from ground tableau for first-order logic with equality, cf. [7]. By duality this signature restriction of the γ rules is compatible with RS.

7 Concluding Remarks

We have presented two proof systems for Peirce logic. Both are tableaux-style proof systems, with the difference that one is a refutation calculus and the other is a calculus for proving validities of relations and sets. It is not difficult to see that the duality between tableau and Rasiowa-Sikorski proof systems generalises quite naturally to other logics, especially first-order logic.

An implementation of the tableau calculus for Peirce logic was developed by Nellas [16]. By the duality result shown in this paper it can also be used as a prover for the Rasiowa-Sikorski calculus for Peirce logic.

Can the presented calculi be used to prove validities of Peirce algebra? We know that Maddux's sequent calculus of relational logic can be used to prove validities in relation algebra [14]. Maddux proved that an equation about relations is true in every relation algebra iff its three-variable translation has a four-variable proof in first-order logic. Because of the known connection between sequent calculi and tableau calculi, we expect this result to carry over to Peirce algebra and proofs or refutations constructed by the systems presented in this paper. This would provide a method to prove validities in Peirce algebra by considering the validity, or satisfiability, of an equation (represented as a suitable Peirce logic formula) and the proof, or refutation, of it in one of our systems. If the proof of a Peirce logic formula corresponding to a validity in Peirce algebra uses at most four variables then the equation would be valid in every Peirce algebra. (Dually for the refutation of a Peirce logic formula.)

References

[1] C. Brink. Boolean modules. *J. Algebra*, 71(2):291–313, 1981. 240

[2] C. Brink, K. Britz, and R. A. Schmidt. Peirce algebras. *Formal Aspects of Computing*, 6(3):339–358, 1994. 239, 240, 241

[3] K. Britz. Relations and programs. Master's thesis, Univ. Stellenbosch, South Africa, 1988. 239

[4] J. Dawson and R. Goré. A mechanised proof system for relation algebra using display logic. In *Proc. JELIA'98*, vol. 1489 of *LNAI*, pp. 264–278. Springer, 1998. 239

[5] M. de Rijke. *Extending Modal Logic*. PhD thesis, Univ. Amsterdam, 1993. 239, 241

[6] I. Düntsch and E. Orlowska. A proof system for contact relation algebras. *J. Philos. Logic*, 29:241–262, 2000. 239

[7] M. Fitting. *First-Order Logic and Automated Theorem Proving*. Texts and Monographs in Computer Science. Springer, 1990. 239, 246, 249

[8] G. Gargov and S. Passy. A note on Boolean modal logic. In P. P. Petkov, editor, *Mathematical Logic: Proc. 1988 Heyting Summerschool*, pp. 299–309, New York, 1990. Plenum Press. 239

[9] R. Goré. Cut-free display calculi for relation algebras. In *Selected Papers of CSL'96*, vol. 1258 of *LNCS*, pp. 198–210. Springer, 1996. 239

[10] M. C. B. Hennessy. A proof-system for the first-order relational calculus. *J. Computer and System Sci.*, 20:96–110, 1980. 239

[11] U. Hustadt and R. A. Schmidt. MSPASS: Modal reasoning by translation and first-order resolution. In R. Dyckhoff, editor, *Automated Reasoning with Analytic Tableaux and Related Methods*, vol. 1847 of *LNAI*, pp. 67–71. Springer, 2000. 240

[12] R. C. Lyndon. The representation of relational algebras. *Ann. Math.*, 51:707–729, 1950. 241

[13] W. MacCaull and E. Orlowska. Correspondence results for relational proof systems with applications to the Lambek calculus. *Studia Logica*, 71:279–304, 2002. 239

[14] R. D. Maddux. A sequent calculus for relation algebras. *Ann. Pure Applied Logic*, 25:73–101, 1983. 239, 250

[15] J. D. Monk. On representable relation algebras. *Michigan Math. J.*, 11:207–210, 1964. 241

[16] K. Nellas. Reasoning about sets and relations: A tableaux-based automated theorem prover for Peirce logic. Master's thesis, Univ. Manchester, UK, 2001. 241, 246, 250

[17] E. Orlowska. Relational formalisation of nonclassical logics. In C. Brink, W. Kahl, and G. Schmidt, editors, *Relational Methods in Computer Science*, Advances in Computing, pp. 90–105. Springer, Wien, 1997. 239, 240

[18] H. Rasiowa and R. Sikorski. *The Mathematics of Metamathematics*. Polish Scientific Publ., Warsaw, 1963. 239

[19] R. A. Schmidt. Algebraic terminological representation. Master's thesis, Univ. Cape Town, South Africa, 1991. Available as Technical Report MPI-I-91-216, Max-Planck-Institut für Informatik, Saarbrücken, Germany. 239

[20] R. A. Schmidt. Relational grammars for knowledge representation. In M. Böttner and W. Thümmel, editors, *Variable-Free Semantics*, vol. 3 of *Artikulation und Sprache*, pp. 162–180. Secolo Verlag, Osnabrück, Germany, 2000. 239

[21] R. A. Schmidt and U. Hustadt. Mechanised reasoning and model generation for extended modal logics. In *Theory and Applications of Relational Structures as Knowledge Instruments*, pp. 38–67. Springer, 2003. To appear. 240, 249

[22] W. Schönfeld. Upper bounds for a proof-search in a sequent calculus for relational equations. *Z. Math. Logik Grundlagen Math.*, 28:239–246, 1982. 239

[23] R. M. Smullyan. *First Order Logic*. Springer, Berlin, 1971. 239, 246

[24] A. Tarski. On the calculus of relations. *J. Symbolic Logic*, 6(3):73–89, 1941. 240

[25] W. W. Wadge. A complete natural deduction system for the relational calculus. Theory of Computation Report 5, Univ. Warwick, 1975. 239

An Institution Isomorphism for Planar Graph Colouring*

Giuseppe Scollo

Università di Verona, Dipartimento di Informatica
Ca' Vignal, Strada Le Grazie 15, I-37134 Verona, Italy
giuseppe.scollo@univr.it
http://profs.sci.univr.it/~scollo

Abstract. Maximal planar graphs with vertex resp. edge colouring are naturally cast as (deceptively similar) institutions. One then tries to embody Tait's equivalence algorithms into morphisms between them, and is lead to a partial redesign of those institutions. This paper aims at introducing a few pragmatic questions which arise in this case study, which also showcases the use of relational concepts and notations in the design of the subject institutions, and gives an outline of a solution to the problem of designing an isomorphism between them.

1 Introduction

Institution morphisms are a lively, albeit controversial subject of debate in the community of researchers who investigate abstract model-theoretic concepts and methods [7] in computing. The original definition for these structure maps [10] was soon to compete with different, variously motivated proposals, such as the "maps", "simulations", "transformations", respectively found in [15, 6, 18], among (several) others. Recent work [11] aims at systematic investigation of properties and interrelations of these notions, that surely is a promising, useful effort.

So far, lesser attention seems to have been attracted by pragmatic questions relating to institution morphisms, whatever sensible kind thereof, such as the understanding of how do those maps affect the design of institutions, meant as formalizations of given logical frameworks. This question is not necessarily to be understood in a "comparative" sense; that is to say, our expectation is that even in straightforward cases where different notions of institution morphism have essentially equivalent instances, it may well happen that institutions designed without taking morphisms into account need to be (partially) redesigned when the problem of mapping (relating, translating, structuring) them comes into play.

* This research has been partially supported by MURST Grant prot. 2001017741 under project "Ragionamento su aggregati e numeri a supporto della programmazione e relative verifiche" at the DMI Department of the University of Catania. Previous versions of Sections 1–6 appear in [19].

R. Berghammer et al. (Eds.): RelMiCS/Kleene-Algebra Ws 2003, LNCS 3051, pp. 252–264, 2004.

The present paper is aimed at presenting a little exercise of this kind. We start with introducing and motivating the exercise idea.

The Four Colour Theorem (4CT) is a paradigmatic case of potential applicability of methods and results that are offspring of research on translations between logical frameworks. Here is why the 4CT offers an interesting case study for translation concepts and methods relating to logical frameworks.

Our starting point is a view of the 4CT as a consistency theorem of finite, ad-hoc logics of graph colouring. The plural form *logics* here is purposeful, since a well-known result by Tait [21, 22] proves the equivalence between the 4CT with *vertex colouring* and the 3CT with *edge colouring*. The latter means proper colouring of edges rather than vertices, where "proper" is spelled out as the condition that adjacent edges, i.e. the border of a same triangular face, must be assigned different colours, whereas adjacent vertices must be assigned different colours by a proper vertex colouring.

Tait's equivalence comes equipped with a constructive proof, whereby algorithms are exhibited that turn any given proper 4-colouring of vertices of any given maximal planar graph into a proper 3-colouring of its edges, and vice versa—see e.g. [8] for an outline of Tait's algorithms. In this paper we use somewhat simpler algorithms for graph colouring conversion, that exploit the nice algebraic properties of the Klein 4-group [1].

So, here's our basic idea for an exercise aimed at testing practical impact of institution morphisms, possibly in different flavours, into institution design in the case study in question: 1) formalize maximal planar graph colouring by two distinct institutions, respectively with vertex colourings and edge colourings as models, and 2) (try to) cast Tait's equivalence into a pair of converse morphisms between the two institutions.

The first part of the exercise already raises institution design questions, e.g. the choice of signature morphisms; on pragmatic grounds, one might like to have such morphisms formalize edge contraction, in view of the relevant role played by this operation in reducibility proofs [5], yet contraction doesn't preserve maximality of planar graphs in all cases, which entails that the Set-valued sentence functor ought to map those morphisms to partial functions. One may take the design decision to formulate just vertex, resp. edge permutations as signature morphisms, since these operations are of practical interest, too. This leads to a straightforward solution of the first part of the exercise, already presented in [19], where we also showcase the use of relational concepts and notations, whereby one gets a pleasing conciseness and elegance in its presentation. Moreover, sentences in the institution with edge colouring have an amazing syntactic representation by Matiyasevitch's polynomials [13], whereby the *number* of proper colourings of any given maximal planar graph is readily found.

The second part of the exercise raises new design questions. Since the solution of the first part was determined without taking mutual interpretability of the two institutions into account, one shouldn't be surprised at finding out that Tait's algorithms prove hard to get embodied into structure-preserving maps between those institutions. Our redesign work in this respect seems quite instructive:

because of space constraints, we only give an outline of the key ingredients of a solution to the problem of designing an isomorphism between the subject institutions, whose technical details and proofs are available in [20].

2 Graph Colouring Preliminaries

The first proof of the 4CT [2, 3, 4] raised controversial discussions due to its combinatorial complexity which, for the analysis of the nearly 2000 graphs involved, required the construction of a program—whose correctness was not proven though. A new proof was obtained by [16], keeping the structure of the previous proof but cutting down to 633 the number of graphs involved. The dream yet remains that a simpler reason for the truth of this theorem may exist. However, our interest in carrying out the present exercise arises in the converse research direction—investigation into pragmatics of institution design rather than search of a new proof of the 4CT.

As customary, we only consider colourings of *maximal* planar graphs, viz. those where no new edge may be added between existing vertices without losing planarity. Maximal planar graphs are also referred to as *triangulations* of the sphere, thanks to the well-known bijection established by stereographic projection between the plane and the surface of the sphere. In the next Sections we shall adopt the following notational conventions.

Notation

$\underline{\mathbf{n}}$: the finite ordinal with n elements, viz. the natural numbers less than n.
$1_{\underline{\mathbf{n}}}$, $1'_{\underline{\mathbf{n}}}$, $0'_{\underline{\mathbf{n}}}$: resp. the *universal*, *identity* and *diversity* binary relations on $\underline{\mathbf{n}}$, thus $0'_{\underline{\mathbf{n}}} = 1_{\underline{\mathbf{n}}} \setminus 1'_{\underline{\mathbf{n}}}$, that is, the Boolean complement of the identity relation on $\underline{\mathbf{n}}$.
$\mathrm{r}^{\smile}$: relation-algebraic converse of r. Consistently, we also let $\mathrm{f}^{\smile}$ denote the inverse of an invertible function f.
r ; s : relation-algebraic composition of binary relations r, s. Consistently, we also let f ; g denote the function composition g∘f.
$\mathrm{T_V}(n)$: the set of $\underline{\mathbf{n+2}}$-labeled $(n+2)$-vertex triangulations of the sphere.
$\mathrm{T_E}(n)$: the set of $\underline{\mathbf{3n}}$-labeled $3n$-edge triangulations of the sphere.

3 Institution Preliminaries

The classic definition of institution, already appearing in the paper introducing this concept [9], will suffice for our purposes. Generalizations of this definition were proposed later [10]. Let's fix some basic notation about categories first.

Notation

$|\mathcal{C}|$ is the set of objects of category $\mathcal{C}$.
$\mathcal{C}(a, b)$ is the set of morphisms from a to b in category $\mathcal{C}$, for $a, b \in |\mathcal{C}|$.

Set is the category of (small[1]) sets with total functions as morphisms.
Cat is the category of (locally small[2]) categories with functors as morphisms.

An *institution* is a 4-tuple $\mathcal{I}$ = (Sig, Sen, Mod, $\models$), with:

(i) Sig a category, whose objects are called *signatures*,
(ii) Sen:Sig→Set a functor, sending each signature Σ to the set Sen(Σ) of Σ-sentences, and each signature morphism $\pi{:}\Sigma_1{\to}\Sigma_2$ to the mapping Sen(π):Sen(Σ_1)→Sen(Σ_2) that translates Σ_1-sentences to Σ_2-sentences,
(iii) Mod:Sig^{op}→Cat a contravariant functor, sending each signature Σ to the category Mod(Σ) of Σ-models, and each signature morphism $\pi{:}\Sigma_1{\to}\Sigma_2$ to the π-reduction functor Mod(π):Mod(Σ_2)→Mod(Σ_1),
(iv) $\models$: a |Sig|-indexed relation $\{\models_\Sigma \subseteq |\mathsf{Mod}(\Sigma)|\times\mathsf{Sen}(\Sigma) \mid \Sigma\in|\mathsf{Sig}|\}$, viz. a satisfaction relation between Σ-models and Σ-sentences for each $\Sigma\in$|Sig|, such that the following *satisfaction condition* holds for all $\Sigma_1,\Sigma_2\in$|Sig|, signature morphisms $\pi\in\mathsf{Sig}(\Sigma_1,\Sigma_2)$, Σ_2-models M and Σ_1-sentences φ:

$$\mathsf{Mod}(\pi)(\mathrm{M}) \models_{\Sigma_1} \varphi \Leftrightarrow \mathrm{M} \models_{\Sigma_2} \mathsf{Sen}(\pi)(\varphi)$$

Notation

A few notational conventions will simplify the presentation. We shall henceforth adopt the abbreviations: $\pi\varphi$ for Sen(π)(φ), and Mπ for Mod(π)(M), where $\pi{:}\Sigma_1{\to}\Sigma_2$ is a signature morphism, φ is a Σ_1-sentence, and M is a Σ_2-model.

When considering different institutions, it proves convenient to decorate the name of each element of the 4-tuple which an institution consists of, by adding the institution name as first subscript.

The original definition of institution morphism proposed in [10] is as follows[3].

Let $\mathcal{I}$ = ($\mathsf{Sig}_\mathcal{I}$, $\mathsf{Sen}_\mathcal{I}$, $\mathsf{Mod}_\mathcal{I}$, $\models_\mathcal{I}$), $\mathcal{I}'$ = ($\mathsf{Sig}_{\mathcal{I}'}$, $\mathsf{Sen}_{\mathcal{I}'}$, $\mathsf{Mod}_{\mathcal{I}'}$, $\models_{\mathcal{I}'}$) be institutions. An *institution morphism* $\mathcal{T}$: $\mathcal{I}\to\mathcal{I}'$ is a 3-tuple $\mathcal{T}$ = (Φ, α, β), with:

(i) Φ: $\mathsf{Sig}_\mathcal{I}\to\mathsf{Sig}_{\mathcal{I}'}$ a functor,
(ii) α: Φ ; $\mathsf{Sen}_{\mathcal{I}'} \to \mathsf{Sen}_\mathcal{I}$ a natural transformation (*sentence transformation*),
(iii) β: $\mathsf{Mod}_\mathcal{I} \to \Phi^{op}$; $\mathsf{Mod}_{\mathcal{I}'}$ a natural transformation (*model transformation*),

such that the following *satisfaction condition* holds, for all signatures $\Sigma\in|\mathsf{Sig}_\mathcal{I}|$, models $\mathrm{M}\in|\mathsf{Mod}_\mathcal{I}(\Sigma)|$ and sentences $\varphi'\in\mathsf{Sen}_{\mathcal{I}'}(\Phi(\Sigma))$:

$$\mathrm{M} \models_{\mathcal{I},\Sigma} \alpha_\Sigma(\varphi') \Leftrightarrow \beta_\Sigma(\mathrm{M}) \models_{\mathcal{I}',\Phi(\Sigma)} \varphi'$$

---

1 i.e., excluding proper classes

2 i.e., those with a small set of morphisms between any two objects

3 modulo a minor notational detail: in the definition of institution presented here, the type of the model functor follows a traditional convention for contravariant functors [12]; this explains the occurrence of the *dual* functor Φ^{op} in the type of the model transformation as presented here, in the definition of institution morphism as well as comorphism. Recall that Φ^{op} and Φ coincide on objects.

Note that model transformation and the signature functor go in the same direction, whereas sentence transformation goes in the opposite direction.

A somewhat dual concept was proposed under the name of "plain map" of institutions in [15], and renamed "institution comorphism" in [11], to emphasize the duality with the original concept. We welcome this change of terminology, and we further refer to an "institution (co)morphism" whenever the difference doesn't matter. The essential difference is in the directions of the natural transformations involved, both of which change.

4 A Vertex Colouring Institution

Syntax will be abstract, exploiting the fact that institutions do not force one to deal with concrete syntax. Signatures are just positive numbers, ranking maximal planar graphs by their size, and we take the *bijective* relabelings of vertices as signature morphisms. This restriction is a design decision, motivated as follows.

Each $n>0$ is the rank of the maximal planar graphs, or triangulations of the sphere, that have $n+2$ vertices. Vertex colouring of such structures require that each vertex be given a unique identity. To this purpose we consider vertices to be uniquely labeled by the elements of finite ordinal $\underline{\mathbf{n+2}}$, for triangulations of rank n. Bijective relabelings are thus just label permutations. The pragmatic question arises as to what purpose could be served by non-bijective maps on finite ordinals. On the one hand, loss of surjectivity appears useless, insofar as it introduces labels in the morphism codomain that are not made use of to label any vertex, according to the morphism image. On the other hand, though, loss of injectivity would seem to be of some use, inasmuch it amounts to identify formerly distinct vertices, thus it could prove useful to formalize edge contraction—whenever an edge connects two such vertices. This operation, however, does not preserve maximality of planar graphs (this may happen when a vertex of degree 3 is opposite to the contracted edge). So, if one admits non-injective relabelings as signature morphisms, then the Set-valued sentence functor, giving the set of vertex-labeled triangulations of rank n for each $n > 0$, ought to map those morphisms to *partial* functions, whereas only total functions are available as morphisms in Set.

Signatures

$|\mathsf{Sig}_{\mathcal{V}}| = \mathbf{N}\setminus\{0\}$
$\mathsf{Sig}_{\mathcal{V}}(n,n) = \{\pi{:}\underline{\mathbf{n+2}}{\rightarrow}\underline{\mathbf{n+2}} \mid \pi \text{ is bijective}\}$
$\mathsf{Sig}_{\mathcal{V}}(m,n) = \emptyset \text{ if } m \neq n$

Sentences

Each $\theta{\in}\mathrm{T}_{\mathrm{V}}(n)$ is represented by the symmetric quotient of a binary relation on vertices, $\epsilon_\theta \overset{def}{=} \eta_\theta/\mathrm{Sym}$, where η_θ is the irreflexive, symmetric edge relation of θ, thus satisfies the relation-algebraic laws $\eta_\theta{\leq}0'_{\underline{\mathbf{n+2}}}$, $\eta_\theta{=}\eta_\theta{}^{\smile}$, while $|\eta_\theta|{=}6n$, but the Sym quotient turns ordered pairs into unordered ones, thus $|\epsilon_\theta|=3n$. As a matter of notation, we write $i\ \epsilon_\theta\ j$ or $\{i,j\}\in\epsilon_\theta$, rather than the more cumbersome $\{(i,j),(j,i)\}\in\epsilon_\theta$, whenever $\{(i,j),(j,i)\}\subseteq\eta_\theta$. We thus define:

$$\mathsf{Sen}_{\mathcal{V}}(n) = \{\epsilon_\theta \mid \theta\in\mathrm{T}_{\mathrm{V}}(n)\}$$

Sentence Translation

If $\pi\in\mathsf{Sig}_{\mathcal{V}}(n,n)$, $\epsilon_\theta\in\mathsf{Sen}_{\mathcal{V}}(n)$, then $\pi\epsilon_\theta\in\mathsf{Sen}_{\mathcal{V}}(n)$, with $(\pi i)\ \pi\epsilon_\theta\ (\pi j) \Leftrightarrow i\ \epsilon_\theta\ j$.

Models

The model functor assigns to each signature $n > 0$ the category of 4-colourings of the $n+2$ vertices, with colour permutations as model morphisms, thus:

$$|\mathsf{Mod}_{\mathcal{V}}(n)| = \underline{\mathbf{4}}^{\underline{\mathbf{n+2}}}$$
$$\forall\mu,\mu'\in|\mathsf{Mod}_{\mathcal{V}}(n)|.\mathsf{Mod}_{\mathcal{V}}(n)(\mu,\mu') = \{\rho\in\underline{\mathbf{4}}^{\underline{\mathbf{4}}}|\rho \text{ is bijective}, \mu' = \mu\,;\,\rho\}$$

Model Reduction

If $\pi{:}\underline{\mathbf{n+2}}\to\underline{\mathbf{n+2}}\in\mathsf{Sig}_{\mathcal{V}}(n,n)$ and $\mu{:}\underline{\mathbf{n+2}}\to\underline{\mathbf{4}}\in|\mathsf{Mod}_{\mathcal{V}}(n)|$, then $\mu\pi\in|\mathsf{Mod}_{\mathcal{V}}(n)|$,with $(\mu\pi)i \stackrel{def}{=} \mu(\pi\ i)$, and $\rho\pi \stackrel{def}{=} \rho$ for all colour permutations $\rho{:}\underline{\mathbf{4}}\leftrightarrow\underline{\mathbf{4}}$. This makes model reduction to be a functor, thus $\rho\pi\in\mathsf{Mod}_{\mathcal{V}}(n)(\mu\pi,\mu'\pi)$ if $\rho\in\mathsf{Mod}_{\mathcal{V}}(n)(\mu,\mu')$, since $(\mu\pi)\,;\,\rho = (\mu\,;\,\rho)\pi$, by an easy check.

Satisfaction

In $\mathcal{V}$, a $n-$model satisfies an $n-$sentence iff it is a proper vertex colouring of that triangulation, that is:

$$\mu \models_{\mathcal{V},n} \epsilon_\theta \text{ iff } \forall i,j\in\underline{\mathbf{n+2}}.\ i\epsilon_\theta j \Rightarrow \mu i{\neq}\mu j$$

A relation-algebraic formulation of this definition may exploit the "oriented" edge relation η_θ from which ϵ_θ is obtained as a quotient, and the view of the 4-colouring map as a binary relation $\mu\subseteq\underline{\mathbf{n+2}}\times\underline{\mathbf{4}}$. Then we get:

$$\mu \models_{\mathcal{V},n} \epsilon_\theta \text{ iff } \mu^{\smile};\eta_\theta;\mu \leq 0'_{\underline{\mathbf{4}}}$$

It is easy to show that this definition complies with the *satisfaction condition*:

$$\mu\pi \models_{\mathcal{V},n} \epsilon_\theta \Leftrightarrow \mu \models_{\mathcal{V},n} \pi\epsilon_\theta$$

therefore $\mathcal{V}$ is an *institution*.

5 An Edge Colouring Institution

Syntax will be somewhat more concrete, inspired by Matiyasevich's polynomial representation of triangulations of the sphere [13]. Signatures remain the same, but we now take the bijective relabelings of *edges* as signature morphisms.

Signatures

$|\mathsf{Sig}_{\mathcal{E}}| = |\mathsf{Sig}_{\mathcal{V}}| = \mathbf{N}\backslash\{0\}$
$\mathsf{Sig}_{\mathcal{E}}(n,n) = \{\pi{:}\underline{\mathbf{3n}}{\rightarrow}\underline{\mathbf{3n}} \mid \pi \text{ is bijective}\}$
$\mathsf{Sig}_{\mathcal{E}}(m,n) = \emptyset$ if $m \neq n$

Sentences

Sentences in $\mathsf{Sen}_{\mathcal{E}}(n)$, ranged over by ψ_ϑ, are represented by Matiyasevich's polynomials in product form:

$$\psi_\vartheta \;=\; \prod_{\mathrm{t}_{ijk}\in\vartheta} (x_i - x_j)(x_j - x_k)(x_k - x_i)$$

where $\vartheta{\in}\mathrm{T_E}(n)$ and t_{ijk} is a triangular face of ϑ having edges labeled i, j, k in clockwise order. We thus define: $\mathsf{Sen}_{\mathcal{E}}(n) = \{\psi_\vartheta \mid \vartheta{\in}\mathrm{T_E}(n)\}$.

Sentence Translation

If $\pi{\in}\mathsf{Sig}_{\mathcal{E}}(n,n)$ and $\psi_\vartheta{\in}\mathsf{Sen}_{\mathcal{E}}(n)$ represented as above, then $\pi\psi_\vartheta{\in}\mathsf{Sen}_{\mathcal{E}}(n)$, with

$$\pi\psi_\vartheta \;=\; \prod_{\mathrm{t}_{ijk}\in\vartheta} (x_{\pi i} - x_{\pi j})(x_{\pi j} - x_{\pi k})(x_{\pi k} - x_{\pi i})$$

Models

The model functor assigns to each signature $n > 0$ the category of 3-colourings of the $3n$ edges, with colour permutations as model morphisms, thus:

$|\mathsf{Mod}_{\mathcal{E}}(n)| = \underline{\mathbf{3}}^{\underline{\mathbf{3n}}}$
$\forall \nu, \nu'{\in}|\mathsf{Mod}_{\mathcal{E}}(n)|.\mathsf{Mod}_{\mathcal{E}}(n)(\nu,\nu') = \{\rho{\in}\underline{\mathbf{3}}^{\underline{\mathbf{3}}} | \rho \text{ is bijective}, \nu' = \nu\,;\,\rho\}$

Model Reduction

If $\pi{:}\underline{\mathbf{3n}}{\rightarrow}\underline{\mathbf{3n}} \in \mathsf{Sig}_{\mathcal{E}}(n,n)$ and $\nu{:}\underline{\mathbf{3n}}{\rightarrow}\underline{\mathbf{3}} \in |\mathsf{Mod}_{\mathcal{E}}(n)|$, then $\nu\pi \in |\mathsf{Mod}_{\mathcal{E}}(n)|$, with $(\nu\pi)i \stackrel{def}{=} \nu(\pi\ i)$, and $\rho\pi \stackrel{def}{=} \rho$ for all colour permutations $\rho{:}\underline{\mathbf{3}}{\leftrightarrow}\underline{\mathbf{3}}$. This makes model reduction to be a functor, thus $\rho\pi{\in}\mathsf{Mod}_{\mathcal{E}}(n)(\nu\pi,\nu'\pi)$ if $\rho{\in}\mathsf{Mod}_{\mathcal{E}}(n)(\nu,\nu')$, since $(\nu\pi)\,;\,\rho = (\nu\,;\,\rho)\pi$, by an easy check.

Satisfaction

In $\mathcal{E}$, a n-model satisfies an n-sentence iff it is a proper edge colouring of that triangulation, that is:

$$\nu \models_{\mathcal{E},n} \psi_\vartheta \text{ iff } \forall i,j{\in}\underline{\mathbf{3n}}.(x_i - x_j) \text{ occurs in } \psi_\vartheta \Rightarrow \nu i{\neq}\nu j$$

A relation-algebraic formulation of this definition may use the binary relation of "occurrence in ψ_ϑ", $\xi_{\psi_\vartheta} \leq 1_{\underline{\mathbf{3n}}}$: $i \; \xi_{\psi_\vartheta} \; j$ iff $(x_i - x_j)$ occurs in ψ_ϑ. Then, by using the view of a 3-colouring map as a binary relation $\nu \subseteq \underline{\mathbf{3n}} \times \underline{\mathbf{3}}$, we get:

$$\nu \models_{\mathcal{E},n} \psi_\vartheta \text{ iff } \nu^{\smile}; \xi_{\psi_\vartheta}; \nu \leq 0'_{\underline{\mathbf{3}}}$$

It is easy to show that this definition complies with the *satisfaction condition*:

$$\nu\pi \models_{\mathcal{E},n} \psi_\vartheta \Leftrightarrow \nu \models_{\mathcal{E},n} \pi\psi_\vartheta$$

therefore $\mathcal{E}$, too, is an *institution.*

6 Tait's Equivalence

A triangulation admits a proper 4-colouring of its vertices if, and only if, it admits a proper 3-colouring of its edges. This is Tait's classical result [21, 22], albeit here stated in graph-theoretic terms rather than, as in its original formulation, in terms of cubic map colourings. The equivalence is shown by exhibiting two algorithms, which we are going to recast in graph-theoretic terms, that for any given triangulation respectively turn any proper 4-colouring of its vertices into a proper 3-colouring of its edges, and vice versa.

We take $\underline{\mathbf{4}}$ as the set of colours for vertex-colouring and $\underline{\mathbf{4}} \backslash \underline{\mathbf{1}}$ that for edge-colouring. Taking the latter rather than $\underline{\mathbf{3}}$ somewhat simplifies the presentation of Tait's algorithms, thanks to the properties of an elegant, algebraic construction which uses the *Klein 4-group*, as provided in [1]. We take $\underline{\mathbf{4}}$ as the group carrier, with 0 as its neutral element. Every element is self-inverse, and the binary group operation $\oplus$ further satisfies $x \oplus y = z$ whenever $\{x, y, z\} = \underline{\mathbf{4}} \backslash \underline{\mathbf{1}}$. This defines $\oplus$, since $0 \oplus x = x \oplus 0 = x \oplus x = 0$ for all $x \in \underline{\mathbf{4}}$, by the previous conditions.

4CT ⇒ 3CT

Let $\mu{:}\underline{\mathbf{n+2}} \rightarrow \underline{\mathbf{4}}$ be a proper 4-colouring of given triangulation $\theta \in \mathrm{T_V}(n)$. For each edge x in θ, let $\mu i \neq \mu j$ be the colours assigned by μ to the vertices connected by x. Then their Klein sum $\mu i \oplus \mu j$ is the colour assigned to edge x.

By the properties of the Klein 4-group, this colour is never 0 insofar as $\mu i \neq \mu j$ (by assumption, μ is a proper colouring of the vertices of θ). Furthermore, any two edges sharing a face get different colours since they share one vertex (thus one addend of the Klein sums yielding their respective colours), whereas the other two vertices they resp. join are coloured differently by μ, as they are the ends of the third edge sharing the same face. We thus have a proper 3-colouring of the edges of θ, with colours out of $\underline{\mathbf{4}} \backslash \underline{\mathbf{1}}$.

3CT ⇒ 4CT

The construction in the converse direction is a bit more complex. Let $\nu{:}\underline{\mathbf{3n}} \rightarrow \underline{\mathbf{4}} \backslash \underline{\mathbf{1}}$ be a proper 3-colouring of given triangulation $\vartheta \in \mathrm{T_E}(n)$. Choose a vertex in the triangulation as start-vertex, and assign it colour 0. Every other vertex is then coloured by the Klein sum of the colours assigned by ν to the edges of any path from the start-vertex to that vertex.

Of course, the specified construction is only sound if Klein summation of the colours assigned by ν proves invariant for all paths joining any given pair of vertices. This holds because (i) every element is self-inverse in the Klein group and (ii) Klein summation of the colours assigned by ν along every circuit turns out to be 0. A proof of this fact is worked out in [1] (pp. 22–23), for the colouring of cubic maps, but it is readily interpreted in our present setting [19, 20].

7 Morphism-Driven Redesign of Institutions

A basic obstacle makes it impossible to embody Tait's algorithms into (whatever kind of) morphism between the $\mathcal{V}$ and $\mathcal{E}$ institutions presented above, and that is: the lack of a non-trivial functorial mapping between their categories of signatures. Although those categories share their objects, their signature morphisms differ, and these prove hard to map. It's seems worthwhile to review the implicit reason for the choice of different signature morphisms in the design of the aforementioned institutions.

The choice of signature morphisms for $\mathcal{V}$ was just the obvious one, as far as abstract syntax for vertex colourings is concerned. Similarly, that for $\mathcal{E}$ was inspired by Matiyasevich's polynomial representation of triangulations, where only the naming of edges matter, thus it seemed fairly natural to take edge renamings as the edge colouring counterpart of vertex renamings for vertex colouring, as far as abstract syntax for edge colourings is concerned. This choice is actually sentence independent, in that it only depends on the rank of the triangulation (since every triangulation of given rank n has the same number of edges, that is $3n$), therefore it was appropriate as a design choice for signature morphisms.

Our "local" design choices of signature morphisms prove no longer appropriate when a wider perspective is taken, that is to say, as soon as one needs to know which vertices are connected by which edges—as it happens to be the case with Tait's algorithms. An outline of a solution to this problem follows; the interested reader is referred to [20], where technical details are fully worked out.

In a first approximation to our institution redesign problem, we try to keep $\mathcal{V}$ unchanged, as well as the sentences in the edge colouring institution.

The unordered edge relation ϵ_θ on vertices shows up in $\mathcal{V}$ as sentence representation. We need to keep this information when moving to Matiyasevich's representation used in $\mathcal{E}$. Here any enumeration of the edges does the job, thus one may choose a particular enumeration that be determined by the vertex labeling only. To this purpose we define the *lexicographic edge-labeling map* $\lambda_\theta{:}\epsilon_\theta{\rightarrow}\underline{\mathbf{3n}}$ for every triangulation $\theta{\in}\mathrm{T_V}(n)$ represented in $\mathcal{V}$ by ϵ_θ, as the unique enumeration of edges which satisfies:

$$\forall\, i, j, i', j'. i\ \epsilon_\theta\ j, i'\ \epsilon_\theta\ j', i < j, i' < j' \Rightarrow$$
$$(\lambda_\theta\{i,j\} < \lambda_\theta\{i',j'\} \Leftrightarrow (i < i' \vee (i = i' \wedge j < j')))$$

The new version of the edge-colouring institution, let it be $\mathcal{E}'$, has the same signature morphisms as $\mathcal{V}$, the same category of signatures thus: $\mathsf{Sig}_{\mathcal{E}'} = \mathsf{Sig}_{\mathcal{V}}$.

For sentences in $\mathcal{E}'$, we keep Matiyasevich's polynomial representation, but enumerating edges by the λ_θ map, thus for $\theta \in \mathrm{T_V}(n)$ we define

$$\psi_\theta = \prod_{t_{ijk} \in \theta} (x_{\lambda_\theta\{i,j\}} - x_{\lambda_\theta\{j,k\}})(x_{\lambda_\theta\{j,k\}} - x_{\lambda_\theta\{k,i\}})(x_{\lambda_\theta\{k,i\}} - x_{\lambda_\theta\{i,j\}})$$

where t_{ijk} is a triangular face having *vertices* labeled i, j, k, in clockwise order. Let $\mathsf{Sen}_{\mathcal{E}'}(n)$ be the set of all such sentences.

The change of signature morphisms just made, requires a straightforward adaptation of sentence translation along them. For each θ, the λ_θ map induces, for every permutation π of $\underline{\mathbf{n+2}}$, a unique permutation $\pi_\theta^\#$ of $\underline{\mathbf{3n}}$ that commutes with the lexicographic edge-labeling map. Now, if $\pi \in \mathsf{Sig}_{\mathcal{E}'}(n, n)$ and $\psi_\theta \in \mathsf{Sen}_{\mathcal{E}'}(n)$ is a Matiyasevich's polynomial as above, then $\pi\psi_\theta \in \mathsf{Sen}_{\mathcal{E}'}(n)$ is defined by

$$\pi\psi_\theta \stackrel{def}{=}$$
$$\prod_{t_{ijk} \in \theta} (x_{\pi_\theta^\# \lambda_\theta\{i,j\}} - x_{\pi_\theta^\# \lambda_\theta\{j,k\}})(x_{\pi_\theta^\# \lambda_\theta\{j,k\}} - x_{\pi_\theta^\# \lambda_\theta\{k,i\}})(x_{\pi_\theta^\# \lambda_\theta\{k,i\}} - x_{\pi_\theta^\# \lambda_\theta\{i,j\}}).$$

The check that $\mathsf{Sen}_{\mathcal{E}'}$ so defined is indeed a functor is straightforward.

The change with the signature morphisms affects the model category more deeply. We drop the assumption, made in the 4CT⇒3CT part of Tait's equivalence proof, that the vertex colouring $\mu : \underline{\mathbf{n+2}} \rightarrow \underline{\mathbf{4}}$ be proper. Then, by Klein sum, we may get edges "coloured 0", meaning that the two vertices have the same colour. One may use this within an edge colouring model to mean that it will not satisfy any sentence having some edge which links a pair of vertices whose label pair is "coloured 0" in that model. In the converse direction, we may only well-define vertex colourings as images of edge colourings in terms of Klein summation of edge colours along paths if the soundness condition is met that enabled us to do so in Tait's equivalence proof. We thus require edge-colouring models to have Klein sum 0 along every triangle. We refer the reader to [20] for further details about $\mathcal{E}'$ because, perhaps surprisingly, this turns out not to be apt to our redesign purpose yet.

The story of our ultimate design problem has to do with the choice of a start-vertex, which is assigned colour 0 in the construction of a 4-colouring of vertices out of a given 3-colouring of edges presented in Section 6. At a first glance, it may seem natural to make that choice appear as a constituent of the model transformation part of an institution (co)morphism between the institutions $\mathcal{V}$ and $\mathcal{E}'$ introduced above. At a closer look, however, such an idea proves troublesome precisely with respect to naturality of the transformation!

We thus introduce yet new versions $\mathcal{V}'$ and $\mathcal{E}''$ of both institutions, that respectively inherit both the category of signatures and the sentence functor from $\mathcal{V}$ and $\mathcal{E}'$, but have slightly modified model functors, defined as follows.

Extended Model Functor for Vertex Colouring

$|\mathsf{Mod}_{\mathcal{V}'}(n)| = \{v = \langle \mu_v, \mathrm{s}_v \rangle \mid \mu_v \in \underline{4}^{\underline{\mathbf{n+2}}}, \mathrm{s}_v \in \underline{\mathbf{n+2}}, \mu_v \mathrm{s}_v = 0\}$

$\forall v, v' \in |\mathsf{Mod}_{\mathcal{V}'}(n)|. \mathsf{Mod}_{\mathcal{V}}(n)(v, v') = \{\sigma \in \underline{4}^{\underline{4}} \mid \sigma \text{ is bijective}, \mu_{v'} = \mu_v ; \sigma\}$

Model reduction, for a given n-model $v = \langle \mu_v, \mathrm{s}_v \rangle$ and a signature morphism $\pi: \underline{\mathbf{n+2}} \rightarrow \underline{\mathbf{n+2}}$, is defined by $v\pi = \langle \mu_v \pi, \mathrm{s}_{v\pi} \rangle$, where $(\mu_v \pi) i = \mu_v(\pi i)$ like in $\mathcal{V}$, and moreover $\mathrm{s}_{v\pi} \stackrel{def}{=} \pi^{\smile} \mathrm{s}_v$. Letting $\sigma\pi = \sigma$ for all n-model morphisms σ and n-signature morphisms π, makes model reduction to be a functor, since $(\mu_v \pi) ; \sigma = (\mu_v ; \sigma)\pi$, as it is straightforward to check.

Extended Model Functor for Edge Colouring

$|\mathsf{Mod}_{\mathcal{E}''}(n)| = \{\varepsilon = \langle \nu_\varepsilon : E_{\mathbf{n+2}} \rightarrow \underline{4}, \mathrm{s}_\varepsilon \rangle \mid$
$\mathrm{s}_\varepsilon \in \underline{\mathbf{n+2}} \wedge (i \neq j \neq k \neq i \Rightarrow \nu_\varepsilon\{i,j\} \oplus \nu_\varepsilon\{j,k\} \oplus \nu_\varepsilon\{k,i\} = 0)\}$

$\forall \varepsilon, \varepsilon' \in |\mathsf{Mod}_{\mathcal{E}''}(n)|.\ \mathsf{Mod}_{\mathcal{E}''}(n)(\varepsilon, \varepsilon') = \mathsf{Mod}_{\mathcal{E}'}(n)(\nu_\varepsilon, \nu_{\varepsilon'}) =$
$\{\rho \in \underline{4}^{\underline{4}}, \mid \rho \text{ is bijective}, \rho 0 = 0, \nu_{\varepsilon'} = \nu_\varepsilon ; \rho\}$

So, the n-model morphisms still are the 0-preserving colour permutations. Model reduction, for a given n-model $\varepsilon = \langle \nu_\varepsilon, \mathrm{s}_\varepsilon \rangle$ and signature morphism $\pi: \underline{\mathbf{n+2}} \rightarrow \underline{\mathbf{n+2}}$, is defined by $\varepsilon\pi = \langle \nu_\varepsilon \pi, \mathrm{s}_{\varepsilon\pi} \rangle$, where $(\nu_\varepsilon \pi)\{i,j\} = \nu_\varepsilon\{\pi i, \pi j\}$ like in $\mathcal{E}'$, and moreover $\mathrm{s}_{\varepsilon\pi} \stackrel{def}{=} \pi^{\smile} \mathrm{s}_\varepsilon$. Letting $\rho\pi = \rho$ for all n-model morphisms ρ and n-signature morphisms π, makes model reduction to be a functor, since $(\nu_\varepsilon \pi) ; \rho = (\nu_\varepsilon ; \rho)\pi$, as it is straightforward to check.

Satisfaction

Satisfaction in $\mathcal{V}'$ and $\mathcal{E}''$ is defined as in $\mathcal{V}$ and $\mathcal{E}'$, respectively, but only using the colouring map part of the model, whence the satisfaction condition is met just as it is in the respective previous versions of those institutions. For $\mathcal{E}''$, according to the chosen sentence representation (same as in in $\mathcal{E}'$) we have

$\nu \models_{\mathcal{E}',n} \psi_\theta$ iff
$\forall i, j, k \in \underline{\mathbf{n+2}}.(x_{\lambda_\theta\{i,j\}} - x_{\lambda_\theta\{j,k\}})$ occurs in $\psi_\theta \Rightarrow 0 \neq \nu\{i,j\} \neq \nu\{j,k\} \neq 0$.

For a relation-algebraic formulation of satisfaction in these institutions see [20].

8 Outline of a Graph Colouring Institution Isomorphism

A final check is in place, viz. that the fine tuning of the model functors just worked out indeed solves the problem which motivated it, that is, it allows one to build an isomorphism between the graph colouring institutions $\mathcal{V}'$ and $\mathcal{E}''$.

As a matter of notational convenience, for each of the three constituents of the isomorphism $\mathcal{T} : \mathcal{V}' \leftrightarrow \mathcal{E}''$, we use the same symbol in either direction, so we need not specify whether $\mathcal{T}$ is an institution morphism or comorphism. Furthermore, if γ is a bijection, we write $x \overset{\gamma}{\leftrightarrow} y$ to mean $y = \gamma x$ and $x = \gamma^{\smile} y$.

Let $\Phi \overset{def}{=} \mathbf{1}_{\mathsf{Sig}_{\mathcal{V}'}}$ be the identity functor on the category of signatures (the same in both institutions), then we are going to show the existence of natural isomorphisms α: $\mathsf{Sen}_{\mathcal{V}'} \leftrightarrow \mathsf{Sen}_{\mathcal{E}''}$ and β: $\mathsf{Mod}_{\mathcal{V}'} \leftrightarrow \mathsf{Mod}_{\mathcal{E}''}$, so that $\mathcal{T} = \langle \Phi, \alpha, \beta \rangle$ is an institution isomorphism $\mathcal{T} : \mathcal{V}' \leftrightarrow \mathcal{E}''$.

For each $n > 0$, the sentence transformation component α_n : $\mathsf{Sen}_{\mathcal{V}'}(n) \leftrightarrow \mathsf{Sen}_{\mathcal{E}''}(n)$ is the bijection whereby $\epsilon_\theta \overset{\alpha_n}{\longleftrightarrow} \psi_\theta$ for each $\theta \in \mathrm{T}_{\mathcal{V}}(n)$, with ϵ_θ as defined in Section 4 and ψ_θ as defined in Section 7.

The model transformation component β_n : $\mathsf{Mod}_{\mathcal{V}'}(n) \leftrightarrow \mathsf{Mod}_{\mathcal{E}''}(n)$ is the invertible functor defined by the following conditions.

Objects: for all $\upsilon = \langle \mu_\upsilon, \mathrm{s}_\upsilon \rangle \in |\mathsf{Mod}_{\mathcal{V}'}(n)|$ and $\varepsilon = \langle \nu_\varepsilon, \mathrm{s}_\varepsilon \rangle \in |\mathsf{Mod}_{\mathcal{E}''}(n)|$,

$$\upsilon \overset{\beta_n}{\longleftrightarrow} \varepsilon \text{ iff } \mathrm{s}_\upsilon = \mathrm{s}_\varepsilon \wedge \forall i, j \in \underline{\mathbf{n+2}}. i \neq j \Rightarrow \nu_\varepsilon\{i, j\} = \mu_\upsilon(i) \oplus \mu_\upsilon(j).$$

Arrows: for all $\upsilon, \upsilon' \in |\mathsf{Mod}_{\mathcal{V}'}(n)|$, $\varepsilon, \varepsilon' \in |\mathsf{Mod}_{\mathcal{E}''}(n)|$, $\sigma \in \mathsf{Mod}_{\mathcal{V}'}(n)(\upsilon, \upsilon')$, and $\rho \in \mathsf{Mod}_{\mathcal{E}''}(n)(\varepsilon, \varepsilon')$,

$$\sigma \overset{\beta_n}{\longleftrightarrow} \rho \text{ iff } \forall x \in \underline{\mathbf{4}}. \rho x = \sigma 0 \oplus \sigma x.$$

A little calculation shows that the clauses given above do indeed define a functor in either direction. For arrows, the map in the $\mathcal{E}'' \rightarrow \mathcal{V}'$ direction is also determined by the condition $\sigma(\mu_\upsilon \mathrm{s}_{\upsilon'}) = 0$. This, together with the conditions given above (on arrows as well as on objects), uniquely determines the σ permutation for a given ρ, thanks to the algebraic properties of the Klein sum. The verification of these facts is left to the reader. Naturality of transformations and the satisfaction condition are proven in [20].

9 Concluding Remarks

The forgetful morphisms $\mathcal{E}'' \rightarrow \mathcal{E}'$ and $\mathcal{V}' \rightarrow \mathcal{V}$, whereby the start-vertex disappears from the model structure, are also comorphisms $\mathcal{E}' \rightarrow \mathcal{E}''$ and $\mathcal{V} \rightarrow \mathcal{V}'$. Our case study seems to suggest a general technique to relate institutions, when some structure translation algorithm between their models is known, but that introduces additional, specific structure for the sole purpose of translation.

References

[1] M. Aigner, *Graphentheorie—Eine Eintwicklung aus dem 4-Farbenproblem*, B. G. Teubner, Stuttgart (1984). 253, 259, 260

[2] K. Appel and W. Haken, Every planar map is four colorable. Part I. Discharging, *Illinois J. Math.* **21** (1977), 429-490. 254

[3] K. Appel, W. Haken and J. Koch, Every planar map is four colorable. Part II. Reducibility, *Illinois J. Math.* **21** (1977), 491–567. 254

[4] K. Appel and W. Haken, Every planar map is four colorable, *Contemporary Math.* **98** (1989). 254

[5] G. D. Birkhoff, The reducibility of maps, *Amer. J. Math.* **35** (1913), 114-128. 253

[6] M. Cerioli, *Relationships between logical formalisms*, Ph. D. Thesis, University of Genova, March 1993. 252

[7] H.-D. Ebbinghaus, Extended logics: the general framework, in: J. Barwise and S. Feferman (Eds.) *Model-Theoretic Logics*, Springer-Verlag, Berlin (1985) 25–76. 252

[8] R. Fritsch and G. Fritsch, *The Four Colour Theorem*, Springer-Verlag, New York (1998). 253

[9] J. A. Goguen and R. M. Burstall, Introducing Institutions, in: E. Clarke and D. Kozen (Eds.), *Proceedings, Logics of Programming Workshop*, Lecture Notes in Computer Science, **164**, Springer (1984) 221–256. 254

[10] J. A. Goguen and R. M. Burstall, Institutions: Abstract model theory for specification and programming, *J. Assoc. Comput. Mach.* **39** (1992) 95–146. 252, 254, 255

[11] J. A. Goguen and G. Roşu, Institution Morphisms, *Formal Aspects of Computing* **13** (2002) 274–307. At URL: `http://www.cs.ucsd.edu/users/goguen/pubs/` 252, 256

[12] S. Mac Lane, *Categories for the Working Mathematician*, Springer-Verlag, New York (1971). 255

[13] Y. Matiyasevich, *A Polynomial Related to Colourings of Triangulation of Sphere*, July 4, 1997. At URL: `http://logic.pdmi.ras.ru/~yumat/Journal/Triangular/triang.htm` 253, 257

[14] Y. Matiyasevich, The Four Colour Theorem as a possible corollary of binomial summation, *Theoretical Computer Science*, **257**(1-2):167–183, 2001.

[15] J. Meseguer, General Logics, in: H.-D. Ebbinghaus *et al.* (Eds.), *Logic Colloquium '87*, North-Holland, Amsterdam (1989) 275–329. 252, 256

[16] N. Robertson, D.P. Sanders, P.D. Seymour and R. Thomas, The four colour theorem, *J. Combin. Theory* Ser. B **70** (1997), 2-44. 254

[17] T. L. Saati, Thirteen colorful variations on Guthrie's four-color conjecture, *American Mathematical Monthly*, **79**(1):2–43 (1972).

[18] A. Salibra and G. Scollo, Interpolation and compactness in categories of pre-institutions, *Mathematical Structures in Computer Science* **6**, (1996) 261–286. 252

[19] G. Scollo, Graph colouring institutions, in: R. Berghammer, B. Möller (Eds.), *7th Seminar RelMiCS, 2nd Workshop Kleene Algebra*, Christian-Albrechts-Universität zu Kiel, Bad Malente, Germany, May 12-17, 2003, pp. 288–297. At URL: `http://www.informatik.uni-kiel.de/~relmics7` 252, 253, 260

[20] G. Scollo, Morphism-driven design of graph colouring institutions, RR 03/2003, University of Verona, Dipartimento di Informatica, March 2003. 254, 260, 261, 262, 263

[21] P. G. Tait, Note on a theorem in the geometry of position, *Trans. Roy. Soc. Edinburgh* **29** (1880), 657-660. printed in *Scientific Papers* **1**, 408–411. 253, 259

[22] P. G. Tait, On Listing's topology, *Phil. Mag. V. Ser.* **17** (1884), 30–46, printed in *Scientific Papers* **2**, 85–98. 253, 259

Decomposing Relations into Orderings

Michael Winter

Department of Computer Science
Brock University, St. Catharines, Ontario, Canada, L2S 3A1
mwinter@cosc.brocku.ca

Abstract. In various applications one is interested in decomposing a given relation in a suitable manner. In this paper we want to study several decompositions of a binary relation R of the form $R = F; E; G^{\smile}$, i.e., into a composition of a partial function F, an ordering E and the converse of a partial function G.

1 Introduction

In applications of computer science the information to handle is often given by a suitable relation. For example, a database is usually considered to be an n-ary relation. In multicriteria decision making a relation is given and one asks for so-called concepts and the corresponding lattice generated by them [4]. Since these relations tend to be very huge one is either interested in representing them efficiently or in computing the desired information with acceptable complexity. In database theory the notions of functional and difunctional dependencies are studied [2, 5] to reduce the size of the relation. Basically, both notions use a decomposition of the relation R into a composition of F and the converse of G where F and G are univalent (partial functions)[1], i.e., $R = F; G^{\smile}$. This is a special case of a decomposition of the form $R = F; E; G^{\smile}$ where F and G are univalent and E is a partial ordering[2]. Furthermore, a concept lattice of a relation R may be defined as such a decomposition where E is a complete lattice.

In this paper we want to study several decompositions of the form above. First, we give a short introduction to the theory of Schröder categories (or heterogeneous relation algebras), a convenient algebraic theory for binary relations. Afterwards we recall some basic properties of difunctional relations. As a generalization of difunctional, rectangular relations and a convenient subset of relations of Ferrers type (see, e.g., [10]) we will define relations of order-shape. They may be represented as mentioned above where F and G are, in addition, surjective. In the last section, we will show that every relation R may be decomposed. It turns out that there are two possibilities. Beside the representation above, R may be decomposed as $R = F; E; G^{\smile}$, where F and G are, in addition, total, i.e., mappings.

[1] In the case of functional dependencies G is the identity.

[2] For functional/difunctional dependencies E may be chosen as the identity.

R. Berghammer et al. (Eds.): RelMiCS/Kleene-Algebra Ws 2003, LNCS 3051, pp. 265–277, 2004.

There are several application of the theory invented in this paper. The notion of a concept lattice of a relation R seems to be isomorphic to the Dedekind-McNeille completion of the ordering E of the decomposition of R. Since E usually provides more structure than R this might lead to some new and faster algorithms for computing the concept lattice of R. Furthermore, in the finite case an ordering E may be represented by its Hasse-diagram H. This relation H usually requires less storage than E. The operations induced via the decomposition $R = F;E;G^\smile$ by the map $E \mapsto H$ and the reflexive-transitive closure may lead to a more efficient representation of relations within a computer. Both application are subject of ongoing research by the author.

For lack of space we have to omit several proofs throughout the paper. The author may communicate them to any interested reader.

2 Schröder Categories

In the remainder of this paper, we use the following notations. To indicate that a morphism R of a category $\mathcal{R}$ has source A and target B we write $R : A \to B$. The collection of all morphisms $R : A \to B$ is denoted by $\mathcal{R}[A, B]$ and the composition of a morphism $R : A \to B$ followed by a morphism $S : B \to C$ by $R; S$. The identity morphism on A is denoted by $\mathbb{I}_A$.

In this section we recall some fundamental properties of Schröder categories [6, 7]. For further details we refer to [1, 3, 10, 11, 12].

Definition 1. *A Schröder category $\mathcal{R}$ is a category satisfying the following:*

1. *For all objects A and B the collection $\mathcal{R}[A, B]$ is a complete Boolean algebra. Meet, join, complement, the induced ordering, the least and the greatest elements are denoted by $\sqcap, \sqcup, \overline{}, \sqsubseteq, \perp\!\!\!\perp_{AB}, \top\!\!\!\top_{AB}$, respectively.*
2. *There is a unary operation $^\smile$ mapping each relation $R : A \to B$ to a relation $R^\smile : B \to A$, called the converse operation.*
3. *The Schröder equivalences $Q; R \sqsubseteq S \Leftrightarrow Q^\smile; \overline{S} \sqsubseteq \overline{R} \Leftrightarrow \overline{S}; R^\smile \sqsubseteq \overline{Q}$ hold for relations $Q : A \to B, R : B \to C$ and $S : A \to C$.*

The standard model of a Schröder category is the category of concrete binary relations between sets. In the case of finite sets such a relation may be represented by a Boolean matrix. The relational connectives correspond to the usual matrix operations.

In the remainder of this paper, we will use some basic properties of Schröder categories without explicit reference. A proof may be found in [1, 3, 10, 11, 12].

The concept of univalent, total, injective, surjective and bijective relations is defined as usual (see, e.g. [10]). In the remainder of the paper we will use lower letters for mappings, i.e., for univalent and total relations.

Definition 2. *Let $R : A \to A$ be a relation. R is called*

1. *reflexive iff $\mathbb{I}_A \sqsubseteq R$,*
2. *irreflexive iff $R \sqsubseteq \overline{\mathbb{I}}_A$,*

3. *transitive iff* $R; R \sqsubseteq R$,
4. *dense iff* $R \sqsubseteq R; R$,
5. *idempotent iff* R *is transitive and dense,*
6. *symmetric iff* $R^{\smile} \sqsubseteq R$,
7. *antisymmetric iff* $R \sqcap R^{\smile} \sqsubseteq \mathbb{I}_A$,
8. *a preordering iff* R *is reflexive and transitive,*
9. *an ordering iff* R *is an antisymmetric preordering,*
10. *a linear ordering iff* R *is an ordering and* $R \sqcup R^{\smile} = \mathbb{T}_{AA}$,
11. *a strict ordering iff* R *is transitive and irreflexive,*
12. *a linear strict ordering iff* R *is a strict ordering and* $R \sqcup R^{\smile} = \overline{\mathbb{I}}_A$.

Suppose $E : A \to A$ and $E' : B \to B$ are orderings. A map $h : A \to B$ is called monotonic iff $E; h \sqsubseteq h; E'$. An isomorphism between E and E' is a bijective map h such that h and $h^{\smile}$ are monotonic.

The left residual $Q/R : C \to A$ of a relation $Q : A \to B$ over $R : C \to B$ is defined by $Q/R := \overline{\overline{R}; Q^{\smile}}$. By the Schröder equivalences this relation may be characterized as the greatest solution X of $X; Q \sqsubseteq R$. Analogously, the right residual $Q \backslash S : B \to D$ of a relation $Q : A \to B$ over $S : A \to D$, defined by $Q \backslash S := \overline{Q^{\smile}; \overline{S}}$ is the greatest solution Y of $Q; Y \sqsubseteq S$. We are also interested in relations which share properties of left and right residuals simultaneously, called symmetric quotients. This construction is defined by

$$\mathrm{syq}(Q, S) := (Q/S) \sqcap (Q^{\smile} \backslash S^{\smile}).$$

The relational description of disjoint unions is the relational sum [11]. This construction corresponds to the categorical product[3].

Definition 3. *Let A and B be objects of a Schröder category $\mathcal{R}$. An object $A+B$ together with two relations $\iota : A \to A+B$ and $\kappa : B \to A+B$ is called a relational sum of A and B iff $\iota; \iota^{\smile} = \mathbb{I}_A$, $\kappa; \kappa^{\smile} = \mathbb{I}_B$, $\iota; \kappa^{\smile} = \perp\!\!\!\perp_{AB}$ and $\iota^{\smile}; \iota \sqcup \kappa^{\smile}; \kappa = \mathbb{I}_{A+B}$. $\mathcal{R}$ has relational sums iff for every pair of objects a relational sum exists.*

A symmetric and idempotent relation R may be considered as a partial equivalence relation. It seems natural to switch to the set of existing equivalence classes of R.

Definition 4. *Let $Q : A \to A$ be a symmetric and idempotent relation. An object B together with a relation $R : B \to A$ is called the splitting of Q (or R splits Q) iff $R; R^{\smile} = \mathbb{I}_B$ and $R^{\smile}; R = Q$. A Schröder category has splittings iff for all symmetric and idempotent relations a splitting exists.*

Relational sums as well as splittings may not exist in a given Schröder category $\mathcal{R}$. But, it is possible to embed $\mathcal{R}$ into a Schröder category $\mathcal{R}^{+}_{SI}$ which offers these constructions [3, 12]. We call $\mathcal{R}^{+}_{SI}$ the completion of $\mathcal{R}$.

Splittings allow us to to switch from preorderings to the ordering of the equivalence classes. The proof of the next lemma may be found in [12].

[3] By the converse operation, a Schröder category is self-dual. Therefore, a product is also a coproduct.

Lemma 1. *Let $P : A \to A$ be a preordering and $f : A \to B$ be the converse of the splitting of $P \sqcap P^{\smile}$. Then $E := f^{\smile}; P; f$ is an ordering with $f; E; f^{\smile} = P$.*

3 Difunctional Relations

A Boolean matrix which may be written (modulo rearranging rows and columns) in block diagonal form may be of special interest. They may be characterized by an algebraic property [8, 10].

Definition 5. *A relation $R : A \to B$ is called difunctional iff $R; R^{\smile}; R \sqsubseteq R$.*

The following theorem was also given in [10].

Theorem 1.

1. *A product $F; G^{\smile}$ is always difunctional, if F and G are univalent.*
2. *The decomposition $R = F; G^{\smile}$ of a difunctional relation $R : A \to B$ in two univalent and surjective relations $F : A \to C$ and $G : B \to C$ is unique up to isomorphism.*
3. *For an arbitrary relation $R : A \to B$ the construct $R^d := R \sqcap \overline{R; \overline{R}^{\smile}; R}$ is difunctional and included in R.*

In general, the relation R^d is not maximal among the difunctional relations included in R. Consider the following relations:

$$R = \begin{pmatrix} 1&1&0 \\ 1&1&1 \\ 0&1&1 \end{pmatrix}, \quad D_1 = \begin{pmatrix} 1&0&0 \\ 0&1&1 \\ 0&1&1 \end{pmatrix}, \quad D_2 = \begin{pmatrix} 1&1&0 \\ 1&1&0 \\ 0&0&1 \end{pmatrix}, \quad R^d = \begin{pmatrix} 1&0&0 \\ 0&0&0 \\ 0&0&1 \end{pmatrix}.$$

D_1 and D_2 are the maximal difunctional relations included in R and R^d is even less than $D_1 \sqcap D_2$.

A decomposition of a difunctional relation into two univalent and surjective relations may not exist. But, under a slight assumption on the corresponding category we are able to prove the existence.

Theorem 2. *Suppose $\mathcal{R}$ has splittings. Then for every difunctional relation $R : A \to B$ there is a decomposition $R = F; G^{\smile}$ with F and G univalent and surjective.*

Proof. First of all, $R; R^{\smile}$ is symmetric and idempotent since R is difunctional. Therefore, there is an object C and a relation $F : A \to C$ such that $F^{\smile}; F = \mathbb{I}_C$ and $F; F^{\smile} = R; R^{\smile}$. Now, we define $G := R^{\smile}; F$ and conclude that

$$\begin{aligned} G^{\smile}; G &= F^{\smile}; R; R^{\smile}; F \\ &= F^{\smile}; F; F^{\smile}; F && F \text{ splits } R; R^{\smile} \\ &= \mathbb{I}_C, && F \text{ univalent and surjective} \\ F; G^{\smile} &= F; F^{\smile}; R \\ &= R; R^{\smile}; R && F \text{ splits } R; R^{\smile} \\ &= R. && R \text{ difunctional} \end{aligned}$$

This finishes the proof. □

Notice again, that the existence of splittings is not a strong assumption since they are guaranteed in the completion of $\mathcal{R}$.

In the next lemma we have summarized some basic properties of the relation R^d.

Lemma 2. *Let $R, S : A \to B, T : B \to C$ and $U : C \to A$ be relations. Then we have*

1. $R; R^{d\smile}; R \sqsubseteq R,$
2. $R \sqcap R^d; R^\smile; R^d = R^d,$
3. $R^d; R^{d\smile}; \overline{R;T} \sqsubseteq \overline{R^d; R^{d\smile}; R; T},$
4. $\overline{U;R}; R^{d\smile}; R^d \sqsubseteq \overline{U; R; R^{d\smile}; R^d},$
5. $R^d; R^{d\smile}; (S \sqcap R) = R^d; R^{d\smile}; S \sqcap R^d; R^{d\smile}; R,$
6. $(S \sqcap R); R^{d\smile}; R^d = S; R^{d\smile}; R^d \sqcap R; R^{d\smile}; R^d,$
7. $R^d \sqsubseteq (R^d; R^{d\smile}; R; R^{d\smile}; R^d)^d,$
8. $(\mathbb{I}_A \sqcap R^d; R^{d\smile}); \mathrm{syq}(R^\smile, R^\smile) = R^d; R^{d\smile}.$

4 Relations of Order-Shape

We are interested in those relation R which may be decomposed into $R = F; E; G^\smile$ with F and G univalent and surjective and E an ordering. Therefore, consider the following ordering E:

$$\begin{pmatrix} 1&0&0&0&1&0&1\\ 0&1&0&0&0&1&0\\ 0&0&1&0&0&0&0\\ 0&0&0&1&0&0&0\\ 0&0&0&0&1&0&1\\ 0&0&0&0&0&1&0\\ 0&0&0&0&0&0&1 \end{pmatrix}$$

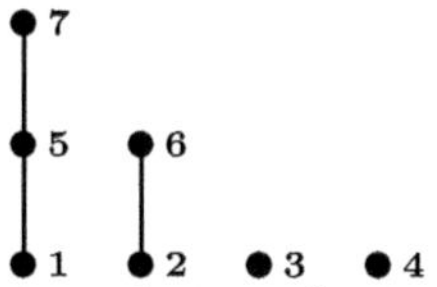

Composing suitable univalent and surjective relations from the left and the right we get the following Boolean matrix:

$$\begin{pmatrix}
1&1&1&0&0&0&0&0&0&0&0&0&0&1&1&0&1&1&0\\
1&1&1&0&0&0&0&0&0&0&0&0&0&1&1&0&1&1&0\\
0&0&0&1&1&0&0&0&0&0&0&0&0&0&0&1&0&0&0\\
0&0&0&1&1&0&0&0&0&0&0&0&0&0&0&1&0&0&0\\
0&0&0&1&1&0&0&0&0&0&0&0&0&0&0&1&0&0&0\\
0&0&0&0&0&0&0&0&0&0&0&0&0&0&0&0&0&0&0\\
0&0&0&0&0&1&1&1&1&0&0&0&0&0&0&0&0&0&0\\
0&0&0&0&0&0&0&0&0&1&1&1&0&0&0&0&0&0&0\\
0&0&0&0&0&0&0&0&0&1&1&1&0&0&0&0&0&0&0\\
0&0&0&0&0&0&0&0&0&0&0&0&0&1&1&0&1&1&0\\
0&0&0&0&0&0&0&0&0&0&0&0&0&0&0&1&0&0&0\\
0&0&0&0&0&0&0&0&0&0&0&0&0&0&0&0&1&1&0
\end{pmatrix}$$

Our aim is now to find an algebraic criterion for a concrete relation R to be representable by a matrix of the shape above (modulo rearranging rows and columns). We have to express that two rows corresponding to the same element in the domain A of E have to be identical. The relation R^d corresponds to the identity on A in the same way that R corresponds to E. Therefore, $R^d; R^{d\smile}$ and $R^{d\smile}; R^d$ are the symmetric and idempotent relations whose equivalence classes are the elements of A. A relation of the shape above has to respect both partial equivalence relations. Consequently, it may be characterized by $R^d; R^{d\smile}; R; R^{d\smile}; R^d = R$. In the following definition we use an apparently weaker property. The proof of equivalence is given in Lemma 3.

Definition 6. *We say that a relation $R : A \to B$ is of order-shape iff* $R \sqsubseteq R^d;R^{d^\smile};R;R^{d^\smile};R^d$.

In the next lemma we have collected some definition variants equivalent to the notion of relations of order-shape.

Lemma 3. *For a relation $R : A \to B$ the following properties are equivalent:*

1. *R is of order-shape,*
2. $R^d;R^{d^\smile};R;R^{d^\smile};R^d = R$,
3. $R;\mathbb{T}_{BC} \sqsubseteq R^d;\mathbb{T}_{BC}$ *and* $\mathbb{T}_{CA};R \sqsubseteq \mathbb{T}_{CA};R^d$ *for all objects C,*
4. $R;\mathbb{T}_{BC} = R^d;\mathbb{T}_{BC}$ *and* $\mathbb{T}_{CA};R = \mathbb{T}_{CA};R^d$ *for all objects C,*
5. $\mathbb{I}_A \sqcap R;R^\smile = \mathbb{I}_A \sqcap R^d;R^{d^\smile}$ *and* $\mathbb{I}_B \sqcap R^\smile;R = \mathbb{I}_B \sqcap R^{d^\smile};R^d$.

In the following lemma we have summarized some properties of decompositions of the form $F;E;G^\smile$.

Lemma 4. *Suppose $R = F;E;G^\smile$ with F and G univalent and surjective and E an ordering. Then we have*

1. $R^d = F;G^\smile$,
2. $F;F^\smile = R^d;R^{d^\smile}$ *and* $G;G^\smile = R^{d^\smile};R^d$,
3. $E = F^\smile;R;G$.

Now, we are ready to generalize Theorem 1 to relations of order-shape.

Theorem 3. 1. *A product $F;E;G^\smile$ is always of order-shape, if F and G are univalent and surjective and E is an ordering.*
2. *The decomposition $R = F;E;G^\smile$ of an order-shaped relation $R : A \to B$ by two univalent and surjective relations $F : A \to C$ and $G : B \to C$ and an ordering E is unique up to isomorphism.*
3. *For an arbitrary relation $R : A \to B$ the construct $R^d;R^{d^\smile};R;R^{d^\smile};R^d$ is of order-shape and included in R.*

Proof. 1. First of all, we conclude

$$
\begin{aligned}
(a)\ (F;E;G^\smile)^{d^\smile};F
&= (G;E^\smile;F^\smile \sqcap \overline{G;E^\smile;F^\smile;\overline{F;E;G^\smile};G;E^\smile;F^\smile});F\\
&= G;E^\smile \sqcap \overline{G;E^\smile;F^\smile;\overline{F;E;G^\smile};G;E^\smile;F^\smile;F}\\
&= G;E^\smile \sqcap \overline{G;E^\smile;\overline{E};E^\smile}\\
&= G;E^\smile \sqcap \overline{G;\overline{E}}\\
&\sqsupseteq G;E^\smile \sqcap G;E\\
&= G;(E^\smile \sqcap E)\\
&= G
\end{aligned}
$$

and analogously (b) $F \sqsubseteq (F;E;G^{\smile})^d;G$. The assertion follows from

$$\begin{aligned}
&(F;E;G^{\smile})^d;(F;E;G^{\smile})^{d\smile};F;E;G^{\smile};(F;E;G^{\smile})^{d\smile};(F;E;G^{\smile})^d \\
&\qquad \sqsupseteq (F;E;G^{\smile})^d;G;E;F^{\smile};(F;E;G^{\smile})^d && (a),(b) \\
&\qquad \sqsupseteq F;E;G^{\smile}. && (a),(b)
\end{aligned}$$

2. Suppose $F' : A \to D$, $G' : B \to D$ and $E' : D \to D$ is another decomposition of R. From Lemma 4 (1) we get $F;G^{\smile} = R^d = F';G'^{\smile}$. We define $h := F^{\smile};F'$ and conclude that

$$\begin{aligned}
h^{\smile};h &= F'^{\smile};F;F^{\smile};F' \\
&= F'^{\smile};F;G^{\smile};G;F^{\smile};F' && G \text{ univalent and surjective} \\
&= F'^{\smile};F';G'^{\smile};G';F'^{\smile};F' && F;G^{\smile} = R^d = F';G'^{\smile} \\
&= \mathbb{I}_D, && F',G' \text{ univalent and surjective} \\
h;h^{\smile} &= F^{\smile};F';F'^{\smile};F \\
&= F^{\smile};F';G'^{\smile};G';F'^{\smile};F && G' \text{ univalent and surjective} \\
&= F^{\smile};F;G^{\smile};G;F^{\smile};F && F;G^{\smile} = R^d = F';G'^{\smile} \\
&= \mathbb{I}_C. && F,G \text{ univalent and surjective}
\end{aligned}$$

Furthermore, h and $h^{\smile}$ are monotonic, i.e.,

$$\begin{aligned}
E;h &= E;F^{\smile};F' \\
&= F^{\smile};F;E;G^{\smile};G;F^{\smile};F' && F,G \text{ univalent and surjective} \\
&= F^{\smile};R;G;F^{\smile};F' \\
&= F^{\smile};R;G';F'^{\smile};F' && F;G^{\smile} = R^d = F';G'^{\smile} \\
&= F^{\smile};R;G' && F' \text{ univalent and surjective} \\
&= F^{\smile};F';E';G'^{\smile};G' \\
&= F^{\smile};F';E' && G' \text{ univalent and surjective} \\
&= h;E', \\
E';h^{\smile} &= h^{\smile};h;E';h^{\smile} && h \text{ bijective map} \\
&= h^{\smile};E;h;h^{\smile} && \text{see above} \\
&= h^{\smile};E. && h \text{ bijective map}
\end{aligned}$$

3. For brevity, let $S := R^d;R^{d\smile};R;R^{d\smile};R^d$. Then $S \sqsubseteq R$ follows immediately from Lemma 2 (1) and $R^d \sqsubseteq R$. Furthermore, we have

$$\begin{aligned}
&S^d;S^{d\smile};S;S^{d\smile};S^d \\
&\qquad \sqsupseteq R^d;R^{d\smile};S;R^{d\smile};R^d && \text{Lemma 2 (7)} \\
&\qquad = R^d;R^{d\smile};R^d;R^{d\smile};R;R^{d\smile};R^d;R^{d\smile};R^d
\end{aligned}$$

$$
\begin{aligned}
&= R^d; R^{d^\smile}; R; R^{d^\smile}; R^d && R^d \text{ difunctional} \\
&= S
\end{aligned}
$$

so that S is of order-shape. □

As for difunctional relations we need a slight assumption on the corresponding Schröder category to prove the existence of a decomposition of a relation of order-shape.

Theorem 4. *Suppose $\mathcal{R}$ has splittings. Then for every relation $R : A \to B$ of order-shape there is a decomposition $R = F; E; G^\smile$ with F and G univalent and surjective and E an ordering.*

Proof. By Theorem 1 the relation R^d is difunctional and by Theorem 2 there are two univalent and surjective relations $F : A \to C$ and $G : B \to C$ such that $F; G^\smile = R^d$. Now, we define $E := F^\smile; R; G$ and conclude that

$$
\begin{aligned}
F; E; G^\smile &= F; F^\smile; R; G; G^\smile \\
&= F; G^\smile; G; F^\smile; R; G; F^\smile; F; G^\smile && F, G \text{ univalent and surjective} \\
&= R^d; R^{d^\smile}; R; R^{d^\smile}; R^d && \text{Definition } F, G \\
&= R. && \text{Lemma 3 (2)}
\end{aligned}
$$

Furthermore, E is an ordering, which follows from

$$
\begin{aligned}
\mathbb{I}_C &= F^\smile; F; G^\smile; G && F, G \text{ univalent and surjective} \\
&= F^\smile; R^d; G && \text{Definition } F, G \\
&\sqsubseteq F^\smile; R; G \\
&= E, \\
E; E &= F^\smile; R; G; F^\smile; R; G \\
&= F^\smile; R; R^{d^\smile}; R; G && \text{Definition } F, G \\
&\sqsubseteq F^\smile; R; G && \text{Lemma 2 (1)} \\
&= E, \\
E \sqcap E^\smile &= F^\smile; R; G \sqcap G^\smile; R^\smile; F \\
&= G^\smile; (G; F^\smile; R; G; F^\smile \sqcap R^\smile); F \\
&= G^\smile; (R^{d^\smile}; R; R^{d^\smile} \sqcap R^\smile); F && \text{Definition } F, G \\
&= G^\smile; R^{d^\smile}; F && \text{Lemma 2 (2)} \\
&= G^\smile; G; F^\smile; F && \text{Definition } F, G \\
&= \mathbb{I}_C. && F, G \text{ univalent and surjective}
\end{aligned}
$$

This finishes the proof. □

In the remainder of this section we want to show that relations of order-shape are a generalization of some known classes of relations.

Definition 7. *A relation $R : A \to B$ is called*

1. *rectangular iff $R;\mathbb{T}_{BA};R \sqsubseteq R$,*
2. *of Ferrers type iff $R;\overline{R}^{\smile};R \sqsubseteq R$,*
3. *of strong Ferrers type iff $R;\overline{R}^{\smile};R \sqsubseteq R^d;R^{d\smile};R;R^{d\smile};R^d$.*

A concrete relation of Ferrers type may be written as a Boolean matrix in staircase block form by suitably rearranging rows and columns [9, 10]. Furthermore, in [10] it was proved that if a relation of Ferrers type is decomposable into an ordering it has to be linear. The next lemma shows that such a decomposition may not exist. Therefore, we have introduced the new notion of relations of strong Ferrers type.

Lemma 5. *Let $R \neq \perp\!\!\!\perp_{AA}$ be a dense linear strict-ordering. Then we have*

1. *$R;\overline{R}^{\smile};R = R$ and hence R is of Ferrers type,*
2. *$R^d = \perp\!\!\!\perp_{AA}$,*
3. *R is not of order-shape.*

Consequently, there are relations of Ferrers type which are not decomposable into a pair of univalent and surjective relations and a linear ordering.

Lemma 6. *Let $R : A \to B$ be a relation. Then we have*

1. *if R is difunctional then R is of order-shape,*
2. *if R is rectangular then R is of order-shape,*
3. *$R : A \to B$ is of strong Ferrers type iff R is of Ferrers type and of order-shape.*

In the next lemma we want to characterize the above classes of relations by properties of the corresponding ordering E.

Lemma 7. *Let $R : A \to B$ be a relation of order-shape and $R = F;E;G^{\smile}$ its decomposition into univalent and surjective relations $F : A \to C$ and $G : B \to C$ and an ordering E. Then we have*

1. *R is difunctional iff $E = \mathbb{I}_C$,*
2. *R is of Ferrers type iff E is a linear ordering,*
3. *R is rectangular iff $E = \mathbb{I}_C = \mathbb{T}_{CC}$.*

The last lemma also shows that the decomposition of a difunctional relation from the last section and the decomposition of a relation of Ferrers type from [10] coincide with the decomposition given in this section.

5 Decomposition of Arbitrary Relations

Suppose $R : A \to B$ has a decomposition of the form $R = F; E; G^{\smile}$ where $F : A \to C$ and $G : B \to C$ are univalent. A sensible property of such a decomposition may be formulated as follows. All elements of C are in the codomain of either F or G, i.e., there are no useless elements (in respect to the decomposition). This requirement may be expressed by $F^{\smile}; F \sqcup G^{\smile}; G = \mathbb{I}_C$. In the remainder of this paper a decomposition $R = F; E; G^{\smile}$ is always a decomposition of R as indicated above. Notice that the decomposition of a relation of order-shape fulfills that requirement since F and G are surjective.

Suppose $\mathcal{R}$ has relational sums and $R : A \to B$ is an arbitrary relation in $\mathcal{R}$. Then we may define $F := \iota, G := \kappa$ and $E := \mathbb{I}_{A+B} \sqcup \iota^{\smile}; R; \kappa$ and conclude that $R = F; E; G^{\smile}$, i.e., that R has a decomposition. But it does not correspond to the decompositions given in the previous sections. Consider the following example:

$$R := \begin{pmatrix} 1 & 1 \\ 1 & 1 \end{pmatrix}, \quad F = \begin{pmatrix} 1 & 0 & 0 & 0 \\ 0 & 1 & 0 & 0 \end{pmatrix}, \quad G = \begin{pmatrix} 0 & 0 & 1 & 0 \\ 0 & 0 & 0 & 1 \end{pmatrix}, \quad E = \begin{pmatrix} 1 & 0 & 1 & 1 \\ 0 & 1 & 1 & 1 \\ 0 & 0 & 1 & 0 \\ 0 & 0 & 0 & 1 \end{pmatrix}$$

R is rectangular but neither F and G are surjective nor $E = \mathbb{I} = \mathbb{T}$. The domain of the ordering E has to many elements (4 instead of 1) such that we should be interested in the "least" decomposition. Such a property is usually expressed by terminal elements in a suitable category.

Definition 8. *Let $R = F; E; G^{\smile}$ and $R = F'; E'; G'^{\smile}$ be two decompositions of R. A monotone map h from E to E' is called a homomorphism from the first to the second decomposition iff $F; h = F'$ and $G; h = G'$.*

It is easy to verify that the decompositions of R together with homomorphisms between them constitute a category. Therefore, a terminal object in this category (if it exists) is unique up to isomorphism. In the remainder of this paper we will call this terminal object the partial decomposition of R. If F and G are, in addition, total we call the corresponding terminal object the total decomposition of R.

First, we want to concentrate on decompositions where F and G are mappings.

Lemma 8. *Let $R : A \to B$ be a relation, and $R = f; E; g^{\smile}$ be a decomposition of R with*

$$f; E; f^{\smile} = R/R, \qquad g; E; g^{\smile} = R\backslash R, \qquad g; E; f^{\smile} = \overline{R^{\smile}; \overline{R}; R^{\smile}}.$$

Then $R = f; E; g^{\smile}$ is a total decomposition of R.

Later on, we will show that the converse implication of the last lemma is also valid.

Theorem 5. *Suppose $\mathcal{R}$ has splittings and relational sums. Then for every relation $R : A \to B$ there is a total decomposition.*

Proof. Let ι and κ be the relational sum of A and B. Then define the following relation: $P := \iota^{\smile}; (R/R); \iota \sqcup \kappa^{\smile}; (R\backslash R); \kappa \sqcup \iota^{\smile}; R; \kappa \sqcup \kappa^{\smile}; \overline{R^{\smile}; \overline{R}; R^{\smile}}; \iota$. Then P is a pre-ordering, which follows from

$$\begin{aligned}
\mathbb{I}_{A+B} &= \iota^{\smile}; \iota \sqcup \kappa^{\smile}; \kappa && \text{Definition sum} \\
&\sqsubseteq \iota^{\smile}; (R/R); \iota \sqcup \kappa^{\smile}; (R\backslash R); \kappa \\
&\sqsubseteq P, \\
P; P &= \iota^{\smile}; (R/R); (R/R); \iota \sqcup \iota^{\smile}; (R/R); R; \kappa && \text{Definition sum} \\
&\quad \sqcup \iota^{\smile}; R; (R\backslash R); \kappa \sqcup \iota^{\smile}; R; \overline{R^{\smile}; \overline{R}; R^{\smile}}; \iota \\
&\quad \sqcup \kappa^{\smile}; (R\backslash R); (R\backslash R); \kappa \sqcup \kappa^{\smile}; (R\backslash R); \overline{R^{\smile}; \overline{R}; R^{\smile}}; \iota \\
&\quad \sqcup \kappa^{\smile}; \overline{R^{\smile}; \overline{R}; R^{\smile}}; (R/R); \iota \sqcup \kappa^{\smile}; \overline{R^{\smile}; \overline{R}; R^{\smile}}; R; \kappa \\
&\sqsubseteq P.
\end{aligned}$$

Furthermore, the definition of a relational sum shows that

$$\iota; P; \iota^{\smile} = R/R, \qquad \kappa; P; \kappa^{\smile} = R\backslash R, \qquad \iota; P; \kappa = R, \qquad \kappa; P; \iota = \overline{R^{\smile}; \overline{R}; R^{\smile}}.$$

Now, let $l : C \to A + B$ be the converse of the splitting of $P \sqcap P^{\smile}$, i.e.,

$$l^{\smile}; l = \mathbb{I}_C \quad \text{and} \quad l; l^{\smile} = P \sqcap P^{\smile}.$$

From Lemma 1 we conclude that $E := l^{\smile}; P; l$ is an ordering with $l; E; l^{\smile} = P$. We define $f := \iota; l$ and $g := \kappa; l$. Then f and g are mappings and we have

$$\begin{aligned}
f^{\smile}; f \sqcup g^{\smile}; g &= l^{\smile}; \iota^{\smile}; \iota; l \sqcup l^{\smile}; \kappa^{\smile}; \kappa; l \\
&= l^{\smile}; (\iota^{\smile}; \iota \sqcup \kappa^{\smile}; \kappa); l \\
&= l^{\smile}; l && \text{Definition sum} \\
&= \mathbb{I}_C, && \text{Definition } l \\
f; E; f^{\smile} &= \iota; l; E; l^{\smile}; \iota^{\smile} \\
&= \iota; P; \iota^{\smile} && \text{Lemma 1} \\
&= R/R. && \text{Definition of } P \text{ and } \iota
\end{aligned}$$

$g; E; g^{\smile} = R\backslash R$, $f; E; g^{\smile} = R$ and $g; E; f^{\smile} = \overline{R^{\smile}; \overline{R}; R^{\smile}}$ are shown analogously. The assertion follows from Lemma 8. □

Now, we are ready to prove the converse implication of Lemma 8.

Theorem 6. *Let $R : A \to B$ be a relation and $R = f; E; g^{\smile}$ be a decomposition of R. Then this decomposition is a total decomposition of R iff*

$$f; E; f^{\smile} = R/R, \qquad g; E; g^{\smile} = R\backslash R, \qquad g; E; f^{\smile} = \overline{R^{\smile}; \overline{R}; R^{\smile}}.$$

Proof. The implication $\Leftarrow$ was already shown in Lemma 8. By the last theorem, in the completion of $\mathcal{R}$ there is decomposition $R = f';E';g'^{\smile}$ of R with the required properties for f', g', E' instead of f, g, E. Since this decomposition is also a total decomposition there is an isomorphism h from E' to E. We conclude that

$$\begin{aligned} f;E;f^{\smile} &= f';h;E;h^{\smile};f' && h \text{ homomorphism} \\ &= f';E';f' && h \text{ homomorphism} \\ &= R/R. && \text{assumption} \end{aligned}$$

The other equations follow analogously. □

In the next lemma we want to show that the total decomposition of a total, surjective relation of order-shape corresponds to the decomposition given in the last section.

Lemma 9. *Let $R : A \rightharpoonup B$ be a relation and $R = f;E;g^{\smile}$ be a total decomposition of R. Then R is total, surjective and of order-shape iff f and g are surjective.*

Now, we want to concentrate on decompositions of the form $R = F;E;G^{\smile}$, where F and G are univalent. Such a partial decomposition is given by a convenient subset of a total decomposition. Therefore, we get the following theorem on the existence of partial decompositions.

Theorem 7. *Suppose $\mathcal{R}$ has splittings and relational sums. Then for every relation $R : A \rightharpoonup B$ there is a partial decomposition.*

Proof sketch: By Theorem 5 there is a total decomposition $R = f;E;g^{\smile}$ of R. Now, let H be the splitting of $f^{\smile};(\mathbb{I}_A \sqcap R;R^{\smile});f \sqcup g^{\smile};(\mathbb{I}_B \sqcap R^{\smile};R);g$ and define $F := (\mathbb{I}_A \sqcap R;R^{\smile});f;H$, $G := (\mathbb{I}_B \sqcap R^{\smile};R);g;H$ and $E' := H^{\smile};E;H$. Then F, G and E' is the required partial decomposition. □

Last but not least, we want to show that the partial decomposition of a relation of order-shape corresponds to the decomposition given in the last section.

Lemma 10. *Let $R : A \rightharpoonup B$ be a relation and $R = F;E;G^{\smile}$ be a partial decomposition of R. Then R is of order-shape iff F and G are surjective.*

Acknowledgement

I would like to thank the anonymous referees for their helpful comments.

References

[1] Chin L. H., Tarski A.: Distributive and modular laws in the arithmetic of relation algebras. University of California Press, Berkley and Los Angeles (1951). 266

[2] Codd E. F.: A relational model of data for large shared data banks. Comm. ACM 13(6) (1970), 377-387. 265

[3] Freyd P., Scedrov A.: Categories, Allegories. North-Holland (1990). 266, 267

[4] Ganter B., Wille R.: Formal Concept Analysis - Mathematical Foundations. Springer (1999) 265

[5] Jaoua A., Belkhiter N., Ounalli H., Moukam T: Databases. *In:* Brink C., Kahl W., Schmidt G. (eds.), Relational Methods in Computer Science, Advances in Computer Science, Springer Vienna (1997), 196-210. 265

[6] Olivier J. P., Serrato D.: Catégories de Dedekind. Morphismes dans les Catégories de Schröder. C. R. Acad. Sci. Paris 290 (1980), 939-941. 266

[7] Olivier J. P., Serrato D.: Squares and Rectangles in Relational Categories - Three Cases: Semilattice, Distributive lattice and Boolean Non-unitary. Fuzzy sets and systems 72 (1995), 167-178. 266

[8] Riguet J.: Quelques propriétés des relations difonctionelles. C. R. Acad. Sci. Paris 230 (1950), 1999-2000. 268

[9] Riguet J.: Les relations de Ferrers. C. R. Acad. Sci. Paris 232 (1951), 1729-1730. 273

[10] Schmidt G., Ströhlein T.: Relationen und Graphen. Springer (1989); English version: Relations and Graphs. Discrete Mathematics for Computer Scientists, EATCS Monographs on Theoret. Comput. Sci., Springer (1993). 265, 266, 268, 273

[11] Schmidt G., Hattensperger C., Winter M.: Heterogeneous Relation Algebras. *In:* Brink C., Kahl W., Schmidt G. (eds.), Relational Methods in Computer Science, Advances in Computer Science, Springer Vienna (1997), 40-54. 266, 267

[12] Winter M.: Strukturtheorie heterogener Relationenalgebren mit Anwendung auf Nichtdetermismus in Programmiersprachen. Dissertationsverlag NG Kopierladen GmbH, München (1998). 266, 267

Author Index

Lecture Notes in Computer Science

For information about Vols. 1–2976

please contact your bookseller or Springer-Verlag

Vol. 3092: J. Eckstein, H. Baumeister (Eds.), Extreme Programming and Agile Processes in Software Engineering. XVI, 358 pages. 2004.

Vol. 3083: W. Emmerich, A.L. Wolf (Eds.), Component Deployment. X, 249 pages. 2004.

Vol. 3076: D. Buell (Ed.), Algorithmic Number Theory. XI, 451 pages. 2004.

Vol. 3074: B. Kuijpers, P. Revesz (Eds.), Constraint Databases and Applications. XII, 181 pages. 2004.

Vol. 3066: S. Tsumoto, R. S lowiński, J. Komorowski, J.W. Grzymala-Busse (Eds.), Rough Sets and Current Trends in Computing. XX, 853 pages. 2004. (Subseries LNAI).

Vol. 3065: A. Lomuscio, D. Nute (Eds.), Deontic Logic. X, 275 pages. 2004. (Subseries LNAI).

Vol. 3064: D. Bienstock, G. Nemhauser (Eds.), Integer Programming and Combinatorial Optimization. XI, 445 pages. 2004.

Vol. 3063: A. Llamosí, A. Strohmeier (Eds.), Reliable Software Technologies - Ada-Europe 2004. XIII, 333 pages. 2004.

Vol. 3062: J.L. Pfaltz, M. Nagl, B. Böhlen (Eds.), Applications of Graph Transformations with Industrial Relevance. XV, 500 pages. 2004.

Vol. 3060: A.Y. Tawfik, S.D. Goodwin (Eds.), Advances in Artificial Intelligence. XIII, 582 pages. 2004. (Subseries LNAI).

Vol. 3059: C.C. Ribeiro, S.L. Martins (Eds.), Experimental and Efficient Algorithms. X, 586 pages. 2004.

Vol. 3058: N. Sebe, M.S. Lew, T.S. Huang (Eds.), Computer Vision in Human-Computer Interaction. X, 233 pages. 2004.

Vol. 3056: H. Dai, R. Srikant, C. Zhang (Eds.), Advances in Knowledge Discovery and Data Mining. XIX, 713 pages. 2004. (Subseries LNAI).

Vol. 3054: I. Crnkovic, J.A. Stafford, H.W. Schmidt, K. Wallnau (Eds.), Component-Based Software Engineering. XI, 311 pages. 2004.

Vol. 3053: C. Bussler, J. Davies, D. Fensel, R. Studer (Eds.), The Semantic Web: Research and Applications. XIII, 490 pages. 2004.

Vol. 3052: W. Zimmermann, B. Thalheim (Eds.), Abstract State Machines 2004. Advances in Theory and Practice. XII, 235 pages. 2004.

Vol. 3051: R. Berghammer, B. Möller, G. Struth (Eds.), Relational and Kleene-Algebraic Methods in Computer Science. X, 279 pages. 2004.

Vol. 3047: F. Oquendo, B. Warboys, R. Morrison (Eds.), Software Architecture. X, 279 pages. 2004.

Vol. 3046: A. Laganà, M.L. Gavrilova, V. Kumar, Y. Mun, C.K. Tan, O. Gervasi (Eds.), Computational Science and Its Applications - ICCSA 2004. LIII, 1016 pages. 2004.

Vol. 3045: A. Laganà, M.L. Gavrilova, V. Kumar, Y. Mun, C.K. Tan, O. Gervasi (Eds.), Computational Science and Its Applications – ICCSA 2004. LIII, 1040 pages. 2004.

Vol. 3044: A. Laganà, M.L. Gavrilova, V. Kumar, Y. Mun, C.K. Tan, O. Gervasi (Eds.), Computational Science and Its Applications – ICCSA 2004. LIII, 1140 pages. 2004.

Vol. 3043: A. Laganà, M.L. Gavrilova, V. Kumar, Y. Mun, C.J.K. Tan, O. Gervasi (Eds.), Computational Science and Its Applications – ICCSA 2004. LIII, 1180 pages. 2004.

Vol. 3042: N. Mitrou, K. Kontovasilis, G.N. Rouskas, I. Iliadis, L. Merakos (Eds.), NETWORKING 2004, Networking Technologies, Services, and Protocols; Performance of Computer and Communication Networks; Mobile and Wireless Communications. XXXIII, 1519 pages. 2004.

Vol. 3039: M. Bubak, G.D.v. Albada, P.M. Sloot, J. Dongarra (Eds.), Computational Science - 2004. LXVI, 1271 pages. 2004.

Vol. 3038: M. Bubak, G.D.v. Albada, P.M. Sloot, J. Dongarra (Eds.), Computational Science - ICCS 2004. LXVI, 1311 pages. 2004.

Vol. 3037: M. Bubak, G.D.v. Albada, P.M. Sloot, J. Dongarra (Eds.), Computational Science - ICCS 2004. LXVI, 745 pages. 2004.

Vol. 3036: M. Bubak, G.D.v. Albada, P.M. Sloot, J. Dongarra (Eds.), Computational Science - ICCS 2004. LXVI, 713 pages. 2004.

Vol. 3035: M.A. Wimmer (Ed.), Knowledge Management in Electronic Government. XII, 326 pages. 2004. (Subseries LNAI).

Vol. 3034: J. Favela, E. Menasalvas, E. Chávez (Eds.), Advances in Web Intelligence. XIII, 227 pages. 2004. (Subseries LNAI).

Vol. 3033: M. Li, X.-H. Sun, Q. Deng, J. Ni (Eds.), Grid and Cooperative Computing. XXXVIII, 1076 pages. 2004.

Vol. 3032: M. Li, X.-H. Sun, Q. Deng, J. Ni (Eds.), Grid and Cooperative Computing. XXXVII, 1112 pages. 2004.

Vol. 3031: A. Butz, A. Krüger, P. Olivier (Eds.), Smart Graphics. X, 165 pages. 2004.

Vol. 3030: P. Giorgini, B. Henderson-Sellers, M. Winikoff (Eds.), Agent-Oriented Information Systems. XIV, 207 pages. 2004. (Subseries LNAI).

Vol. 3029: B. Orchard, C. Yang, M. Ali (Eds.), Innovations in Applied Artificial Intelligence. XXI, 1272 pages. 2004. (Subseries LNAI).

Vol. 3028: D. Neuenschwander, Probabilistic and Statistical Methods in Cryptology. X, 158 pages. 2004.

Vol. 3027: C. Cachin, J. Camenisch (Eds.), Advances in Cryptology - EUROCRYPT 2004. XI, 628 pages. 2004.

Vol. 3026: C. Ramamoorthy, R. Lee, K.W. Lee (Eds.), Software Engineering Research and Applications. XV, 377 pages. 2004.

Vol. 3025: G.A. Vouros, T. Panayiotopoulos (Eds.), Methods and Applications of Artificial Intelligence. XV, 546 pages. 2004. (Subseries LNAI).

Vol. 3024: T. Pajdla, J. Matas (Eds.), Computer Vision - ECCV 2004. XXVIII, 621 pages. 2004.

Vol. 3023: T. Pajdla, J. Matas (Eds.), Computer Vision - ECCV 2004. XXVIII, 611 pages. 2004.

Vol. 3022: T. Pajdla, J. Matas (Eds.), Computer Vision - ECCV 2004. XXVIII, 621 pages. 2004.

Vol. 3021: T. Pajdla, J. Matas (Eds.), Computer Vision - ECCV 2004. XXVIII, 633 pages. 2004.

Vol. 3019: R. Wyrzykowski, J. Dongarra, M. Paprzycki, J. Wasniewski (Eds.), Parallel Processing and Applied Mathematics. XIX, 1174 pages. 2004.

Vol. 3016: C. Lengauer, D. Batory, C. Consel, M. Odersky (Eds.), Domain-Specific Program Generation. XII, 325 pages. 2004.

Vol. 3015: C. Barakat, I. Pratt (Eds.), Passive and Active Network Measurement. XI, 300 pages. 2004.

Vol. 3014: F. van der Linden (Ed.), Software Product-Family Engineering. IX, 486 pages. 2004.

Vol. 3012: K. Kurumatani, S.-H. Chen, A. Ohuchi (Eds.), Multi-Agnets for Mass User Support. X, 217 pages. 2004. (Subseries LNAI).

Vol. 3011: J.-C. Régin, M. Rueher (Eds.), Integration of AI and OR Techniques in Constraint Programming for Combinatorial Optimization Problems. XI, 415 pages. 2004.

Vol. 3010: K.R. Apt, F. Fages, F. Rossi, P. Szeredi, J. Váncza (Eds.), Recent Advances in Constraints. VIII, 285 pages. 2004. (Subseries LNAI).

Vol. 3009: F. Bomarius, H. Iida (Eds.), Product Focused Software Process Improvement. XIV, 584 pages. 2004.

Vol. 3008: S. Heuel, Uncertain Projective Geometry. XVII, 205 pages. 2004.

Vol. 3007: J.X. Yu, X. Lin, H. Lu, Y. Zhang (Eds.), Advanced Web Technologies and Applications. XXII, 936 pages. 2004.

Vol. 3006: M. Matsui, R. Zuccherato (Eds.), Selected Areas in Cryptography. XI, 361 pages. 2004.

Vol. 3005: G.R. Raidl, S. Cagnoni, J. Branke, D.W. Corne, R. Drechsler, Y. Jin, C.G. Johnson, P. Machado, E. Marchiori, F. Rothlauf, G.D. Smith, G. Squillero (Eds.), Applications of Evolutionary Computing. XVII, 562 pages. 2004.

Vol. 3004: J. Gottlieb, G.R. Raidl (Eds.), Evolutionary Computation in Combinatorial Optimization. X, 241 pages. 2004.

Vol. 3003: M. Keijzer, U.-M. O'Reilly, S.M. Lucas, E. Costa, T. Soule (Eds.), Genetic Programming. XI, 410 pages. 2004.

Vol. 3002: D.L. Hicks (Ed.), Metainformatics. X, 213 pages. 2004.

Vol. 3001: A. Ferscha, F. Mattern (Eds.), Pervasive Computing. XVII, 358 pages. 2004.

Vol. 2999: E.A. Boiten, J. Derrick, G. Smith (Eds.), In grated Formal Methods. XI, 541 pages. 2004.

Vol. 2998: Y. Kameyama, P.J. Stuckey (Eds.), Functio and Logic Programming. X, 307 pages. 2004.

Vol. 2997: S. McDonald, J. Tait (Eds.), Advances in Inf mation Retrieval. XIII, 427 pages. 2004.

Vol. 2996: V. Diekert, M. Habib (Eds.), STACS 2004. X 658 pages. 2004.

Vol. 2995: C. Jensen, S. Poslad, T. Dimitrakos (Eds.), Tr Management. XIII, 377 pages. 2004.

Vol. 2994: E. Rahm (Ed.), Data Integration in the Li Sciences. X, 221 pages. 2004. (Subseries LNBI).

Vol. 2993: R. Alur, G.J. Pappas (Eds.), Hybrid System Computation and Control. XII, 674 pages. 2004.

Vol. 2992: E. Bertino, S. Christodoulakis, D. Plexousaki V. Christophides, M. Koubarakis, K. Böhm, E. Ferra (Eds.), Advances in Database Technology - EDBT 200 XVIII, 877 pages. 2004.

Vol. 2991: R. Alt, A. Frommer, R.B. Kearfott, W. Luth (Eds.), Numerical Software with Result Verification. 315 pages. 2004.

Vol. 2990: J. Leite, A. Omicini, L. Sterling, P. Torro (Eds.), Declarative Agent Languages and Techniques. XI 281 pages. 2004. (Subseries LNAI).

Vol. 2989: S. Graf, L. Mounier (Eds.), Model Checkin Software. X, 309 pages. 2004.

Vol. 2988: K. Jensen, A. Podelski (Eds.), Tools and Algo rithms for the Construction and Analysis of Systems. XIV 608 pages. 2004.

Vol. 2987: I. Walukiewicz (Ed.), Foundations of Softwar Science and Computation Structures. XIII, 529 pages 2004.

Vol. 2986: D. Schmidt (Ed.), Programming Languages and Systems. XII, 417 pages. 2004.

Vol. 2985: E. Duesterwald (Ed.), Compiler Construction. X, 313 pages. 2004.

Vol. 2984: M. Wermelinger, T. Margaria-Steffen (Eds.), Fundamental Approaches to Software Engineering. XII, 389 pages. 2004.

Vol. 2983: S. Istrail, M.S. Waterman, A. Clark (Eds.), Computational Methods for SNPs and Haplotype Inference. IX, 153 pages. 2004. (Subseries LNBI).

Vol. 2982: N. Wakamiya, M. Solarski, J. Sterbenz (Eds.), Active Networks. XI, 308 pages. 2004.

Vol. 2981: C. Müller-Schloer, T. Ungerer, B. Bauer (Eds.), Organic and Pervasive Computing – ARCS 2004. XI, 339 pages. 2004.

Vol. 2980: A. Blackwell, K. Marriott, A. Shimojima (Eds.), Diagrammatic Representation and Inference. XV, 448 pages. 2004. (Subseries LNAI).

Vol. 2979: I. Stoica, Stateless Core: A Scalable Approach for Quality of Service in the Internet. XVI, 219 pages. 2004.

Vol. 2978: R. Groz, R.M. Hierons (Eds.), Testing of Communicating Systems. XII, 225 pages. 2004.

Vol. 2977: G. Di Marzo Serugendo, A. Karageorgos, O.F. Rana, F. Zambonelli (Eds.), Engineering Self-Organising Systems. X, 299 pages. 2004. (Subseries LNAI).

GPSR Compliance

The European Union's (EU) General Product Safety Regulation (GPSR) is a set of rules that requires consumer products to be safe and our obligations to ensure this.

If you have any concerns about our products, you can contact us on ProductSafety@springernature.com

In case Publisher is established outside the EU, the EU authorized representative is:

Springer Nature Customer Service Center GmbH
Europaplatz 3
69115 Heidelberg, Germany

Batch number: 09490862

Printed by Printforce, the Netherlands